HIT TO KILL

HIT TO KILL

The New Battle Over Shielding America From Missile Attack

BRADLEY GRAHAM

 Public Affairs *New York*

Copyright © 2001, 2003 by Bradley Graham.
Published in the United States by PublicAffairs™,
a member of the Perseus Books Group.
All rights reserved.
Printed in the United States of America.
No part of this book may be reproduced in any manner whatsoever without written
permission except in the case of brief quotations embodied in critical articles and
reviews. For information, address PublicAffairs, 250 West 57th Street, Suite 1321,
New York, NY 10107. PublicAffairs books are available at special discounts for
bulk purchases in the U.S. by corporations, institutions, and other organizations.
For more information, please contact the Special Markets Department at the
Perseus Books Group, 11 Cambridge Center, Cambridge, MA 02142, or call
(617) 252–5298.

Book design by Barbara Werden.
Set in 10.6-point Sabon

Library of Congress Cataloging-in-Publication data
Graham, Bradley.
Hit to kill: the new battle over shielding America from missile attack /
Bradley Graham.
p. cm.
Includes bibliographical references and index.
ISBN 1-58648-209-2 (PBK.)
1. Ballistic missile defenses—United States. I. Title.
UG743 .G695 2001
358.1'7182'0973—dc21
2001048152

10 9 8 7 6 5 4 3 2 1

To Lissa
And to Wynne, Cole, and Max

Contents

Cast of Characters

Madeleine Albright	Secretary of State under Clinton
Steve Andreasen	NSC director for defense policy and arms control
Robert Bell	NSC special assistant to the President for defense policy and arms control
Samuel "Sandy" Berger	National security adviser to Clinton
Joseph Biden	Democratic senator from Delaware
James Bodner	Deputy undersecretary of Defense for policy
George W. Bush	President of the United States
Stephen Cambone	Senior aide to Rumsfeld
Bill Carpenter	Raytheon executive who managed the kill vehicle program
Richard Cheney	Vice president of the United States under George W. Bush
Bill Clinton	President of the United States
Thad Cochran	Republican senator from Mississippi

William Cohen	Secretary of Defense under Clinton, 1997–2001
Philip Coyle	Pentagon director of operational testing and evaluation
Keith Englander	Technical director of the National Missile Defense program
Leon Fuerth	National security adviser to Vice President Gore
Jacques Gansler	Undersecretary of Defense for acquisition
Richard Garwin	Physicist
Al Gore	Vice president of the United States under Clinton, Democratic presidential candidate in 2000
Mort Halperin	Director of State Department's policy planning staff
John Hamre	Deputy secretary of Defense under Clinton
Jesse Helms	Republican senator from North Carolina
John Holum	Senior adviser for arms control and international security at the State Department
Igor Ivanov	Foreign minister of Russia
Sergei Ivanov	National security adviser to Putin, now minister of Defense
Lieutenant General Ronald Kadish	Director of the Ballistic Missile Defense Organization
Carl Levin	Democratic senator from Michigan
Jack Lew	Director of Office of Management and Budget
Georgi Mamedov	Deputy foreign minister of Russia

Hans Mark	Deputy undersecretary of Defense for research and engineering
Sylvia Matthews	Deputy director of Office of Management and Budget
Ron Meyer	Deputy program manager of Raytheon's kill vehicle program
Admiral Richard Mies	Commander-in-chief of U.S. Strategic Command
Frank Miller	Principal deputy assistant secretary of Defense for strategy and threat reduction
Major General Willie Nance, Jr.	Director of the National Missile Defense program
John Peller	Boeing program manager for National Missile Defense
William Perry	Secretary of Defense under Clinton, 1994–1997
John Podesta	Chief of staff to Clinton
Theodore Postol	MIT professor
Colin Powell	Secretary of State under George W. Bush
Vladimir Putin	President of Russia
General Joseph Ralston	Vice chairman of the Joint Chiefs of Staff
Donald Rumsfeld	Secretary of Defense under George W. Bush
General Henry "Hugh" Shelton	Chairman of the Joint Chiefs of Staff
Walter Slocombe	Undersecretary of Defense for policy
Robert Soule	Director of Office of Program Analysis and Evaluation

James Steinberg	Deputy national security adviser to Clinton
Strobe Talbott	Deputy secretary of State
Edward "Ted" Warner	Assistant secretary of Defense for strategy and threat reduction
Larry Welch	President of Institute for Defense Analyses
Curt Weldon	Republican congressman from Pennsylvania
Paul Wolfowitz	Deputy secretary of Defense under George W. Bush

Chronology

1944	Germany launches V-2 missile attacks against Britain.
1957	Soviet Union tests its first ICBM and launches *Sputnik* satellite.
1962	Army's Zeus missile interceptor, fired from Kwajalein Atoll, passes within two kilometers of mock warhead shot from California, or close enough to destroy it with a nuclear blast.
1963	Development of Nike-X system initiated with plan to use a combination of short-range Sprint and extended-range Spartan missiles.
1964	Soviet Union first detected developing Galosh antimissile system around Moscow.
1966	China fires its first nuclear-armed missile.
1967	Defense Secretary Robert McNamara announces plan to proceed with Sentinel system to protect U.S. cities from Chinese attack.
1969	President Richard Nixon opts for Safeguard system, relocating planned defensive sites out of metropolitan areas and centering them on offensive U.S. missile silos.

1972	Anti-Ballistic Missile Treaty signed, along with the first Strategic Arms Limitation Treaty (SALT I).
1974	ABM Treaty protocol reduces the number of permitted missile defense sites to one each in the United States and the Soviet Union.
1975	Safeguard starts operating in North Dakota but closes after several months.
1982	U.S. Army tests HOE, a prototype hit-to-kill antimissile system.
1983	President Ronald Reagan calls for research on an antimissile shield that will render nuclear weapons "impotent and obsolete."
1984	Strategic Defense Initiative Organization established.
1989	The Berlin Wall comes down.
1991	George Bush scales back notion for an antimissile system, proposing GPALS to guard the United States and allies against limited attack. U.S. and allied forces evict Iraqi troops from Kuwait in the Persian Gulf War. The first Strategic Arms Reduction Treaty (START I) is concluded in July, a month before a coup attempt against Soviet Premier Mikhail Gorbachev precipitates the breakup of the Soviet Union. Congress passes the Missile Defense Act calling for deployment of an ABM Treaty–compliant defense by 1996.
1993	Bush and Gorbachev sign START II. Bill Clinton takes office, and within four months his Defense secretary declares the "end of the Star Wars era," shifting emphasis to development of battlefield antimissile systems.
1994	The Republican Party's "Contract with America" calls for renewed commitment to national missile defense.

1995 A National Intelligence Estimate concludes that no new missile threat to the continental United States will emerge in the next fifteen years.

1996 Senator Robert Dole and House Speaker Newt Gingrich introduce a "Defend America" bill, which goes nowhere after its price tag is pegged at up to $60 billion. The Pentagon unveils its "three-plus-three" plan calling for three years of further development of a national antimissile system followed by three years of construction, but putting off a deployment decision until after the first three-year period.

1997 Clinton and Russian President Boris Yeltsin agree in Helsinki on the outline of a START III accord for reducing nuclear arsenals to between 2,000 and 2,500 warheads.

March 1998 The Welch panel warns of a "rush to failure" in missile defense programs.

April 1998 Boeing is chosen as the LSI for the National Missile Defense program. Pakistan conducts its first test of the Ghauri medium-range missile.

May 1998 Pakistan and India set off underground nuclear tests.

July 1998 The Rumsfeld commission warns that North Korea and Iran could develop missiles capable of striking U.S. territory within five years, and with little or no warning. Iran for the first time flies its Shahab 3 medium-range missile.

August 1998 North Korea fires a three-stage Taepodong I missile, attempting to put a satellite into orbit.

October 1998 Eight GOP congressional leaders write Clinton declaring the ABM Treaty null and void because of the dissolution of the Soviet Union.

December 1998 Clinton's top national security advisers recommend increased funding for national missile defense and the start of discussions with Russia about amending the ABM Treaty, along with talks on a START III accord. The Pentagon shifts the planned deployment date from 2003 to 2005. Raytheon receives a contract to develop the kill vehicle. U.S. warplanes bomb Iraq. Clinton is impeached by the House of Representatives.

January 1999 — Secretary of State Madeleine Albright in Moscow informs Russian officials of U.S. missile defense plans. Senate begins impeachment trial of Clinton; trial ends in acquittal in mid-February.

March 1999 — Senate passes a bill calling for deployment of a national missile defense "as soon as technologically possible." House passes a similar measure. Clinton declares his opposition to abrogating the ABM Treaty. U.S. warplanes begin an eleven-week bombing campaign to evict Yugoslav forces from Kosovo.

July 1999 — Clinton signs the National Missile Defense Act but asserts that any deployment decision will be based on considerations of the actual threat, the cost and operational effectiveness of the planned system, and the impact that fielding it would have on arms control.

August 1999 — Clinton approves a system architecture that would start with the deployment of one hundred interceptors in Alaska. He authorizes negotiations with Russia to amend the ABM Treaty and reach a START III agreement. A decision on whether to begin construction of the antimissile system is set for June 2000.

September 1999 — North Korea agrees to suspend further missile testing. A second Welch panel report characterizes the missile defense program as plagued still by inadequate testing, spare parts shortages, and management lapses.

October 1999 — In its first intercept attempt, the kill vehicle rams into a target missile over the central Pacific Ocean. The

Senate rejects ratification of the Comprehensive Test Ban Treaty.

November 1999
Deputy Secretary of State Strobe Talbott runs into a chorus of European opposition to the U.S. missile defense plan at a NATO meeting in Brussels.

January 2000
In its second intercept attempt, the kill vehicle misses its target, owing to an obstructed coolant line.
Vladimir Putin succeeds Yeltsin as president of Russia.

February 2000
A delegation of high-level U.S. officials travels to Beijing to discuss missile defense plans.

March 2000
A former TRW employee alleges faked tests and computer evaluations in developing software for a prototype kill vehicle no longer being considered.

April 2000
The Union of Concerned Scientists and MIT release a report challenging the effectiveness of the planned antimissile system against countermeasures.

May 2000
Other scientific groups and former Clinton administration officials urge Clinton not to go forward with the planned system. The Joint Chiefs of Staff make clear their discomfort at trying for a quick deal with Russia on deeper cuts in nuclear arms. George W. Bush gives a speech appealing for a larger missile defense program coupled with an overhaul in U.S. nuclear weapons strategy involving significant reductions, possibly taken unilaterally.

June 2000
Clinton meets Putin in Moscow, but the Russians show no interest in a deal involving national missile defense.

July 2000
A third intercept attempt ends in a miss when the kill vehicle fails to separate from its booster.

August 2000
Clinton rejects an appeal from Defense Secretary William Cohen to authorize site preparation in Alaska for construction of an X-band radar and

decides to defer a deployment decision to the next administration.

October 2000   Albright travels to North Korea amid indications that Kim Jong Il is ready to curtail missile production and exports in exchange for foreign economic assistance. Kim wants Clinton to come to North Korea to seal a deal, but U.S. officials have difficulty pinning down all the details of a possible accord.

December 2000   Bush emerges as president after a disputed vote. Clinton decides against going to North Korea.

January 2001   In confirmation hearings, Donald Rumsfeld refers to the ABM Treaty as "ancient history" and a "strait-jacket," but Colin Powell asserts there will be "a long way to go" and "a lot of conversations" with the Russians and European and Asian allies before the United States considers walking away from the treaty.

February 2001   Defense Secretary Rumsfeld in Europe underscores the Bush administration's determination to proceed with an antimissile system, describing missile defense as nothing less than a moral imperative. Russia floats its own proposal for developing a mobile, European-based, theater missile defense system, using Russian technology.

March 2001   Bush declares he has no intention of picking up quickly where Clinton left off in missile talks with North Korea. At the same time, he quietly decides to pass up an option to proceed with construction activities on Shemya Island, Alaska, in 2000, thereby avoiding an early confrontation over the ABM Treaty.

May 2001   In a speech at the National Defense University, Bush calls for moving beyond the ABM Treaty to a new strategic framework based on a cooperative relationship with Russia. He says the Pentagon will test various land-, sea-, and air-based missile defense systems and "will evaluate what works and what does not." Democratic lawmakers charge that Bush's enthusiasm for missile defense risks undermining U.S. security and busting the federal budget.

June 2001   Bush meets Putin in Slovenia, and although the Russian leader warns against unilateral U.S. action, the two men display a surprising degree of camaraderie. In the Senate, control suddenly shifts to the Democrats after James Jeffords of Vermont leaves the Republican Party to become an independent. Bush announces a willingness to resume talks with North Korea but on a broader agenda.

July 2001   In another intercept attempt—the first in a year—the kill vehicle hits a mock warhead target 144 miles above the central Pacific Ocean. Bush and Putin, meeting in Italy, announce a new set of arms control talks to address both defensive and offensive weapons.

November 2001   Bush tells Putin that the United States will unilaterally reduce its nuclear arsenal to 1,700 to 2,200 warheads. Putin reciprocates by announcing his intention to cut the Russian arsenal to between 1,500 and 2,200. But he also insists that the arms cuts be codified in a formal arms control treaty.

December 2001   Bush gives Russia formal notice on December 13 that the United States will be withdrawing from the ABM Treaty. "I have concluded the ABM Treaty hinders our government's ability to develop ways to protect our people from future terrorist or rogue state missile attacks," the president says.

January 2002   Rumsfeld issues a memo establishing new authority for the Pentagon's missile defense team, granting it an extraordinary exemption from the planning and reporting requirements normally applied to major acquisition programs. He also elevates the team to full agency rank, changing its name from the Ballistic Missile Defense Organization to the Missile Defense Agency. "The special nature of missile defense development, operations and support calls for nonstandard approaches to both acquisition and requirements generation," Rumsfeld says.

May 2002   Bush, in Moscow, joins Putin in signing the Strategic Offensive Reductions Treaty committing the United States and Russia to reducing their nuclear arsenals to between 1,700 and 2,200 warheads each by the end of 2012.

June 2002   ABM Treaty ends with U.S. withdrawal. The treaty had lasted thirty years and eighteen days. Two days after its demise, the United States breaks ground at Fort Greely, Alaska, for the construction of six interceptor missile silos and other military facilities.

December 2002   Bush announces plan to begin fielding a rudimentary system for defending the United States against missile attack, with an initial ten land-based interceptors to go in Alaska and California in 2004, and ten more at Fort Greely in 2005. Defense officials point to a series of successful flight tests—four intercepts in five attempts under Bush, five out of eight overall since 1999—as evidence the "hit-to-kill" concept is sound. But the Pentagon decides to cancel three intercept attempts initially scheduled for 2003 in order to concentrate on development of a new booster, which is months behind schedule.

Introduction

In the late 1990s, military crews in Colorado's Cheyenne Mountain, the nation's nerve center dedicated to warning of nuclear attack, stopped showing visitors simulated trajectories of a Russian or Chinese missile launch against the United States. Instead, they switched to a North Korean scenario. The change marked an important shift in U.S. focus away from possible nuclear war with Russia or China and toward the growing possibility that resourceful Third World nations—principally North Korea, Iran, Iraq, or Libya—would menace the United States with increasingly powerful rockets.

The North Korean simulation, which remains the featured attraction at the North American Aerospace Defense (NORAD) facility today, plays out on a map of the Northern Hemisphere. The map appears on a large overhead screen in the mountain's command center, a cramped room with a high ceiling and banks of computer monitors. The room is situated deep in a warren of tunnels carved out of solid granite, a surreal subterranean complex built in the 1960s that retains its cold war feel, with four-foot-thick, twenty-five-ton steel doors sealing access to the outside world, hot-line phones to Washington, and a war-room clock still displaying the time in Moscow.

As the mock attack gets under way, a red, doughnut-like symbol glows on a spot in North Korea, indicating the launch of a single missile a minute earlier. The lapse of a minute reflects the time required for infrared sensors on U.S. satellites 22,300 miles above the Earth to detect the burning rocket motors. A box to the right reports the time and place of launch and type of missile. Seconds later, a fan-like shape pops onto the screen, indicating the missile's general direction out over the Pacific toward Alaska. Then an early-warning radar in Alaska picks up the missile, and a tiny, upside-down triangle shines in the vicinity of Anchorage—the projected point of impact. The screen also shows the estimated time of impact.

Watching even a simulated missile arc its way rapidly to an American city is an unnerving experience for many visitors, whose reactions have not varied much over the years, whether the pretend perpetrator was Russia or North Korea. Inevitably, someone asks when in the tracking sequence would U.S. forces knock down the deadly intruder. Sometimes the briefer waits an expectant moment before responding, sometimes not. But always the answer is the same: "They wouldn't, because they can't." And the visitors go away in disbelief.

Ronald Reagan certainly did. He too was told during a tour of the mountain in July 1979 that while the United States could closely track a missile, the mighty U.S. military could do nothing to stop one. The powerful impression this left on the future president about America's vulnerability has become part of the lore of what compelled Reagan to undertake the Strategic Defense Initiative, popularly known as "Star Wars," and assume the role of evangelist-in-chief of the missile defense movement.

Reagan was not the first U.S. president to dream of erecting a national umbrella against missile attack. Lyndon Johnson proposed building such a defense system, and Richard Nixon actually initiated construction of an antimissile weapon in North Dakota that was completed under Gerald Ford. In the autumn of 1975, a system of missile interceptors tipped with nuclear explosives went operational and stood guard over fields of offensive missile silos. But the system was shut down after a few months for technical and cost reasons.

For a time the idea of a nationwide antimissile defense appeared to

fade with Reagan and with the end of the cold war.
develop a combination of space-based interceptors an
satellites died of its own weight. The technology was ...
even if it had been, the sheer scale of such a project would have made it
prohibitively expensive. The first President George Bush put forward a
scaled-back—though still very ambitious—notion of a global defense
aimed at blocking limited attacks or accidental launches, and he entered
discussions with a new, democratically elected Russian government
headed by Boris Yeltsin, who was excited about the prospect of some
kind of cooperative missile defense arrangement. But the Bush plan had
little time to get off the ground, either technologically or diplomatically.

Taking over in 1993, Bill Clinton abandoned the talks with Russia
on possible joint defenses and reduced work on a national antimissile
weapon to a low-priority research and development program. He concentrated instead on designing shorter-range systems to defend soldiers
against such menaces as Iraqi Scud missiles, which had been fired at
U.S. troops during the 1991 Persian Gulf War.

And that was where the national missile defense story rested
through much of the 1990s.

Now the issue is back, with a Republican president again leading the
push for antimissile protection and critics at home and abroad warning
that the United States has embarked on a foolhardy quest that threatens
to run up the national debt while falling short of expectations.
Questions about how much a nationwide missile defense would cost,
whether it can ever be made to work reliably, and what need really
exists for it have re-ignited the political and scientific debate.

Overshadowing the debate is the lingering impact of the worst terrorist strike in U.S. history. The nightmare of suicide terrorists hijacking
commercial airliners and flying them into the World Trade Center towers and the Pentagon has awakened America to a heightened sense of its
own vulnerability, galvanizing support for expanded measures to
defend the homeland. Critics of missile defense noted that no antimissile system would have stopped the attacks of September 11, 2001,
pressing their argument that low-tech terrorism poses a much graver

threat to U.S. security than long-range missiles from hostile Third World nations. But the coordination and skill that went into the unforeseen terrorist feat have given everyone pause and prompted rethinking of other scenarios previously considered unlikely. Missile defense proponents have felt fortified in their resolve to put up a system as soon as possible. After all, they argue, if terrorists could sow such destruction by turning planes of peaceful passengers into guided missiles, imagine what they or small hostile nations could do with actual ballistic missiles armed with nuclear devices or biological or chemical warfare agents.

In this context, a look back at how missile defense re-emerged over the last few years as a dominant political controversy remains as relevant as ever. The story is complex and contentious. It is a tale of disputed intelligence assessments and reactive political decisions, of hurried technical development and embarrassing misfires, of dated old world treaties and ill-defined new world orders.

Clinton himself had been deeply skeptical of both the need for a defense against long-range missiles and of the ability of defense contractors to build a dependable and affordable system. He had regarded Reagan's Star Wars initiative as a senseless endeavor that ended up wasting billions of dollars. But his thinking began to change in 1998. Startled by a North Korean missile launch that summer, and under pressure from a Republican-led Congress and his own Republican defense secretary to intensify development of a national missile defense, he agreed to boost funding for a limited system, one that would use land-based missiles to intercept enemy warheads in space. Soon after, the Pentagon adopted a compressed development schedule of about seven years which, while helped by work already done on some system components, still amounted to only about half the time normally allotted for a major new weapons program. Having thus embarked on a direct collision course with the 1972 Anti-Ballistic Missile (ABM) Treaty, which bans nationwide antimissile systems, U.S. officials started pressing a resistant Russia for changes in the accord to allow territorial protection against at least small-scale attacks. Even Congress, which had been bitterly split between Republicans and Democrats over missile defense, went on record in 1999 overwhelmingly favoring deployment of a limited shield "as soon as technologically possible," while also endorsing renewed arms reduction talks with Russia.

But Clinton's attempt at a middle way, balancing concern about a Third World missile threat against a desire to maintain the traditional arms control framework, proved unworkable. It encountered stiff foreign resistance, stinging scientific criticism, and poor technological performance. It could not transcend intrinsic tensions between the Defense and State Departments. And it was ultimately stranded by the unraveling of the fragile political consensus that had existed for a limited antimissile defense.

Most of Clinton's top national security advisers ended up feeling they had been pushed into a course of action with which they never grew comfortable. Although they certainly had their own share of diplomatic and programmatic miscalculations to blame, they also were let down on a number of fronts. The intelligence community got mired at the outset in a controversy over its ability to estimate the missile threat to the United States. Boeing, the lead contractor, fumbled early management of the program, committing an ethical lapse that compromised development of a key element of the proposed system, then neglected to put its best managers in charge or to impose adequate quality controls. The military chiefs never were enthusiastic about the program and resisted offering deeper cuts in nuclear weapons to gain Russian acceptance of national missile defense. And the Russians chose not to make a deal with Clinton in his final months in office.

In the end, Clinton put the program on hold, deferring to his successor a decision on whether to deploy a system. George W. Bush has taken a more aggressive approach, making clear his intention to proceed with deployment of a very limited antimissile system in Alaska and California, although just how the system expands and improves will depend on a broadened program of experimentation with land-, sea-, air- and space-based alternatives. The reasons put forward by the new president to justify his renewed drive echo those offered by advocates stretching back half a century. They reflect not just a concern about the threat of missile attack but also a commonsensical—and even moral—notion that a good defense is as necessary as a good offense. Also behind the push is an abiding faith in the ability of American scientists and engineers to achieve what has so far been unachievable. Still, Bush's initiative will require spending billions more dollars on prototype systems before establishing whether his faith is well placed.

America's on-again, off-again argument with itself—and with the rest of the world—over whether to build a defense against intercontinental-range ballistic missiles (ICBMs) is in a class by itself. No other proposed U.S. weapon system has fueled such sustained debate—not any jet fighter, combat ship, or armored vehicle. The United States already has sunk more than $120 billion over fifty years into pursuit of a nationwide antimissile shield, amounting to the longest-running, most expensive military research and development effort ever. The spending continues at the rate of more than $6 billion a year—more than $9 billion if work on five shorter-range antimissile systems is counted as well. The undertaking involves some of the nation's largest defense contractors and some of the Pentagon's most talented program managers.

After so much investment and so much debate over so many years, it is little wonder that most Americans—like most of those who visit Cheyenne Mountain—are surprised to discover that the Pentagon still has no way of shooting down an intercontinental ballistic missile heading toward the United States. Opinion surveys routinely show that about two-thirds of all Americans think such a missile defense system already exists.

The reasons for the contentiousness derive partly from the gravity of what is at stake—namely, survival in an age of nuclear bombs. Any move that might alter the current strategic balance necessarily carries tremendous significance. The horror of atomic attack, once a dominant national nightmare, is not remarked on much these days. But the gruesome consequences of what a single twenty-megaton bomb would do, falling on a major metropolitan area, were calculated with comprehensive precision nearly forty years ago by the Physicians for Social Responsibility. Writing in the *New England Journal of Medicine*, the group described the fireball that would form, reaching out for two miles in every direction from the point of detonation, or "ground zero." Out to four miles, the blast would produce pressures of twenty-five pounds per square inch and winds in excess of 650 mph. Even bomb shelters deep underground would be crushed. Out to a distance of sixteen miles, the heat would still be intense enough to ignite all easily flammable materials—houses, paper, cloth, leaves, gasoline—starting hundreds of thousands of fires. Fanned by blast winds still in excess of 100 mph, these fires would merge into a giant firestorm more than thirty miles across and covering 800 square miles. As for casualty estimates, out of a population of 2.8 million—roughly the size of the San Diego metro area today—about one million would die within minutes and another 500,000 would suffer serious burns from fires, stab wounds from flying debris, ruptured ear drums and collapsed lungs from tremendous pressure waves, blindness from light flashes and other major injuries.

Then, too, nuclear devices are no longer alone in their ability to wipe out entire populations. A host of chemical and biological agents also turn up nowadays on lists of weapons of mass destruction. The grave consequences of these potential Armageddons draw no argument. Rather, the battle comes over how urgent a threat they pose and how best to prevent them—whether by erecting missile defenses or relying on more traditional means of diplomacy, arms control agreements, technology controls, economic sanctions, pre-emptive strikes, and warnings of military retaliation.

During the cold war, the nuclear peace was kept not by virtue of any defense but strictly by the strength of offsetting offenses. The thinking was that as long as one nuclear superpower could threaten another with annihilation, neither would be inclined to launch an attack. This doctrine of mutual assured destruction, or MAD, argued against the construction of missile defenses, which were viewed as inherently destabilizing. If one superpower were to deploy an antimissile system, the reasoning went, the rational response would be for its adversary to build more offensive weapons in order to overwhelm the other's defenses. This in turn would lead to a spiral of ever more elaborate missile shields and ever more massive arsenals. That was the rationale behind prohibiting nationwide anitmissile systems under the U.S.-Soviet ABM Treaty.

Although to many there was something absurd in the notion of a mutual suicide pact, MAD seemed especially suited to a world dominated by two nuclear superpowers, the United States and the Soviet Union, each with leaders who, it was assumed, would behave rationally, even in times of great crisis. But with the end of the cold war, is it reasonable to expect leaders of smaller countries to behave rationally if they obtain ocean-spanning missiles? Is MAD still plausible in a world where eight nations have tested nuclear weapons and at least twenty-five countries either possess or are in the process of acquiring weapons capable of inflicting mass casualties and the ballistic missiles to deliver them? It is this concern that has prompted the Bush administration to frame a new theory of deterrence, one based on missile defenses as well as threatening offenses.

Opponents of missile defense contend that there is little to suggest that small-country despots would behave any less rationally than superpower dictators. Faced with the credible threat of devastating retaliation, no so-called rogue state would risk a first strike against the United States any more than Moscow would. So why upset the strategic balance by adopting a new approach to deterrence?

Apart from the theoretical argument over whether national antimissile defenses would be more or less stabilizing, there is the technical challenge of building such a weapon. The history of development attempts is marked by repeated underestimations of the breadth of the effort required. "If you go back to the beginning, about half a dozen things needed to be invented to make missile defense work," said Larry Welch, the retired Air Force general who has been hired by the Pentagon several times in recent years to assess the missile defense program. "I don't mean half a dozen things where significant advances had to be made, but half a dozen things that had never been done before."

Nonetheless, proponents express an unshakable belief in America's innovative abilities, particularly given steadfast political will and adequate federal financing, which so often have been lacking. After all, this is the nation that built the first atomic bomb and went to the moon. Indeed, as a difficult technological challenge, the quest for a national missile defense has frequently been likened to the 1940s Manhattan Project, which developed the atomic bomb, and the effort in the 1960s to place a man on the moon. These earlier programs eventually succeeded after some initial failures, but they also differed in important ways from the situation with missile defense. For one, compelled by the pressures of World War II and the cold war, they enjoyed widespread political and scientific support. For another, they were essentially offensive in nature. They did not confront anything like the decoys and other deceptive measures designed to thwart missile defense weapons. Going to the moon was certainly no easy task, but think about trying to do it amid concerns that the moon might shoot back.

Those who have labored for years attempting to develop a workable national missile defense insist that the technology has arrived for at least a limited system, one capable of knocking a couple dozen or so warheads out of the sky using ground- or sea-launched missiles. And in fact, there is no disputing that the computers needed to relay tracking information from satellites and ground radars to these interceptors have advanced considerably. So have the sensors required to discriminate between warheads and decoys in space. Building a national antimissile system, proponents frequently say, is no longer a scientific challenge, merely an engineering one. The real factor holding the United States back, they argue, was an outdated adherence to an ABM Treaty that placed burdensome constraints on development and created artificial distinctions between national and shorter-range battlefield systems. The Bush administration has eliminated those constraints and distinctions.

But prototype interceptors have routinely failed to achieve success under benign conditions, let alone under the stressful circumstances of combat. Over the previous two decades, only sixteen of thirty-six flight tests involving a variety of high altitude "kill vehicles" have scored hits, although to be fair, no program has been allowed a sustained run of trials.

It would be helpful if the scientific community could provide a dispassionate, neutral assessment of the practicability of missile defense.

But unfortunately, scientists have been as divided over the issue as politicians. Many technical experts remain skeptical of any quest for a leak-proof defense, insisting that no antimissile system would operate reliably enough to give American leaders confidence in it. An attacker, they say, could overwhelm or fool a defensive system by adding more warheads or obscuring them in a cloud of decoys.

Plans in the 1960s and 1970s had called for interceptors to carry nuclear weapons that would soar up and explode near incoming warheads. The kind of system that has received the most development in recent years has no explosives, nuclear or otherwise. It depends instead on an interceptor ramming squarely into its target, a concept dubbed "hit-to-kill." But this approach requires precision that is measured in centimeters and microseconds, compared to the hundreds of meters or even kilometers allowed for the old nuclear-tipped weapons.

And that is just in the final homing phase of the mission. To get to the endgame will require the integration of a number of other critical components—including land- or sea-based radars, space-based tracking satellites, and a network of computers for relaying massive amounts of information—all pushing the state of the art in their own areas. It is a coordinating task of unprecedented scale and complexity.

There also is the question of affordability. Spending on missile defense under Bush's proposed plan would account for less than 3 percent of the defense budget. This may seem like a relatively small amount to ensure against the incalculable expense of a nuclear holocaust, but the impact on the budget becomes substantial once the project moves beyond testing to procurement. The complete land-based plan outlined by the Clinton administration was priced as high as $60 billion, and Bush's notion of adding sea- and air-based elements on top of that to produce a layered defense could run the cost over $100 billion. In today's dollars, that would amount to four times the cost of the Manhattan Project, and it would approach the cost of the Apollo moon shots, the most expensive government science project ever. At a time when the military service chiefs talk of needing tens of billions of dollars more a year just to sustain conventional forces and everyday operations, not to mention modernizing for the future—another Bush priority—the prospect of financing a national missile defense has many wondering where all the money will be found.

Something else makes the top military brass, as well as many Democratic lawmakers, especially wary of pursuing a national shield: the rogue-nation missile threat ranks well below their concerns about terrorism and other forms of Third World attack. No Third World state has actually produced a long-range missile yet. Intelligence estimates have repeatedly concluded that Americans have more to fear from suitcase bombs, explosive-laden trucks, or anthrax in the subway than from large missiles, which are more costly, less reliable, and less accurate. Missiles also leave easy-to-detect home addresses. Prestige and intimidation may have more to do with why rogue states are engaged in acquiring them than any actual plan to use them against the United States.

Ultimately, the passions aroused by missile defense, and especially the enduring nature of the controversy, cannot be explained simply by differences over deterrence theory, scientific capabilities, or budgetary trade-offs. The issue has taken on a transcendent, symbolic significance. It has become a proxy between the political left and right, a kind of litmus test for how best to keep America militarily strong and secure—whether, in broad terms, the defense of the nation is better served by arms advances or arms control, by military buildups or diplomatic building blocks, by unilateral initiatives or compromise accords. Framed in this way, the argument has aroused a fervor akin to clashes over theology. There is an almost religious ferocity to the intense partisan political wrangling, and religious terms are often invoked. Proponents talk of the morality of erecting a national defense. Opponents speak of the violated sanctity of the now-defunct ABM Treaty.

Of course, the human yearning for invulnerability is as old as Greek mythology and the aegis cloak of Zeus. But the particular ambition to shield a nation against ballistic missiles has been a distinctly American story. For most of their history, Americans have felt protected from the world and its conflicts by vast oceans. In that light, missile defense can be seen as an attempt to return America to a time when it felt less vulnerable to the actions of distant dictators.

Like Ronald Reagan after his NORAD visit, George W. Bush evinces a certain disbelief at a U.S. strategy that for decades left the nation open to missile attack. And also like Reagan when he declared his intention to render nuclear weapons "impotent and obsolete," Bush and his advisers have launched their own missile defense quest on a note of hyperbole. Their planned defense is more limited than Reagan's, but their associated moves to discard cold war treaties and establish a new strategic framework have been no less ambitious. To the degree that the rest of the world still appears wary of the argument that such doctrinal change and weapons deployment are so urgently necessary, they adopt a pose of perseverance and confidence.

It is far from certain that Bush's plan will succeed where those of his predecessors have not. At home, the president must find a way to sustain the considerable expense of building and expanding an elaborate antimissile system. Abroad, he must overcome continued suspicion on the part of Russia, China and NATO allies that construction of an antimissile system reflects an underlying drive for U.S. military dominance and strategic hegemony. And technologically, he has to show that the designs can work. As daunting as this task seems, in practice it may be even harder than it looks—as the story in the following pages suggests.

I

Threat

Chapter 1

BACK TO THE FUTURE

In the autumn of 1944, the terror and destruction of German V-2 rockets, traveling faster than the speed of sound and slamming one-ton-explosive loads into British neighborhoods, marked the dawn of the missile age. At the end of the war, the Allies learned of Nazi plans to build a larger, two-stage rocket that might have been able to span the Atlantic Ocean, enabling Germany to make good on its intention of striking the United States. This revelation prompted Americans to question whether they could ever feel secure from missile attack.

Several U.S. military studies recommended the immediate development of an antimissile system, but a General Electric report in 1945 concluded that such a defense was beyond the scope of contemporary technology. General Dwight D. Eisenhower, an early missile defense skeptic, scoffed at the idea of shooting down missiles, comparing the challenge to "hitting a bullet with a bullet." Then, in 1957, the United States observed the test of a Soviet intercontinental ballistic missile. Two months later, the Soviets launched *Sputnik*, the first man-made satellite. Together, these events showed that the Soviets could build missiles with enough range to cripple U.S. bomber fleets in a surprise attack. Intelligence estimates at the time predicted that the Soviets would deploy more than five hundred such missiles by the end of 1962.

Antimissile programs then took on a new urgency. The Army seized the lead, developing the Nike-Zeus project as an expansion of its Nike surface-to-air missiles—an anti-aircraft system initiated in 1945. Entirely ground-based, the plan involved dish-type radars for detecting enemy warheads and guiding interceptor missiles to them. The interceptors, armed with atomic devices, were to get close enough to the targets to destroy them in space with nuclear explosions.

No sooner had the Army introduced its concept than others started picking the plan apart, finding technical and operational faults. Some of the concerns were unique to the proposed use of nuclear missiles to shoot at other nuclear missiles. For instance, government review groups argued that nuclear blasts from interceptors could destroy the system's own radars and warned that the Soviets might even choose to explode nuclear weapons high in the atmosphere to blind the radars. Other concerns included doubts about the system's ability to guide the interceptors close enough to destroy their targets and worries that the Soviets could easily overwhelm the system by firing many missiles or confuse it by employing decoys along with active warheads.

Service rivalries came into play as well. While the Army had based its concept on shooting down missiles in their last minutes of flight, providing a point defense of military facilities, the Air Force favored an alternative concept centered on intercepting enemy missiles shortly after launch in their boost phase. The Pentagon's Advanced Research Projects Agency (ARPA) was exploring futuristic technologies for just such an approach under a program called BAMBI (Ballistic Missile Boost Intercept). It came up with a number of concepts for defenses in space, one of which involved housing interceptor missiles in large vehicles that would be stationed in orbit over ICBM sites. Critical of the Army's approach, the Air Force urged the Joint Chiefs not to deploy Nike-Zeus because it could be easily deceived, would cost too much, and might create a false sense of security. Besides, the Air Force argued, offensive retaliation—an Air Force mission—was a better defense.

The Army stood alone in its insistence that Nike-Zeus was effective and had growth potential; the reservations and doubts of higher authorities prevented the Army from proceeding with production. Even though funding for research and development continued to flow into work on antimissile systems, the Eisenhower and Kennedy administrations withheld any decision on deployment.

Subsequent designs modified Nike-Zeus in important ways to correct some of its shortcomings. The follow-on Nike-X system, initiated in 1963, used a layered defense of two missiles to address the risk of being overwhelmed. Under the revised plan, Spartan, an extended-range Zeus missile for interceptions in space, would take the first shots, and Sprint, a short-range missile for low-altitude intercepts, would attack any warheads that had penetrated the first layer.

This system also introduced phased array radar, a new kind of radar that was meant to reduce the vulnerability of earlier radars to direct attack. In contrast to previous mechanically steered, fragile, dish-type radars, the new versions used electronically steered radars housed in structures designed to withstand nuclear blasts. In addition, these radars could scan a much wider area of the sky and handle a larger number of targets than the Nike-Zeus models.

The technical superiority of the Nike-X system strengthened the Army's case for deployment. So did developments in the Soviet Union. In 1964 the United States detected initial construction of an antimissile system around Moscow. The same year the Soviets paraded what they claimed were antimissile interceptors through Red Square during the celebration of the October Revolution.

The intensity of the Soviet antimissile effort helped the U.S. Army rally the other military services to support a U.S. deployment. In 1965 the Joint Chiefs of Staff unanimously recommended that Defense Secretary Robert McNamara request funds for some initial Nike-X components. But just as the military chiefs appeared to be moving toward embracing missile defense, McNamara and the Pentagon's civilian leadership found themselves moving away from the concept as part of a rethinking of U.S. nuclear strategy. The Pentagon's missile defense efforts thus became enmeshed in an emerging body of thought about strategic nuclear deterrence.

McNamara had initially focused strategic planning on destroying Soviet nuclear forces in the event of war, but by the mid-1960s he had come around to the idea that no attainable level of force was sufficient to strike the Soviets and preclude a devastating retaliatory blow, particularly since the Soviets kept building more weapons. His central concern became finding a way to deter the Soviets from nuclear war. To figure out how much force was enough—and impose some fiscal constraint on service requirements—McNamara adopted a new standard for procurement, based on what he called the capability of "assured destruction." He defined this as the capability to destroy a certain percentage of the enemy's population and industrial capacity. This shift in strategic doctrine drew criticism from conservatives who viewed it as capitulation to the Soviets. But a growing and increasingly vocal group of private experts—mostly scientists and former government officials—also was contending that mutual deterrence could be maintained if each side developed a secure, second-strike force and simply left its population vulnerable to annihilation. Missile defense had no place in such a strategy because, so the reasoning went, it would spark a new arms race as each superpower sought to compensate for the other's defense.

McNamara thus joined the argument against missile defense. But President Lyndon Johnson was being lobbied to support an antimissile system, not only by the Joint Chiefs but also by the Senate and House Armed Services Committees. Johnson was concerned as well that the Republicans would pound him about an "ABM gap" in the 1968 elections. In December 1966, McNamara offered Johnson a compromise: seek funds for long-lead items on missile defense but delay a deployment decision while querying the Soviets about negotiations to limit such systems.

Johnson spent the first part of 1967 playing for time, hoping to work out a deal with Moscow that might make missile defense unnecessary. But the Soviets were not interested. When Johnson met Soviet Premier Aleksei Kosygin in Glassboro, New Jersey, in June, McNamara took the opportunity to argue that missile defense threatened strategic stability and must therefore be limited. Kosygin disagreed with the notion that missile defense was destabilizing and said to McNamara, "When I have trouble sleeping nights, it's because of your offensive missiles, not your defensive missiles." The Soviets would not agree to further discussions. With a three-to-one disadvantage in strategic offensive arms, the Soviets had little incentive to bargain over missile defense.

At the same time, the Chinese were beginning to loom as a new threat, having fired their first nuclear-armed missile in 1966. A week before the Glassboro summit, they surprised both the United States and the Soviet Union by announcing the detonation of a hydrogen bomb. If the notion of defending against massive Soviet attack still seemed too much of a reach for U.S. technology, the prospect of a limited defense against the small number of Chinese missiles held some promise.

This is the direction McNamara ultimately took. In September 1967, in a speech in San Francisco, the Defense secretary moved the United States for the first time toward deploying a national missile defense. But it was a heavily hedged and ambivalent step. In the speech, McNamara began by actually making an impassioned case against missile defense, emphasizing that attempts to defend against a large-scale Soviet strike would just fuel the arms race. At the end, however, he announced the decision to proceed with a "thin" system called Sentinel to protect U.S. cities not from Soviet attack but from a much smaller Chinese threat.

Sentinel, an outgrowth of the Nike-X program, envisioned a layered defense using the Spartan and Sprint missiles, ground-based radars, and a multiple site command-and-control system. The plan called for deploying seven hundred interceptors to defend a handful of U.S. population centers around the country.

The Sentinel decision represented a political compromise—an attempt to balance conflicting strategic, technical, and diplomatic considerations. With China beginning to test nuclear devices and missiles, the threat was real and clearly a major motivating factor to do something. Experts were convinced that an antimissile weapon could be built to defend against a limited and relatively unsophisticated attack. Congress went along initially with the Pentagon's technical judgments.

But the anti-Chinese rationale was less a coherent strategic approach than an attempt to appease the pro–missile defense forces while minimizing any provocation to the Soviet Union. If McNamara could not prevent missile defense outright, he could at least keep it limited. Johnson too saw in the Sentinel plan a way of mollifying critics with something while still trying to entice the Soviets into an arms control deal.

By the time action had to be taken to implement the Sentinel deployment, public opposition to missile defense was becoming a factor for the first time. Critics were coalescing into an organized movement of academics, scientists, and former government officials, publishing articles in science and foreign policy journals and pressing their arguments at arms control conferences, in the corridors of power, and in the halls of universities and laboratories. Antimissile systems were portrayed as more complex, less reliable, and considerably more expensive than the missiles they were designed to defeat. Even a limited system, it was noted, would have to provide a nearly perfect defense, since penetration by just one warhead would be a disaster. Enhancing America's offensive capabilities, it was argued, would be cheaper than erecting a defense. In a 1962 article in *Scientific American*, Herbert York, a former Pentagon director of research and engineering, and Jerome Wiesner, President John Kennedy's science adviser, argued that developing defenses would merely spur the Soviets to a new cycle of weapons building and thus intensify the arms race.

The role of scientists is particularly noteworthy. As early as 1964, the Federation of American Scientists (FAS), a nationwide organization of about twenty-five hundred scientists and engineers concerned about the impact of science on national and international affairs, opposed any missile defense deployment. A 1968 article in *Scientific American* by Hans Bethe, a Nobel laureate professor of physics at Cornell University and member of the President's Scientific Advisory Committee, and Richard Garwin, a research scientist at IBM, outlined in public for the first time the technical vulnerabilities of ballistic missile defenses. Their article cited concerns about high-altitude detonations blinding radars on the ground and the prospect of decoys or multiple warheads overwhelming the system.

Members of Congress began to seek the advice of the scientific community, and by the spring of 1969 scientists opposed to missile defense were testifying before congressional committees. This was a new phenomenon; previously, only administration witnesses had testified on defense matters. It was during this period that the core arguments against missile defense solidified and began to take root throughout the military establishment and on Capitol Hill. But the scientific community was itself split. A number of respected experts also made the case for proceeding with a limited antimissile system—among them, Freeman Dyson of the Institute for Advanced Study and Alvin Weinberg of the Oak Ridge National Laboratory. They argued that a strong defense could undercut the value of ICBMs and end the arms race. Even in the absence of a 100 percent effective defense, Dyson wrote in the *Bulletin of the Atomic Scientists*, some benefit would come from saving most of a population. Another prominent supporter was Albert Wohlstetter, a researcher at the RAND Corporation and the Stanford Research Institute, who insisted that defensive systems were necessary to ensure that enough offensive missiles would survive a Soviet first strike to retaliate. He accused opponents of distorting operations research and data; they responded with accusations of contradictory statements, changing rationales, and selective use of intelligence information by members of the administration.

The technical debate left the impression that for every expert declaring that missile defense would not work, another was ready to argue that it would. What finally aroused the general public, though, was the

Army's move in the final year of the Johnson administration to start buying land for the missile defense sites. Opponents warned that cities near defensive missile sites would become "megaton magnets" for the Soviet Union. They also fanned fears by saying that the nuclear warheads of the Spartan and Sprint interceptors might detonate at low altitude during an attack, or accidentally in peacetime, thereby destroying the very cities they were intended to protect. In the face of such heightened public concern, congressional backers began to rethink their commitment. Compounding matters, the Sentinel controversy was occurring against the backdrop of growing opposition to the Vietnam War, which cast clouds of general suspicion across all military programs.

Soon after taking office in 1969, President Richard Nixon decided to deploy Sentinel equipment in a new configuration, relocating defensive sites out of metropolitan areas and basing them around offensive U.S. strategic missile silos. He called the reoriented system Safeguard. With this shift from population defense to silo defense, the Nixon administration hoped to dampen public opposition. At the time, U.S. officials were also increasingly concerned about the vulnerability of U.S. missiles to attack as a result of moves by the Soviet Union to put multiple, independently targetable warheads atop its huge SS-9 missiles.

Still, the Safeguard system proved as controversial as its predecessor, and the debate churned on. In August 1969, Congress narrowly approved funding to begin production of Safeguard, with Vice President Spiro Agnew breaking a fifty-fifty tie vote in the Senate. Over the next two years, the program retained its precarious grip on survival on the strength of its perceived value as a bargaining chip in the talks with the Soviet Union on limiting offensive nuclear weapons that began in November 1969. Already by the early 1970s, contemporary chroniclers were referring to national missile defense as "the most costly, complex and controversial weapon system ever developed by the United States."

In 1972, through negotiations known as the Strategic Arms Limitation Talks (SALT), the United States and Soviet Union agreed to a five-year freeze on strategic launchers and concluded the ABM Treaty restricting each superpower to two antimissile system complexes. In 1974 a protocol was added reducing the number of permitted sites to one each and a maximum of one hundred interceptors. Consistent with their respective antimissile programs at the time, the United States chose to protect a missile base near Grand Forks, North Dakota, while the Soviets retained the Galosh system they had built around Moscow in the 1960s.

The provisions of the ABM Treaty were tailored to prevent either country from deploying a full territorial defense or laying the groundwork for such a defense. To that end, the treaty placed tight restrictions on the development of new types of antimissile weapons, forbidding the testing or deployment of antimissile systems—or even components— that were mobile and land-based, or based in space, at sea, or in the air. The treaty marked a conceptual turning point in the nuclear relationship between the two superpowers. It signified an acknowledgment of deterrence based on mutual vulnerability. By entering into the treaty, the Americans and Soviets seemed to agree that the best way to avoid a massive nuclear attack by the other side was to remain defenseless against one. Understandably, missile defense enthusiasts found such reasoning absurd. Donald Brennan, a Hudson Institute analyst who had been working on missile defense issues, bitterly attacked the notion. He took McNamara's term "assured destruction" and the phrase "mutual deterrence" and combined them into what he called "the concept of mutual assured destruction," thus coining the enduring acronym MAD, which he said appropriately described the official U.S. posture.

It did not take long after the ABM Treaty process reduced the number of allowable sites down to one for even the single U.S. Safeguard facility in North Dakota to start looking expendable. From its inception, Safeguard had faced the same technical criticisms as Sentinel— chiefly that the system was extremely vulnerable to countermeasures and a determined Soviet first strike. Such limitations might have been acceptable in the short term while Safeguard served as a bargaining chip to persuade the Soviets to accept reductions in strategic forces. But after the ABM Treaty and the SALT I accord, there was less justification for keeping Safeguard at all.

Shortly after Safeguard started operating in October 1975, Congress canceled funding for the system, citing its expense and likely ineffectiveness. Operations were halted in February 1976. From start to finish, the program absorbed $5.5 billion, excluding the cost of developing and building the nuclear warheads. By the end, nuclear-tipped interceptors had lost favor as the way to defend against missile attack because of their technical and political liabilities. For one thing, nuclear explosions interfered with the operation of the radar systems that were supposed to control the battle between defending missiles and incoming warheads. For another, the prospect of nuclear blasts even high overhead unnerved populations on the ground.

So the Army shifted its research and development to an alternative approach that avoided explosive devices and relied instead on the kinetic energy of a direct collision to obliterate a target. Such an approach would require significant advances in two main areas. One was optical sensors to overcome the problems that radars had with distinguishing among decoys, boosters, warheads, and debris. The other area was parallel processing by computers at speeds fast enough to interpret the sensor data, incorporate it with radar tracking information, and compute targeting instructions for an interceptor.

By combining the improved capabilities of infrared sensors with small, high-capacity computers, the Army produced interceptors that worked on the principle of kinetic kill. Dubbed "hit-to-kill" vehicles, they represented the first major revolution in ballistic missile defense since the United States began research in the 1940s. This technology was ready for demonstration in 1982, when the Army began what it called its Homing Overlay Experiment, or HOE. In these tests, an experimental vehicle was launched from the Kwajalein missile range in the Marshall Islands using a modified Minuteman rocket. Once in space, the vehicle separated from its booster and homed in on a target missile that had been fired from an Air Force base in California. HOE succeeded in scoring a hit after three failures, but the credibility of the test was called into question years later when investigators at the Congressional General Accounting Office reported that the chances of intercepting the target warhead had been increased by heating it before launch and instructing it to fly sideways, thereby exposing a greater surface area to the interceptor's sensors. In any case, HOE was far too heavy and expensive for operational purposes.

Major advances in the development of lasers also occurred in the 1970s as the Pentagon explored ways of using this technology to shoot down aircraft or missiles. By the early 1980s, these efforts had focused on high-energy lasers based in space in order to overcome the scattering and spread of laser beams caused by the atmosphere. The construction of large mirrors posed a challenge for the evolution of laser systems, as did pointing and tracking with high precision. But of all the technical advances during this period, the promise of directed-energy weapons contributed most to generating renewed interest in deploying an antimissile system.

At the end of the 1970s, a precipitous change in U.S.-Soviet relations resulting from the Soviet invasion of Afghanistan and the failure of the United States to ratify a SALT II agreement set the conditions for reigniting the missile defense debate. The election of Ronald Reagan provided the spark. But it did not come immediately. While Reagan wasted little time launching the largest peacetime military buildup in U.S. history, the strategic modernization program that he presented in October 1981 contained no provision for an antimissile system. A Defense Science Board panel had reviewed the status of various missile defense technologies earlier in the year and concluded that none was on the horizon.

Two years would pass before Reagan, who had developed an interest in missile defense in the 1970s, would return to the idea and give it fresh impetus. In the meantime, several prominent missile defense advocates—among them, Senator Malcolm Wallop, a conservative Wyoming Republican; retired Lieutenant General Daniel Graham, a former Defense Intelligence Agency chief; and Edward Teller, a prominent physicist who had fathered the hydrogen bomb and was pursuing a nuclear-powered X-ray laser—pushed their own strategic defense concepts with Congress and the Pentagon as well as with the White House. To help lobby Reagan and other senior administration officials, Graham and Karl Bendetson, a business executive and former undersecretary of the Army, enlisted several of the president's longtime friends and supporters to form a group called High Frontier that promoted the idea of a space-based defense using kinetic energy weapons.

By early 1983, the military chiefs were themselves open to taking another look at missile defense. They had become frustrated over their failure to devise a basing scheme for the MX missile that would reduce its vulnerability to Soviet attack. When the Navy's chief of staff, Admiral James Watkins, suggested developing an antimissile system, the other service chiefs agreed to present it as one of a number of options to Reagan at a meeting in early March. At the time, Reagan was searching for some new military initiative, something dramatic, that might help break the momentum of a swelling nuclear freeze movement. The president's hard-line, anti-Soviet policies had by then given rise to the largest antinuclear movement in cold war history.

Working almost in secret with a few aides, Reagan drafted an insert to a speech on March 23, 1983, that turned what was to have been a routine defense budget pitch into a historic appeal. Reagan called on the scientific community to apply its great talents to the cause of world peace and to give the United States the means of rendering nuclear weapons "impotent and obsolete." While acknowledging the likelihood that "it will take years, probably decades of effort on many fronts" to build an effective defense, he said he was ordering the start of "a comprehensive and intensive" research and development program to that end. The initiative caught many in the president's own defense bureaucracy by surprise and upset some leading members of his administration, including Secretary of State George Shultz. But it set in motion the largest military research program ever undertaken by the United States and came to dominate discourse and deliberations about defense policy through much of the decade that followed. A panel under the direction of former NASA chief James Fletcher mapped out a technical concept involving five defensive layers. A second panel—chaired by Fred Hoffman of the California think tank Panheuristics—was directed to study the political and strategic ramifications of the policy. Both groups reported to the White House in the autumn of 1983 with relatively optimistic assessments. In early 1984, Reagan issued a presidential directive and a proposed budget to begin what officially was called the Strategic Defense Initiative (SDI)—and what opponents pejoratively labeled "Star Wars," a name taken from the hit science fiction movie of the time.

A large-scale effort spanning the military services, national laboratories, and private industry then burgeoned. Defense Secretary Caspar Weinberger chartered the Strategic Defense Initiative Organization (SDIO) in April 1984 to bypass the Pentagon's bureaucracy and manage the program. Reagan's initiative had prompted a storm of discussion. One count found about one thousand books, monographs, articles, and government publications written about the issue in the first three years after Reagan's speech. Support was limited at the outset, but since 1946 opinion polls had shown that the vast majority of Americans believed in the ability of scientists to develop a missile defense if they put their minds to it.

Proponents of missile defense spent much of the period decrying an expanding Soviet military threat. Moscow's land-based missiles had exceeded U.S. ones in size and explosive yield, they said, and a new Typhoon class of nuclear-powered submarines roamed the seas, with missiles bearing multiple warheads. By the time of Reagan's landslide re-election in November 1984, all of his top officials had lined up behind the Star Wars concept, although for different reasons. Shultz and other moderates in the administration hoped to use SDI as a bargaining chip and compel the Soviets to accept reductions in offensive weapons. Weinberger, his aide Richard Perle, and their fellow hardliners saw SDI as a way to dispose of the ABM Treaty, move away from arms control accords, and defend U.S. missile silos.

All these rationales, though, fell considerably short of Reagan's initial vision of an antimissile system that would render nuclear weapons obsolete. Indeed, critics insisted that the technology just did not exist to ensure a leak-proof defense of the entire U.S. population against a massive attack. A study by the American Physical Society cast doubt on the near-term prospects of laser weapons. Another study by the congressional Office of Technology Assessment, which had full access to classified information, explored the range of technologies that SDIO was considering and found all of them inadequate to serve as a basis for a major change in American strategy. Making many of the same arguments that were leveled against Safeguard in the 1970s, critics in the arms control community complained that SDI would destabilize relations with the Soviets and lead to a sharp increase in the Soviet missile force. They also declared SDI was unaffordable, estimating the cost at hundreds of billions of dollars. The Union of Concerned Scientists organized a boycott of strategic defense research.

For the first several years, the Pentagon's SDIO focused its efforts on space-based lasers and particle beams before concluding, as many critics had asserted, that such technologies were far from mature. As an alternative, missile defense planners put more emphasis on kinetic energy weapons, including space-based interceptors that would be fired from orbiting platforms and ground-based missiles that would launch "kill vehicles" to destroy warheads either in space or high in the atmosphere.

Even so, an influential study in 1986 by the Defense Appropriations subcommittee staff found that the ability to discriminate between warheads and decoys was still beyond the scope of existing technology and the computerized requirement for battle management could not be met. At the same time, SDI budgets were rising, with the administration requesting $5.4 billion for fiscal 1987, a jump of 77 percent over the year before. Even some moderate Republicans had begun to worry about the price tag. In an emotional split with the administration, William Cohen, then a Republican senator from Maine and a member of the Armed Services Committee, backed a Democratic move in 1986 to cut requested SDI funding by $1.5 billion and shift the savings into research on conventional arms. The measure, known as the Balanced Technology Initiative, reflected heightened congressional sentiment that SDI was absorbing funds far out of proportion to its relative contribution to U.S. security.

Worried that missile defense would not survive the Reagan presidency if it remained a research program, SDIO outlined an initial Phase One architecture in mid–1987 and proponents pushed for an early deployment decision. The plan called for hundreds of space-based battle stations and hundreds or thousands of ground-based interceptors, with deployment projected to start in 1994. The system, officials said, would protect a limited number of military installations, not cities. But the plan was still very sketchy, and internal Pentagon auditors pegged its cost at $146 billion. While SDIO whittled the price to $69 billion in a modified Phase One plan a year later, the frequent scuttling of prior plans and bold announcements of new directions was taking its toll on the program. As Reagan's tenure drew to a close, his missile defense initiative, which had stirred such a broad national debate, continued to suffer a seeming lack of coherence and viability. At the same time, the administration's profound skepticism about the value of negotiating with the Soviet Union had given way in 1987 to agreement on a treaty banning U.S. and Soviet intermediate-range missiles in Europe. And while Reagan's resistance to limits on missile defense still posed an obstruction to a bigger deal on reducing strategic weapons—the year before in Rejkjavik, Iceland, it had blocked the most comprehensive and radical proposal ever considered to cut U.S. and Soviet nuclear arsenals—the climate in U.S.-Soviet relations was clearly warming. Reagan's elaborate vision of an all-encompassing antimissile system was receding into history.

At the start of the administration of President George Bush in 1989, a new concept emerged from Lawrence Livermore Laboratory that promised a more affordable national missile defense: a swarm of small, space-based interceptors, each with its own sensors and flight control systems that could steer it into a collision with an enemy warhead. Said to be relatively inexpensive to manufacture and launch, these tiny weapons, named Brilliant Pebbles, were seen as a means of achieving efficiency in missile defense. Yet another Phase One plan put forward in 1990 showed a system made up of 4,600 pebbles plus 1,000 to 2,000 ground-based interceptors.

With the Berlin Wall coming down and the prospect of an imminent U.S.-Soviet arms reduction accord, new rationales began to emerge for an antimissile system that focused more on guarding against accidental launches or rogue nation attacks. In his 1991 State of the Union address, Bush dropped any pretense of trying to blunt a Soviet first strike involving thousands of warheads and proposed a still-slimmer system aimed at protecting the United States—and to some extent its allies—against an accidental launch by Moscow of up to 200 warheads. This approach, called Global Protection Against Limited Strikes, or GPALS, envisioned about 1,000 Brilliant Pebbles and 500 to 1,000 ground- or sea-based interceptors with non-nuclear, hit-to-kill technology. The Pentagon estimated the cost at about $40 billion.

But the Bush program, while a substantially scaled-down version of Reagan's original notion, still amounted to an elaborate array of largely untested components. Arms control advocates objected most to the Brilliant Pebbles component, arguing that such weapons in space constituted a direct assault on the ABM Treaty. Conversely, strong supporters of missile defense saw Bush's plan as a weak compromise that would only delay fielding of a robust capability. It was thus not surprising that GPALS immediately ran into trouble on Capitol Hill as Democrats pressed to reorient the program toward an emphasis on ground-based systems.

In May, two Republican senators—William Cohen of Maine and John Warner of Virginia—issued a white paper that they called "a basis for consensus." It proposed deferring a decision on space-based interceptors "for the time being" while moving ahead with a system of 700 to 1,200 ground-based interceptors at five to seven sites "in the interest of providing the nation as soon as possible with at least some protection against ballistic missiles." The paper also urged the Bush administration to begin talks with Moscow on amending the ABM Treaty to allow for such deployments. Senator Sam Nunn, the Georgia Democrat who was his party's leading voice on defense, moved quickly to ally with the two Republican senators and form a center-right consensus around a draft missile defense bill. Three years earlier, Nunn had introduced his own architecture, a much reduced version of Reagan's Phase One that also had focused on defending against a mistakenly fired nuclear missile or an unauthorized launch. He had called it ALPS, for Accidental Launch Protection System, and now was eager to incorporate its elements in the new legislation. To gain the support of Republicans, the senators agreed to highlight Brilliant Pebbles as a potential follow-on deployment. To meet the concerns of Democrats, the bill left ambiguous the prospect of missile defense deployments beyond the initial treaty-compliant system.

While some Democratic opposition remained, critics stood little chance of winning. The United States had just fought the Persian Gulf War, and scenes of Iraqi Scud missile attacks against U.S. forces in Saudi Arabia were still fresh in the minds of lawmakers. The missile attacks had brought home the deadly reality of missile proliferation, and the perceived success of the Patriot system in countering the Scuds had created an impression of effective defense. The Patriot, originally designed as an air-defense system, had been rushed into service during the war to defend against the short-range Scuds by exploding interceptors in their path. More than a year would pass before the claims of near-perfect intercept rates advanced by the Pentagon and the Patriot's manufacturer, Raytheon, were disproven. A 1992 General Accounting Office review eventually concluded that, at best, the Patriots hit only four of the forty-five Scud warheads engaged as a result of the weapon's limitations and the inherent tendency of the Scuds to break up in flight. But in the months immediately after the war, when Congress was still debating the Missile Defense Act, arguments questioning the workability of antimissile systems held less sway.

Revelations that emerged after the Gulf War about Iraq's pursuit of a nuclear program also served to underscore the limitations of international nonproliferation efforts in ensuring U.S. security. And beyond Third World threats, there was growing concern about Moscow's control over its own strategic forces. Adding a sense of urgency, a coup attempt against Soviet President Mikhail Gorbachev in mid-August 1991 confronted U.S. authorities with the new reality that the Soviet empire was collapsing, which might leave the huge Soviet nuclear arsenal spread among smaller independent powers and pose new problems for preventing the sale or theft—or unauthorized launch—of nuclear devices. Representative Les Aspin, the Wisconsin Democrat who chaired the House Armed Services Committee, issued a white paper entitled "A New Kind of Threat: Nuclear Weapons in an Uncertain Soviet Union" and called for a new approach to superpower relations, although he remained more wary than his Senate colleagues about setting a firm date for deployment of even a treaty-compliant national antimissile system and pushing for treaty amendments to permit an expanded system.

Just the previous month, Bush and Gorbachev had concluded the Strategic Arms Reduction Treaty (START), agreeing to slash nuclear arsenals to 6,000 warheads in the first strategic weapons accord in more than a decade. In September, Bush announced that the United States would take several unilateral actions: eliminate all ground-launched tactical nuclear weapons; remove tactical nuclear weapons from ships and submarines; take all Minuteman II missiles off alert; terminate development of mobile systems for the MX and Midgetman ICBMs; and cancel the short-range SRAM-II missile program. He asked the Soviets to reciprocate with similar moves. In the area of missile defense, he said proposals would be forthcoming to develop shared early-warning capabilities.

At first it appeared that Bush's initiative would undercut his own missile defense efforts on Capitol Hill. It started a Democratic stampede for more defense cuts and fortified a House effort to gut Brilliant Pebbles. But then in October, Gorbachev provided new impetus to the U.S. missile defense plan. Urging a further 50 percent reduction in strategic weapons and declaring a one-year moratorium on nuclear testing, the Soviet leader also proposed that the superpowers "study the possibility of creating joint systems to avert nuclear missile attacks with ground- and space-based elements." The invitation for joint action represented a major break with Moscow's past policy of seeking to terminate SDI, even if Soviet authorities, like American ones, remained divided over just how to proceed. In any case, Congress voted overwhelmingly in November to pass the Missile Defense Act of 1991, and Bush signed it into law. The measure called for deployment of a treaty-compliant defense by "the earliest possible date allowed by the availability of appropriate technology or by fiscal year 1996." It also urged the president to "pursue immediate discussions with the Soviet Union" to amend the treaty to allow more ground-based interceptor sites as well as space-based sensors. It referred to the 1996 deployment as just the "initial step toward deployment" of a "highly-effective defense of the United States against limited attacks of ballistic missiles." Although the legislation cut funding for Brilliant Pebbles to just about 60 percent of what the White House had requested, it increased overall SDI funding by more than a third over the previous year. What was more significant, the commitment to deployment had finally been made. A new, if fragile, consensus had formed again for the first time since the late 1960s. Political support for a limited missile defense was enjoying its highest level in twenty years.

Like the Sentinel decision, the Missile Defense Act represented an attempt at compromise on an inherently polarizing issue. With the breakup of the Soviet Union, scenarios involving accidental or unauthorized launches had become a new focus of concern and the main justification for a missile defense system. But many lawmakers went along with the idea only on the understanding the ABM Treaty would somehow hold. Then, in January 1992, momentum toward some kind of joint U.S.-Russian antimissile effort appeared to pick up speed following Gorbachev's replacement by Boris Yeltsin. In a speech to the United Nations Security Council, Yeltsin proposed deploying and jointly operating "a global system of protection of the world community, based on a revised American SDI and advanced technologies developed by the Russian military-industrial complex." Yeltsin discussed his proposal with Bush the following day, and the two leaders finalized their strategic arms and defense proposals at a June 17 summit in Washington. On the offensive side, they signed a document outlining provisions of what would become START II. This committed the two countries to eliminate all multiple-warhead, intercontinental-range ballistic missiles and reduce their strategic arsenals to between 3,000 and 3,500 warheads by 2003. On defenses, they issued a joint statement committing their countries to work together with allies and other interested states to develop the concept for a Global Protection System. To this end, they set up a high-level group to explore the establishment of an early-warning center and other information-sharing arrangements as well as possible technical cooperation. Talks began between U.S. special envoy Dennis Ross and veteran Russian diplomat Georgi Mamedov.

Even the technology appeared to be maturing. In January 1991, just as the Gulf War was starting, an experimental ground-based interceptor built by Lockheed hit a mock warhead that was surrounded in space by balloon decoys. The Exoatmospheric Reentry-Vehicle Interceptor Subsystem, or ERIS, had its roots in the earlier HOE program. But ERIS reflected substantial advances over previous interceptors. Its kill vehicle, for instance, weighed only 73 pounds—a huge reduction from HOE's 544-pound version—and it employed an infrared, reimaging telescope and scene-matching technology to home in on the mock warhead.

Washington's interest in missile defense, however, was not to last. As with the Sentinel decision in the late 1960s, it soon became clear that the Missile Defense Act of 1991 had represented only a temporary truce between the warring groups on missile defense. Within a year, memories of the Gulf War had faded and an air of scandal had enveloped the Pentagon's missile defense effort. General Accounting Office investigators reported that GPALS still had no stable architecture, Brilliant Pebbles was still unproven, and "tremendous technical challenges" lay ahead. Cost estimates for the ground-based portion of the GPALS plan had doubled, and a second ERIS test, conducted in March 1992, failed because of technical errors. Further, the GAO found that the Army had misrepresented the initial ERIS test the year before, claiming that the missile had met its key goal of discriminating between a warhead and decoys when it had not.

With the cold war behind them and the U.S. economy in the throes of a recession, Americans turned their attention to domestic issues and elected Bill Clinton.

Taking office in 1993, the Clinton administration conducted a bottom-up review of the defense budget, then promptly canceled Pentagon plans to issue contracts for development of a national missile defense system. On May 13, Defense Secretary Les Aspin, in declaring the "end of the Star Wars era," described a revamped approach that he said would shift away from a "crash program" to deploy "space-based weapons." The new plan would emphasize development of theater systems to defend U.S. forces against shorter-range battlefield missiles, while national missile defense would be relegated to a back-burner research and development effort. While the Bush administration had put only 20 percent of SDIO's budget into theater defenses, Aspin reversed the proportion, directing 20 percent into national missile defense. The Army and Navy were both working on two types of systems: a point, or "lower-tier," defense to intercept short-range missiles, and an "upper-tier" defense to combat missiles of greater ranges higher in the atmosphere. And the Air Force was developing an aircraft-mounted laser designed to intercept enemy missiles shortly after launch in their boost phase. Emblematic of the new focus on these programs, Aspin changed the name of the managing Pentagon agency from the Strategic Defense Initiative Organization to the Ballistic Missile Defense Organization (BMDO).

A month earlier, the Russians had put forward a new proposal, hoping to further the cooperative effort begun under Bush. They were interested in establishing an early-warning center that would make use of existing U.S. and Russian satellites and a verification regime for the testing and use of missile defense systems. Just how they expected to reconcile some of these measures with the ABM Treaty was unclear. Some Russian statements suggested that pursuit of a mutual global defense system would not violate the treaty. In any case, the Clinton administration was not interested. Having slashed spending on national missile defense, it did little to pursue the prospect of U.S.-Russian cooperation in this area and dropped the Ross-Mamedov diplomatic initiative that had begun in the final months of the Bush administration.

The new U.S. administration did, however, begin what turned into a four-year negotiation with the Russians over clarifying the demarcation line under the ABM Treaty between theater and strategic defenses. American officials were interested in establishing the line high enough so that the upper-tier Army and Navy systems could be tested and deployed.

In the meantime, national missile defense activities took on the characteristics of a technology readiness program like that pursued in the late 1970s—that is, experimental work on sensors, radars, computer processing algorithms, and other aspects critical to any future system. But absent a political commitment to develop or deploy a system, the research lacked direction and had less appeal for the best and brightest engineers and executives in the aerospace business. The effort languished.

Its comeback began not with a military event but a political one. The Republican sweep of both houses of Congress in 1994 set the stage for a new battle over missile defense. Republicans had not controlled the House of Representatives in forty years, and Democrats had maintained comfortable majorities in both chambers since 1986. The Republican victory, which had been led by the right, elevated Jesse Helms of North Carolina to chair the Senate Foreign Relations Committee and Strom Thurmond of South Carolina to head the Senate Armed Services Committee. In the House, Speaker Newt Gingrich of Georgia turned his preelection manifesto, the "Contract with America," into his party's agenda. One of its planks called for "renewing America's commitment to an effective national missile defense by requiring the Defense Department to deploy antiballistic missile systems capable of defending the United States against ballistic missile attack."

Conservative activists had lobbied hard for this revived emphasis on missile defense and sought to make the most of it. The Heritage Foundation assembled a team of former elected representatives, government officials, and military officers who produced a study in mid-1995 saying that a sea-based national antimissile system could be built for an additional expenditure of about $7 billion above the $19 billion budgeted for missile defense through 2001. Another forceful proponent, Frank Gaffney, who had held a mid-level defense policy job in the Pentagon under Ronald Reagan and now ran the Center for Security Policy, conducted polls indicating that most Americans, after having heard about the Star Wars initiative for much of the 1980s, assumed the United States already had a defense against missile attack. Many in these surveys grew deeply concerned when told that the government had left them vulnerable. Tirelessly pushing this fact, Gaffney sought to persuade Republican leaders that missile defense could be a winning issue in the 1996 presidential election against Bill Clinton.

The Republicans bought this premise. They saw in missile defense an issue they could use to hammer on the Democrats and make them look soft on defense. Drawing strong favor among conservatives, missile defense presented a natural wedge issue between Republicans and Democrats. Once it was included in the Contract with America, Republicans quickly embraced deployment of a system as a way of reasserting their traditional appeal for a strong military.

Wasting little time, congressional Republicans sought in the spring of 1995 to write provisions into the 1996 defense authorization bill that would order deployment of a ground-based, multi-site system by 2003. The Senate version reached the floor first, where it promptly ran into a Democratic filibuster. Facing the prospect that the whole authorization bill would remain stalled over the missile defense issue, four senators worked through the summer crafting a compromise.

The four—Republicans John Warner and Bill Cohen and Democrats Sam Nunn and Carl Levin—formed the moderate core of the Armed Services Committee. They were longtime friends and respectful of one another. The compromise they worked out would have committed the United States only to "develop," not "deploy," a national antimissile system. The deal enabled the bill to pass the Senate, but it was rejected in conference negotiations by House representatives, who did not want any part of a bipartisan agreement with the administration. The conference restored the original language, which subsequently drew a presidential veto. Ultimately, in order to get a defense authorization bill passed, lawmakers stripped out the language mandating deployment and settled for doubling the administration's budget request for national missile defense from $371 million to $745 million.

The experience proved bruising. Even with a majority in both the Senate and House, Republican leaders could not bulldoze through new legislation making missile defense the law of the land. The Democratic point men on the issue in the Senate—Nunn and Levin—had demonstrated an ability to force compromise. As a practical matter, then, the Republicans decided to keep national missile defense out of future military authorization bills and handle it in separate legislation. That way, authorization bills could not be held hostage by Democratic filibusters.

In March 1996, Senate Majority Leader Robert Dole, who had emerged as Clinton's Republican challenger in the presidential race, introduced a "Defend America" bill, separate from the defense authorization bill and focused entirely on deploying a national antimissile system. Gingrich sponsored an identical bill in the House. The lawmakers were bent on provoking a confrontation during the presidential campaign, counting on defense of the homeland to carry an emotional appeal.

Conservative columnists, talk-show hosts, and editors rallied behind the measure. The *Wall Street Journal*, for instance, promoted the legislation on June 20 in a special editorial section, which featured calls for deployment from three former secretaries of Defense—Caspar Weinberger, James Schlesinger, and Donald Rumsfeld—along with the nuclear scientist Edward Teller and former CIA Director James Woolsey. Dole delivered several major speeches devoted to missile defense. In the run-up to the election, more hearings were held, speeches given, and columns published on missile defense than on any other defense issue.

But the measure was never acted on, in large part because the Congressional Budget Office (CBO) issued a report in July 1996 estimating that the Dole-Gingrich plan would cost up to $60 billion to deploy. The CBO figure was issued as the bill was about to be introduced for consideration on the House floor. It so spooked Republican House members, especially freshman deficit hawks, that Republican leaders decided after a turbulent party caucus to pull the legislation from floor debate. Dole barely mentioned the topic again, even in direct debates with Clinton, and the issue did not make a difference in the campaigns.

The Clinton administration, however, had taken the Republican challenge on missile defense seriously enough earlier in the year to adopt a compromise plan that helped defuse the issue's political impact. In the spring of 1996, the Pentagon had announced a somewhat stepped-up effort to develop a missile defense system, providing for the development and demonstration of a system within three or four years—that is, by the year 2000—that could be deployed, if the threat warranted, within the following three years—that is, by 2003. Dubbed the "three-plus-three" plan, administration officials portrayed it as a hedge against an emerging long-range missile threat. But it also conveniently allowed Clinton to blunt the Republican attack in his reelection campaign.

In recommending such a crash effort, Lieutenant General Malcolm O'Neill, then the head of BMDO, recognized there was little margin for error, but he considered the schedule doable. He counted on taking advantage of the Army's existing work on kill vehicles and tracking radars, and he also figured on using a variant of the Air Force's Minuteman rocket as the booster. He testified publicly that the only way to get the job done in the time allotted would be to forego lengthy bidding procedures and organize a national consortium of defense contractors to build the system.

But many experts in missile defense and military contracting doubted that a weapon system as complex and untested as national missile defense could be assembled in six years. Lieutenant General Lester Lyles, who took over from O'Neill in 1997, was among the doubters. He commissioned an outside review of all the Pentagon's antimissile system programs, both theater and national. Most advanced by that time were two Army programs—a more sophisticated Patriot system and the Theater High-Altitude Area Defense, or THAAD—but they had suffered delays and cost overruns, and THAAD had experienced a series of embarassing test failures.

Heading the review was a retired Air Force chief of staff, Larry Welch, who had become president of the Institute for Defense Analyses. Although skeptical of national missile defense, Welch had kept his personal views to himself and generally benefitted from a reputation for providing unbiased, precise judgments about Pentagon projects. To assess the missile defense programs, he assembled a sixteen-member panel of civilian and retired military officers who had worked on earlier missile, aeronautical, and naval projects. Their seventy-six-page report issued in early 1998 ripped into the Pentagon's whole effort, saying it was marred by poor planning, insufficient testing, and political pressure to hasten the inauguration of defensive systems. It said that decisions by officials to accept abbreviated timetables and minimal numbers of flight tests had raised the risk of flops, delays, and cost overruns and could end in a "rush to failure."

GOP defense experts on Capitol Hill tried to dismiss the panel's findings, saying the authors of the report failed to appreciate how the urgent need for missile defenses justified unconventional methods and more inventive development programs than those for other weapons systems. But proponents clearly had been thrown on the defensive by Welch's findings. The Clinton administration was also prompted to face the reality that its three-plus-three program was essentially unworkable and in desperate need of more testing, backup parts, and greater attention to quality control.

As a general rule, Welch prided himself on avoiding headline-grabbing language and often edited out catchy phrases in his reports. But somehow he had allowed "rush to failure" into print. And it was that phrase that kept pounding at missile defense advocates. Months later, Welch acknowledged that he had authored the phrase himself and had neglected to delete it. "I go to some lengths when I write reports to try to ensure that there aren't any sound bites," he said, with a trace of a grin. "That one was a great failure on my part."

Chapter 2

THE ROGUES ARE COMING

They were nine men on a special mission and had asked to see George Tenet, keeper of the keys to U.S. intelligence secrets. They were upset. Their initial round of briefings from intelligence officials had been superficial, providing little more than what was already available in the open press about the threat of missile attack against the United States. Clearly, there was more to the story than what they had been told. After all, these were men who knew what the government was capable of knowing.

One of them even, several years earlier, had served in Tenet's job—director of central intelligence, overseer of the CIA. Another was a former secretary of Defense and White House chief of staff. Two were retired four-star generals—one had commanded U.S. nuclear forces, the other had led the Air Force. The remaining five, all Ph.D.s, also had experience handling state secrets in the sensitive government jobs they once held.

Brought together in January 1998, they had accepted appointments on a blue-ribbon commission chartered by Congress to provide an independent assessment of one of the most politically disputed intelligence issues in Washington: the speed at which Third World countries were developing missiles capable of striking the United States. Now, a month later, speaking one by one around the table in a basement conference room at CIA headquarters outside Washington, D.C., the commissioners told Tenet they wanted greater access to U.S. intelligence information.

After hearing the complaints of the first three commissioners, Tenet tried to short-circuit the rest. "I got it, I got it," he declared. "I don't need to hear any more."

"Yes, you do," replied Donald Rumsfeld, the commission chairman. "Just so you can get the full flavor of what we've been through. This won't take too long, but I really would like each commissioner to take a minute."

So Tenet stayed at the table, chewing—as was his custom—an unlit cigar, absorbing the critical jabs.

"There's more information in *Time* magazine," one panel member complained. "They're giving us stuff they give to congressmen, for God's sake!" quipped another.

Rumsfeld knew he had Tenet's attention. The congressional statute that established the commission had stipulated that the executive branch provide whatever information the group required to do its job. This was not happening. Some influential lawmakers would not be pleased. Tenet did not need this kind of headache.

The CIA chief promised to lift the veil. Indeed, later that afternoon he summoned senior aides to his office and ordered changes in the intelligence community's approach to the commission. Panel members saw an immediate improvement in their access. But the deeper they delved into the evidence of missile development programs in North Korea, Iran, Iraq, and other countries, the more troubled the commissioners grew about the adequacy of the U.S. government's own assessments.

If there was a single moment in the late 1990s when events began to turn in favor of those advocating a national missile defense, this was it.

The meeting with Tenet opened the way for Rumsfeld and his commission to conduct a searing reassessment of the ballistic missile threat to the United States.

In the absence of a foreboding Soviet empire, missile defense advocates needed a compelling new threat to revive interest in their cause. They had hoped to find it in evidence that a small group of menacing Third World authoritarian regimes, designated "rogue" states by U.S. officials, were intently pursuing development of long-range missiles. The specter of these countries—most notably North Korea and Iran—targeting American cities presented a new set of defensive concerns.

But the missile defense crowd had one problem: U.S. intelligence agencies saw little urgency to the threat. This became clear in 1995. With the Republicans newly in control of Congress and pushing to accelerate work on a national antimissile system, the intelligence agencies issued a National Intelligence Estimate (NIE) on the ballistic missile threat that appeared to undercut the GOP initiative. An NIE is the intelligence community's most authoritative projection of future developments in a particular subject area. It is a consensus judgment by the nation's top spy agencies, including the CIA, the Defense Intelligence Agency, and the State Department's Bureau of Intelligence and Research.

The 1995 NIE concluded that no country outside the five major nuclear powers—the United States, Russia, China, Britain, and France—"will develop or otherwise acquire a ballistic missile in the next 15 years that could threaten the contiguous 48 states and Canada."

Republicans were stupefied, not only by the projection that the threat was as long as fifteen years away but also by the certainty with which the estimate was stated. By contrast, an NIE just two years earlier had hedged, noting that uncertainties clouded the community's ability to project beyond ten years. So how could the fifteen-year estimate be stated with such assurance now?

Moreover, the estimate appeared to make a distinction between the threat to Alaska and Hawaii and the threat to the forty-eight continental states. It said that while a missile under development by North

Korea—the Taepodong II—might have a range sufficient to reach Alaska or the westernmost Hawaiian Islands, there was no missile that North Korea or any other Third World adversary was likely to acquire in the next fifteen years that could reach the U.S. mainland. The distinction was curious. Did it mean that only the two outlying states would need protecting, or that a national system was unnecessary, since only the two noncontiguous states faced any kind of missile threat?

And there was more that drew Republican anger, especially on the issue of the assistance with missile development that Third World states might receive from Russia or China. Missile defense advocates had taken foreign assistance as a given—a factor sure to enable countries to acquire long-range missiles more quickly and with little warning to U.S. monitors. But the intelligence community had a different view. The NIE called the prospect of Russia or China aiding Third World missile efforts a "wild card" and appeared to play down the possibility of foreign assistance, observing that international agreements—notably, the Missile Technology Control Regime, begun in 1987 to coordinate export policies and curtail sensitive sales—had significantly limited transfers of missile technology and would continue to do so. The NIE also asserted confidently that the United States could count on detecting any homegrown Third World program for building an intercontinental-range ballistic missile. It predicted a minimum of five years' warning between the time when a country started testing and the time when it could actually deploy an ICBM.

Republicans charged that the intelligence community's findings conflicted with the views of experts inside and outside the intelligence community. They accused the administration of distorting the intelligence process and soft-pedaling the threat to help justify its resistance to a national missile defense system. The appearance of bias and politicization of the intelligence process was reinforced when some details of the NIE were reported first to Democratic senators in a letter from the CIA's director of congressional relations, in what looked like a move timed to refute language in a defense authorization bill supporting an antimissile system. In a dramatic display of congressional ire, Representative Curt Weldon, a Pennsylvania Republican leading the House campaign for a national system, stormed out of a discussion about the NIE with a senior intelligence analyst.

The CIA denied that the assessment had been in any way influenced by political pressure. Richard Cooper, chairman of the National Intelligence Council, which produced the NIE, acknowledged in congressional testimony that government analysts had changed their minds on several points since the 1993 findings. For one thing, they had come to believe that development of the Taepodong would be slower than previously projected. But in general, he maintained, the 1995 judgments were "largely consistent" with the 1993 report.

Republicans challenged the new estimate on virtually every major point. They remained convinced that their campaign for a national antimissile system could strike a popular chord with the American public—and provide an electoral advantage in the 1996 presidential race. They mandated two reviews of the NIE. One, completed a year later by Robert Gates, former CIA director in the first Bush administration, found no evidence of political shading of the NIE and actually endorsed its basic conclusions while faulting its methodology for having been rushed and incomplete. This left much of the Republican case for proceeding with missile defense riding on the outcome of the second review, which was put in the hands of an outside commission chartered to make an independent assessment of the ballistic missile threat.

More than a year passed before agreement was reached among lawmakers and administration officials on a membership list for the commission. In that period, leadership of the intelligence community was in some turmoil: John Deutch resigned in late 1996 following a nineteen-month tenure, and the nomination of Anthony Lake, President Clinton's former national security adviser, ran into fierce Republican opposition, prompting Lake to withdraw. George Tenet, a former White House and congressional staff member who had served as deputy director since 1995, was named in the spring and approved in the summer. But the White House had appeared in no great hurry to see the commission convened. Finally, in January 1998, the group got down to business, meeting in the Old Executive Office Building next to the White House, in a conference room assigned to the director of central intelligence. The betting, even within the group itself, was that the deliberations would prove contentious—and that consensus would be unachievable.

Only three of the nine members had been sponsored by Democratic lawmakers: Richard Garwin, a physicist who had served on various governmental advisory boards and as far back as the 1960s had been among the first in the scientific community to speak out against missile defense; Barry Blechman, a founder of the Henry Stimson Center, a public policy research organization; and retired General Lee Butler, who had overseen the nation's nuclear forces as commander of U.S. Strategic Command in the early 1990s and then, three years after leaving the military, surprisingly joined the ranks of those calling for abolition of all nuclear weapons.

The Republican choices included Rumsfeld, a White House chief of staff and Defense secretary under President Gerald Ford; Paul Wolfowitz, who had served as the Pentagon's policy chief under President Bush and then became dean of the Paul H. Nitze School of Advanced International Studies at Johns Hopkins University; William Schneider Jr., a consultant who had been undersecretary of State for security assistance during President Reagan's tenure and chairman of the president's advisory committee on arms control; William Graham, the White House science and technology director under President Reagan and now head of a national security research firm; R. James Woolsey, a lawyer who had served as Clinton's first director of central intelligence for two years before resigning amid strained relations with the White House; and retired General Larry Welch, the former Air Force chief of staff who was heading a technical review of the Pentagon's missile defense programs.

As chairman, Rumsfeld was determined that the commission come as close as possible to issuing a unanimous report. Experienced in Washington's ways, he recognized that the greater the consensus, the greater the commission's impact would be and the more its findings could rise above the partisan bickering that had marked discussion of the government's own threat assessments. To reduce the prospect of splintering into warring factions, Rumsfeld insisted at the outset that the group stay focused on assessing the missile threat and avoid any run at the question of whether the United States should respond by erecting a national missile defense. The members could not entirely foreclose the possibility of some dissenting views. But they agreed to raise the bar for what it would take to get even a dissenting footnote in the final report. They adopted the "rule of two": if a member wanted to lodge an objection in a footnote, he would have to persuade at least one other commissioner to go along with it.

Even so, the chances of a complete consensus were considered slim. The members themselves went into the process expecting that they would end up divided on some points. Those who had been picked by Democrats—Blechman, Butler, and Garwin—were especially sensitive about not ending up party to any effort that would attack the administration's policy.

Such blue-ribbon panels were notorious in Washington for leaving much of the work to staff members. But from the start, Rumsfeld made clear that his commission would be different. He set a brisk pace, scheduling two or more all-day sessions a week through the winter and spring. He pressed relentlessly for information and placed much of the responsibility for drafting a final report on the panel members themselves. Just what drove Rumsfeld to embrace his mission so aggressively was a source of some curiosity among those involved with the commission. A onetime naval aviator, college wrestler, and congressman before serving in the Ford administration, he had been out of government nearly a quarter of a century and was successfully ensconced in the business world heading a pharmaceutical firm. Because he was living in

Chicago, his work on the commission would require time-consuming commutes to Washington. But the intelligence world seemed to have a particular draw for him. "I think he was determined to push the commission and delve into details because he was fascinated by the issue," said an intelligence official who worked closely with Rumsfeld. "He just wanted to know."

The commission's first briefings came from government intelligence analysts who described missile development efforts by North Korea and Iran and missile programs in Russia and China. These initial sessions were the ones that most disappointed the commissioners. They were filled with familiar and all-too-general descriptions—the standard duty briefs. So in mid-February the commissioners met with Tenet and appealed for greater access to what the intelligence community knew.

The meeting proved a transforming event. Not only did it open spigots of secret information, but it also brought the commissioners closer together. They had shared a sense of frustration—indignation even—over the treatment initially accorded them, and this experience had forged a common bond. Their sense of unity only intensified as the commissioners dove deeper into what the intelligence community knew and did not know about the foreign missile threat. For what quickly became apparent was how poorly staffed, inefficiently structured, and ill-equipped the government's intelligence agencies were to assess the threat. The commissioners had found a common foe. It was ignorance.

Among the most striking—and in the commission's view, handicapping—aspects of the intelligence process was the elaborate compartmentalization of information. Time and again, analysts briefing the commission on one topic would have to leave the room before other analysts could speak. This led to some absurd situations. At the end of one two-hour briefing, for instance, a midlevel manager informed the commissioners, once the briefers themselves had left, that most of what the group had just heard was incorrect. He said this was because the briefers did not have access to the information that was about to be provided to the commissioners. In another case, the commission summoned analysts monitoring North Korea, Iraq, and several other countries with the hope of doing a comparative analysis, only to find that the analysts lacked clearances to know what the others knew and so could not conduct a comparative look across countries.

Because the commissioners themselves were able to move between intelligence compartments, they came to know more than many of the analysts addressing them. "We'd have highly cleared people coming in briefing us who we knew were quite wrong, not because they were dumb but because they had not been exposed," Barry Blechman recounted months later.

The commissioners found that those CIA analysts specifically assigned to monitor missile trade among countries tended to approach the matter more as a law enforcement problem than a strategic one. These analysts focused on trying to determine who was supplying what to whom rather than on assessing the scope, pace, and direction of individual national missile development programs, the motivations behind them, and their potential growth paths.

The commissioners also identified substantial shortfalls in the education and experience of the analysts. Trained in nonscientific and nontechnical disciplines, many analysts appeared to lack the technical, cultural and language skills that would benefit their assignments.

But what most disturbed the panel were the methodologies employed. There was, simply put, a seemingly pervasive reluctance to extrapolate or hypothesize beyond known, proven facts. A case in point came during a session about Iran. William Graham wanted to know what Iran might be doing with the Russian RD-214 engine, which Iran was reported to have acquired. The engine had powered a medium-range Soviet ballistic missile, the SS-4, and served as the first stage of the SL-7 space-launch vehicle. "The answer I got back was, 'We haven't seen the Iranians do anything with it,'" Graham recalled. "And so I asked, 'What do you think their interest might be?' And they said, 'We haven't seen them do anything with it.' And I asked, 'Have you done any analysis about what they might do with it?' And they said, 'Until we see them do something with it, we're not going to consider it.'"

This exchange reinforced a view of the intelligence community as all too narrow-minded. It was what Graham one day described to his fellow commissioners as a disturbing tendency by too many analysts to interpret the absence of evidence as evidence of absence. "It was part of the intelligence community's culture that before they would say something is being done, they wanted to have all kinds of evidence," Graham said months later. "If they didn't have such evidence, they would say there's no indication that it's happening, which is true. The trouble is, there's also no indication that it's not happening."

Only part of the lack of depth and dimension the commissioners were seeing in the government's assessments could be attributed to inadequate training or experience. There also seemed an almost universal reticence, a caution, against saying much about what was known or suspected. Some of this, the commissioners believed, could be understood perhaps as a heightened concern for secrecy in the wake of the espionage scandal involving Aldrich Ames, the CIA officer sentenced to life in prison in 1994 for spying for Moscow. But much of it appeared to be learned behavior. By nature, many analysts seemed noncommittal, unwilling to express what they thought, worried they might be held to account if they did.

The commissioners were inclined to fault the institution for this tendency more than the individuals. They regarded the entire intelligence system as conditioned to give the political decisionmakers the kind of news they wanted and to avoid, delay, or fudge reports that might go against policy and prove publicly embarrassing. Answers to such questions as whether the Iranians were still obtaining Russian technology useful in ICBM development—despite U.S. efforts to block such transfers—fell into the category of undesired news.

Rumsfeld jotted down Graham's quip about the absence of evidence not being evidence of absence. The commission chairman had a habit of keeping a list of pithy phrases and sayings he came across.

In his previous government service at the White House and the Pentagon, he had compiled what he called "Rumsfeld's Rules," a ten-page set of reflections and quotations gathered during his career. For the commission, he started a new list, which came to be labeled "Brilliant Pebbles," a play on the name of the space-based antimissile system under study in the late 1980s and early 1990s.

A section of the list was devoted just to witticisms uttered by Bill Schneider, whose asides often captured the dim view and, at times, downright despair of the commissioners regarding the quality of some of the official briefers. Some examples:

"That fellow has a knowledge board defect. He must have purchased his chip from the low bidder."

"That man has had one year's experience fifty times."

"He's the type that would read the Bible, looking for a loophole."

At one point, so disgusted was Rumsfeld with a briefing that he simply walked out of the room, leaving three fellow commissioners to endure the session. When Rumsfeld returned for lunch, Schneider had this to say about the briefer: "That man is truly a waste of food."

That remark too made the list.

To get at just how well the intelligence community had detected key events in North Korean and Iranian missile development, the commission asked for timeline charts showing when certain events had occurred and when U.S. analysts became aware of them. The result was a series of large, multicolored charts, each stretching several feet across. The presentations—covering four or five incidents in all—amounted to a kind of audit of U.S. intelligence assessments. They showed a startling pattern of lagging estimates, ranging from a few years to more than a decade.

Being starkly reminded of their failings in this way was not a pleasant experience for the analysts involved. During one presentation, a briefer appeared truly testy in responding to the commissioners. Recalled one participant: "This prompted Welch finally to ask, 'You're not happy to be here, are you?' And the analyst said, 'No, I'm not. I'm ticked off that I have to come down and brief a bunch of wacko missile defense advocates.'"

Rumsfeld, insulted by the remark, was so enraged that he stalked out of the room. "Once again, this solidified the commissioners," the participant said. "It was them against the system that was trying to give them the mushroom treatment—that is, put them in a dark room, feed them manure, and then eat them. They didn't like it."

The presentations reinforced for panel members the idea that U.S. officials would probably learn of some missile advances by rogue states only years after the fact. To drive home this point, Rumsfeld brought in a copy of Roberta Wohlstetter's book about Japan's attack on Pearl Harbor and passed out copies of the foreword by the economist Thomas C. Schelling, one of the major contributors to cold war deterrence theory. The foreword said in part:

It is not true that we were caught napping at the time of Pearl Harbor. Rarely has a government been more expectant. We just expected wrong. And it was not our warning that was most at fault, but our strategic analysis.... There is a tendency in our planning to confuse the unfamiliar with the improbable. The contingency we have not considered seriously looks strange; what looks strange is thought improbable; what is improbable need not be considered seriously.... The danger is not that we shall read the signals and indicators with too little skill; the danger is in a poverty of expectations—a routine obsession with a few dangers that may be familiar rather than likely. Alliance diplomacy, inter-service bargaining, appropriations hearings and public discussion all seem to need to focus on a few vivid and oversimplified dangers. The planner should think in subtler and more variegated terms and allow for a wider range of contingencies.

"Afterward, everyone sort of looked at one another and said, 'Boy, that sounds familiar,'" recalled Stephen Cambone, the commission's staff director.

A central focus of the commission's assessment was what could be done using Soviet Scud missile technology. This was the basis of North Korea's missile program, which had begun in the early 1980s with the acquisition of a batch of Scuds from Egypt. Then, by enlarging the Scud B's diameter about 50 percent, North Korea came up in the early 1990s with the Nodong, whose range of about 1,000 to 1,300 kilometers— roughly the distance between Pyongyang and Tokyo—is three to four times greater than the Scud B.

It did not take long for the Nodong, in whole or in part, to end up in several other Third World countries as North Korea energetically promoted its sale. The missile served as a foundation for development of Iran's Shahab 3 and Pakistan's Ghauri missiles.

The question for the commission was whether North Korea could build on the Nodong and develop the advanced propulsion, guidance, and multi-stage separation systems necessary for longer-range flight. U.S. intelligence analysts tended to doubt that the relatively primitive missile technology embodied in the single-stage, three-decade-old Soviet Scud—slender steel cylinder, fixed fins, liquid fuel—could serve as much of a stepping-stone to a long-range missile. The single-stage Scud put out a thrust of 30,000 pounds, propelling the rocket 300 to 600 kilometers at a speed of 1 to 2 kilometers per second. In contrast, America's three-stage Minuteman III, with 202,600 pounds of thrust, could cross nearly 10,000 kilometers at a speed of 7 kilometers per second.

While state-of-the-art U.S. and Russian missiles had onboard computers and gyros to monitor and adjust their position in space, the North Korean models had no such internal guidance. Their engines were timed to burn on low-performance fuels for preassigned amounts of time before shutting off. Just overcoming the demands of staging alone, given the issues of timing and physical stresses involved, appeared nearly overwhelming using such crude technologies. And the missile was just the delivery vehicle; designing a warhead to withstand the heat and other stresses of reentry would involve a whole set of additional technical challenges.

During two working sessions in April 1998, the commissioners heard U.S. government analysts list all the reasons a Third World country would be unlikely anytime soon to produce a long-distance missile. But North Korea already had proven itself quite resourceful, and there was evidence that the country had a multi-stage missile—the Taepodong—under development.

Besides, enough once-sensitive technical information about how to build a missile had become available on the open market to teach the North Koreans whatever they did not know. Underscoring this point, Dick Garwin showed up at a commission meeting one day with a defense contractor's advertisement featuring the Mark 12A reentry vehicle, used on the Minuteman missile. The ad contained information about the hot temperatures at which nose-cone materials are dissipated while reentering the atmosphere from space.

To learn more about just how far and how fast some foreign states might be able to go with the kind of missile technology they already had, the commission ventured outside the government's intelligence network and consulted several aerospace firms, including Boeing and Lockheed Martin.

The notion that Scuds could serve as building blocks for long-range missile development was driven home to the commissioners in a May 7 briefing by Lockheed Martin engineers who had studied the matter and whose company had a long-standing interest in missile defense. During the presentation, Rumsfeld scribbled a sentence on a piece of paper. He often was writing out propositions for the rest of the commission to discuss along the way, laying the foundation for the final report.

"Let me read something to you and see if this is what you're saying," he said to Lockheed's briefers. "Using Scud technology, a country could test-fly a long-range missile within about five years. Is that what you're

saying?" The Lockheed experts concurred, and after some revising by
Rumsfeld and other commissioners, the group ended with a sentence
that became one of its central conclusions: "With the external help now
readily available, a nation with a well-developed, Scud-based ballistic
missile infrastructure would be able to achieve first flight of a long-
range missile, up to and including intercontinental ballistic missile
range, within about five years of deciding to do so."

The statement conveyed the urgency of the threat. It also flew in the
face of the intelligence community's earlier assertion that the United
States was unlikely to face a new long-range missile threat in the next
fifteen years.

In the middle of July, the panel, known formally as the Commission to
Assess the Ballistic Missile Threat to the United States, released its findings and recommendations to Congress. Challenging official U.S. intelligence estimates, the commission said that North Korea and Iran could
develop weapons capable of striking U.S. territory within five years—
and do so with little or no warning. "The threat to the United States
posed by these emerging capabilities is broader, more mature, and
evolving more rapidly than has been reported in estimates and reports
by the intelligence community," the commissioners concluded.

The report went on to highlight what it called the declining ability of
intelligence agencies to assess U.S. vulnerability to ballistic missiles. It
cited shortcomings in the way analysts assessed information as well as a
shortage of satellites and spies to track missile proliferation. And it said
that various tactics—launching missiles from ships, for instance, or basing them in other countries—could further reduce warning times for
U.S. officials. "The intelligence community's ability to provide timely
and accurate estimates of ballistic missile threats to the United States is
eroding," the report said.

At a news conference on Capitol Hill, Rumsfeld attributed his
group's contrasting assessment of the threat to the fact that the commissioners had gained broader access to highly classified information

than was available to most analysts in the compartmentalized intelligence world. He also said the commission had taken "a somewhat different approach" than the intelligence community, weighing information "as senior decisionmakers would." By that he meant they had connected facts more freely, drawn inferences, and applied a more sweeping view.

In particular, the commissioners had placed greater weight on the likelihood that rogue states would receive outside technical assistance from other countries, notably Russia and China. Some Third World nations even had formed their own networks for trading technical information and missile parts. A test of a Ghauri missile in Pakistan, for instance, could feed the North Koreans valuable information on development of their own Nodong.

In effect, the commissioners concluded that to talk about indigenous ballistic missile development was simply no longer relevant. After years that had seen German scientists move to Russia and develop the Scud, which the Russians then gave to the Egyptians, who in turn passed the missile to North Korea, whose engineers ended up building and marketing a version of their own to Pakistan and a handful of other countries—clearly, there was no such thing as indigenous development.

It rankled the commissioners that the 1995 NIE had labeled technical assistance by the haves to the have-nots a "wild card." They read that phrase as suggesting such assistance was somehow outside the mainstream and could be discounted. But in fact it was very real. "Foreign assistance is not a wild card. It is a fact," the report asserted, bluntly taking issue with the 1995 intelligence community report.

The commissioners also stressed that rogue states should not be expected to follow the same missile development patterns that had marked earlier U.S. and Soviet programs. The methodology used by the superpowers had involved extensive testing, and deployment came many years after the first flight trials. Now rogue states were showing little desire to achieve the accuracy that the United States and Soviet Union had required. They did not care as much about safety. They did not need volumes of test data. They were perfectly capable of using technologies, approaches, or equipment that the United States would have rejected as too primitive decades before. North Korea, for one, had shown an extraordinary willingness with its Nodong to test-fly it only once—in 1993—and then deploy it.

Moreover, the commission said, U.S. officials should no longer count on advance warning of foreign missile deployments. It noted that rogue nations had become increasingly able to conceal important elements of their programs. Again, in the case of the Nodong, the missile "was operationally deployed long before the U.S. government recognized that fact," the report observed.

And if the U.S. intelligence community's record of anticipating North Korean developments was poor, its reading of Iranian moves was even worse, according to the commissioners. The panel noted that Iran's ballistic missile program, once heavily dependent on North Korean assistance, had received increased Russian help, ranging from rocket engines to missile designs. Besides, Iran's infrastructure was more advanced than North Korea's. Should Iran decide to go for a long-range missile similar to North Korea's Taepodong, the commission figured, the missile could be ready to fly within five years. In fact, Iran's missile effort might someday overtake North Korea's.

The intelligence community's seeming inability to make precise forecasts of foreign missile development had significant policy implications. But deciding how to characterize those implications provoked one of the greatest controversies in the group. Some members, among them Bill Graham and Jim Woolsey, wanted report language that was openly critical of the administration's policy, which had been based on the presumption that the intelligence community could provide several years of warning before a rogue state acquired an operational long-range missile. Other commissioners, especially Barry Blechman, were wary of coming down too hard on the administration.

In the end, the commissioners simply noted the shrinking warning time for threatening missile deployments, then added, "Therefore, we unanimously recommend that U.S. analyses, practices, and policies that depend on expectations of extended warning of deployment be reviewed and, as appropriate, revised to reflect the reality of an environment in which there may be little or no warning." It was the closest the panel came to a policy recommendation—and it was unanimous.

In fact, the entire report was issued without a single dissenting footnote, just as Rumsfeld had hoped. The final drafting days saw commissioners struggling to accommodate potential objections from any one member. Hours were spent refining language to avoid a footnote, a level of fine-tuning that led Bill Schneider to offer yet another witticism: "This crowd would edit a stop sign."

One especially contentious issue concerned what to say about the prospect that a Third World country might try to launch a ballistic missile from a merchant ship. The point was that this kind of alternative launch mode would not require a long-range rocket and could dramatically shorten the warning time for U.S. officials. Blechman considered the whole idea unrealistic because of the logistics involved, and he saw no reason to include it in the report. But another Democrat-appointed member, Dick Garwin, sided with the Republican technologists Graham and Schneider in arguing that it could be done. After all, they noted, the Russians had launched Scuds from surface ships, and the United States had fired a Polaris from a ship as early as 1962. So the sea-borne threat got a passing reference in the report, with Blechman laboring over compromise language right up to the end—even from a hospital bed where he was recuperating from minor surgery.

It was the closest the commission came to including a footnote. And it illustrated a central tenet of Rumsfeld's approach—namely, that if there is a disagreement, an incessant focus on the facts through briefing after briefing can resolve it.

By speaking in a unanimous voice, the commissioners received a serious hearing in a national debate that had become riven with partisan bickering. While earlier attempts at assessing the threat, including the government's own effort, had been discounted as politically tinged, the Rumsfeld group's effort emerged as perhaps the single most authoritative view.

But its consensus came at some cost in specificity. Its conclusions dealt essentially with what *could* happen, not with what was *most likely* to happen. The commissioners shied away from placing probabilities on their threat scenarios. If they had gone down that road, chances are their judgments would indeed have diverged.

The commissioners also studiously avoided drawing any conclusions as a group about what their findings meant for national missile defense. Here too they certainly would have ended up badly split. Their warning about a growing missile threat constituted a persuasive argument for more rigorous and comprehensive intelligence assessments. But it did not necessarily amount to a summons to intensify development of a national antimissile system.

Still, the commission's report was seized upon by congressional Republicans as further justification for building an antimissile weapon. House Speaker Newt Gingrich hailed the panel's study as "the most important warning about our national security since the end of the cold war."

For its part, the intelligence community, thrown on the defensive and caught in a political thicket, chose to swallow its pride and assume a public stance of general deference toward the Rumsfeld commission's findings. In a letter to Congress, Tenet attempted some defense of the government's earlier threat estimates, saying they had been "supported by the available evidence" and "well tested" in debate. But for the most part, the CIA director chose not to dispute the Rumsfeld report in public. "We need to keep challenging our assumptions," he acknowledged in the letter, and went on to say that the panel's work already had prompted government analysts to review "how we characterize uncertainties and alternative scenarios."

In private, intelligence officials were deeply offended by the report. They expressed indignation at the commission's general assertion that analysts were too narrow in their approaches and were missing the basic picture. "I take issue with the accusation that we take absence of evidence as evidence of absence," one senior intelligence officer said two years later. "The commissioners must recognize that when you're briefing a panel like theirs, you're going to stick to your evidence and talk about what you know. When we write reports, we do go out on speculation branches and sort these issues out."

Some government analysts also belittled the commission's report as simplistic, an indulgence in all the "what coulds." They said the report lacked any sophisticated attempt to assign probabilities to the various possible foreign missile threats. Almost anything *could* happen. For an assessment to be of value to policymakers, however, it should make some judgment about what was *likely* to happen, they said.

Most preposterous in this regard, one senior analyst said, was the commission's assertion that a country with a "well-developed, Scud-based ballistic missile infrastructure" could, if it set its mind to it, get to the point of flight-testing a long-range missile within five years. "Take Congo or Fiji, just grab a country," said the analyst. "Even if they had Scud technology, they lack the economic and scientific support for a missile program. So the first point is, you really need to go country by country, you just can't make blanket statements but must consider technology base, financing, and so on."

The commission's findings also were denounced as exaggerated and misleading by defense experts opposed to a national antimissile system. Among the most outspoken critics, Joseph Cirincione, director of the Non-Proliferation Project at the Carnegie Endowment for International Peace, argued for months afterward that the long-range missile threat to the United States had actually declined by about 50 percent since the mid-1980s as a result of reductions in Russia's nuclear arsenal. He noted that most of the nations possessing ballistic missiles had only short-range capabilities. And while more were acquiring medium-range missiles, the number of Third World states with ballistic missile programs had dropped following the termination of efforts in Brazil, Argentina, Egypt, and South Africa.

But the commission was not finished with the intelligence community. In October, three months after its report appeared amid much fanfare, the group quietly delivered a "side letter" to Congress and to

Tenet, laying out its concerns about how the intelligence agencies were going about the business of making assessments. It presented a searing portrait, describing an intelligence community buffeted by budget cuts, spy scandals, and excessive turnover. It also cited a decline in scientific and engineering competence, a highly charged political atmosphere, and overcompartmentalization of functions and information.

The letter reported that the amount of intelligence resources devoted specifically to monitoring ballistic missile development, as well as weapons of mass destruction generally, had "declined significantly" in the previous five years, despite the prominence of the proliferation issue. Exacerbating matters, the letter said, was the emphasis on intelligence to support real-time military, diplomatic, and counterdrug operations. This left less time and fewer resources for longer-term strategic issues such as missile proliferation. "Treating the threat as one of a hundred or more high-priority issues, all of which are placed on a back burner with each crisis and contingency that comes along, will not improve the capability of the intelligence community to provide actionable warning," the letter said.

The letter also criticized the community for focusing on proliferation more as a law enforcement issue than as a strategic puzzle. It urged that analysts be given broader access to information "wherever it may be held" in the intelligence community. And it called on analysts to consider not only what they knew but what they did not know in venturing beyond the hard evidence in their assessments. For many of the commissioners, this critique of the intelligence community carried greater lasting significance than the commission's final position on the missile threat. While the threat assessments remained open to argument, the deterioration of U.S. intelligence was irrefutable.

All in all, the Rumsfeld group's work was bad news for efforts by the Clinton administration to forestall the Republican push on missile defense. Not only had the commission called into serious question earlier official projections that the Third World threat was a decade or so away, but it also had bluntly challenged the intelligence community's ability to make such projections. The ground on which the administration had stood against national missile defense was eroding. "We'd been having this debate with Capitol Hill for a long time, where we had kept pointing to the NIE that had said the threat was fifteen years away, and suddenly with the Rumsfeld commission report, the NIE offered no protection against this ever-stronger push from the Hill to go farther and faster," one senior administration official recalled.

Clinton and his aides could take some small comfort in the argument that no one really knew how fast the missile danger might materialize. It was all still largely conjecture, still mostly guesswork and speculation about North Korea and Iran, countries that were notoriously difficult for U.S. analysts to read. No dramatic event had yet occurred to substantiate that any rogue state was really as close to developing a long-range missile as the Rumsfeld commission had suggested.

Although that too was about to change.

Chapter 3

WAKEUP CALL

Making clear the urgency of the matter, Air Force General Joseph Ralston got Bob Walpole on the phone at the CIA. The general wanted the senior CIA official to find a satellite that North Korea was claiming to have sent into orbit. It would be North Korea's first satellite, and although U.S. authorities had detected a North Korean missile launch five days earlier, there had been no report of any such cargo.

U.S. intelligence analysts had characterized the North Korean launch on August 31, 1998, as a test of a two-stage missile—and not an entirely successful one at that. And North Korea itself had said nothing publicly about the launch all week.

But now the official North Korean news agency was asserting that what had gone up was a three-stage rocket bearing a satellite, and the satellite was said to be circling Earth every two hours and forty-five minutes, transmitting "the immortal revolutionary hymns" of Kim Jong Il, the North Korean leader, and his deceased father, Kim Il Sung. Lending credence to the astonishing claim, the Russians were reporting that they indeed had spotted the North Korean satellite.

Ralston, vice chairman of the Joint Chiefs of Staff, sounded exasperated. He wanted to know whether U.S. intelligence had missed something. "The Russians can find a satellite," Ralston told Walpole. "Why can't we?"

Walpole, a career intelligence official, had taken over as the CIA's national intelligence officer for strategic and missile programs several months earlier. It was one of the hot-seat jobs at the agency. A renewed Republican campaign to build a national missile defense system had intensified pressure on U.S. intelligence assessments of the missile threat to the United States. In this regard, North Korea was public enemy number one, since, among the rogue states, it had advanced the furthest toward developing a missile capable of reaching U.S. territory. But missile defense proponents suspected the CIA of playing down the missile threat and underestimating how long it would take North Korea and other rogue states to acquire the technology for ocean-spanning missiles. In this atmosphere, the CIA could not afford to bungle the intelligence on North Korea's latest launch.

So U.S. Space Command crews at Cheyenne Mountain in Colorado spent an exhausting Labor Day weekend scouring the skies in search of a North Korean satellite. They never found one and concluded that the North Koreans had nothing in orbit. At the same time, analysts reviewing the launch data did confirm that North Korea had tried to put a satellite up on August 31 using a three-stage rocket—a remarkable development in itself.

Needle-thin and primitive by Western standards, the North Korean missile had flown farther than any fired by a Third World country before. Taking off from a rural stretch of countryside along North Korea's eastern shore, it had dropped its first stage into the Sea of Japan, then soared over the northern tip of the main Japanese island of Honshu and out above the Pacific Ocean, its second stage falling into the water about 1600 kilometers from the launch site and its third stage flying on briefly before burning up with a debris trail that extended more than

4,000 kilometers. Although it never came close to U.S. territory, the missile provided the first concrete evidence that an impoverished, militant state like North Korea had worked out the technology for multi-stage rocketry and thus was coming closer to producing intercontinental missiles capable of threatening American soil.

That U.S. intelligence had missed the satellite-launch aspect was all the more surprising because analysts had known the test was coming. American spy satellites had taken pictures of the launchpad's scaffolding being readied and of missile elements being moved into place at the launch site. The imagery had been viewed at the Pentagon by Defense Secretary Bill Cohen during morning intelligence briefings attended also by General Henry "Hugh" Shelton, chairman of the Joint Chiefs, in the super-secure National Military Command Center located in a window-less expanse within the middle floors of the Pentagon. One of Cohen's assistants had even sent a note to the Republican staff director of the Senate Armed Services Committee advising him of the North Korean preparations.

But the third stage, covered by a shroud, had not been visible on the launchpad. Analysts assumed that the shroud was shielding some kind of mock warhead. In fact, it was enclosing a small, solid-fuel kick-motor and a tiny, rudimentary broadcasting device—what one U.S. military intelligence official later likened to a combination Walkman and cell phone.

Until then, the biggest ballistic missile that North Korea had tested was the single-stage, 45-foot-tall Nodong with a range of about 1300 kilometers. That had been five years earlier. The prototype of a longer-range missile, dubbed the Taepodong I, was spotted by U.S. analysts the next year when put on conspicuous display just north of Pyongyang. Since then, the intelligence community had been predicting that the first flight test of a Taepodong I could come at any time. But U.S. experts presumed the missile had only a two-stage configuration, about 75 feet tall. That is what they looked for and thought they saw when scanning the sketchy preliminary data from the August 31 launch.

U.S. defense satellites with infrared sensors easily detected the

Taepodong I's burning engine after liftoff. Based on a quick read of the trajectory and an assumption that North Korea was firing a two-stage missile, U.S. analysts reported that the missile had faltered, failing to reach its expected altitude. And that remained the official U.S. view for several days. Still, the fact that North Korea had gone ahead and tested a two-stage missile was very disturbing. Officials in Washington swiftly condemned the test, saying it represented a dangerous breakthrough in North Korea's efforts to build longer-range missiles.

The extent of the military threat posed by North Korea's missile program was questionable. Even its single-stage rockets were considered quite inaccurate, and there is much more involved in shooting an intercontinental ballistic missile than simply piling one stage on another. But the launch provoked considerable concern abroad because it suggested that North Korea could, for the first time, reach any part of Japan and all U.S. forces in the region—the 37,000 American troops in South Korea and the 44,000 in Japan. And since North Korea was believed to have at least one or two nuclear weapons and huge stocks of chemical weapons, the launch also aroused fears about the payload that such a missile might carry.

Additionally, the missile flight occurred at a time of mounting Republican criticism over the Clinton administration's handling of North Korea. Just two weeks earlier, a Defense Intelligence Agency report had surfaced suggesting the North Koreans were constructing an underground complex at Kumchangri for housing a nuclear reactor and plutonium reprocessing operation suitable for making nuclear bombs. Under a 1994 accord, North Korea had agreed to freeze plutonium production at its Yongbyon nuclear reprocessing center in exchange for U.S. and other foreign help in building two light-water nuclear power plants, plus the promise of yearly 500,000-ton oil deliveries and the eventual lifting of American economic sanctions. Although the DIA report later proved erroneous, word at the time that North Korea might be violating at least the spirit of its previous deal had prompted Republican lawmakers to hammer the administration for coddling the Stalinist state and deceiving Congress about the extent of North Korean compliance. House Republican leaders were demanding an independent review of the evidence on North Korea. Worried about losing control of the process, U.S. officials had gathered for talks in New York with North Korea, and the administration was preparing to name its own special envoy. But the missile launch threatened to put the whole effort at risk. Incensed by the launch, congressional leaders responded with renewed threats to cut off all food aid and other assistance to the North Koreans.

In Tokyo too there were denunciations and sanctions. The idea of a North Korean rocket whizzing overhead gave the Japanese a new and more immediate sense of their vulnerability to enemy missiles. Japan halted food aid to North Korea and put off talks to establish diplomatic ties. It also announced suspension of its part of the financing for the two nuclear reactors in North Korea being built in conjunction with the United States and South Korea.

Meanwhile, behind the scenes, Japanese authorities were pressing the United States for more intelligence about the North Korean launch. The Americans kept shrugging, saying they had little more to report. But some Japanese suspected Washington of holding back, and there was some reason for their suspicion.

It had taken U.S. officials nearly five years to correct the record on what Japan had been told about the last time North Korea fired a missile—the Nodong launch in 1993. At the time, the United States reported that the missile had fallen short into the Sea of Japan and declared the test a failure. But in early 1998, U.S. authorities quietly informed the Japanese that the test was not a flop after all. A secret review of some previously unprocessed satellite data had revealed that the Nodong had in fact flown over Japan. The revised U.S. assessment stunned the Japanese, because of what it suggested not only about the capability of North Korean missiles to strike Japan but also about the ability of the United States to provide accurate, immediate intelligence.

Now North Korea had shot another missile into the air, and the Americans were again appearing slow to determine where exactly it had gone. U.S. monitors had tracked the flight over Japan, reporting several pieces of the missile falling well to the east in the Pacific. But radar operators on a Japanese destroyer had picked up indications that the missile had traveled even farther.

Information about a foreign missile launch does not arrive in Washington as a complete, clean picture. Rather, the data come in increments from satellites, ship radars, ground monitoring stations, and other sources. The initial reports, based on infrared satellite readings, are inherently fragmentary, while tapes of radar data, which pour in later, require time-consuming processing and analysis. None of the sources produces a neat visual image of the missile arcing across the sky. The missile's precise track must be painstakingly pieced together from myriad radar hits pinged from multiple angles.

Additionally, the intelligence community was accustomed to having time to sort out data about foreign missile launches. "The system was set up during the cold war mainly to monitor Soviet launches for compliance with arms control agreements," Walpole explained. "In those days, no one was asking for an immediate response." In the case of the Taepodong I flight, however, the intelligence community was being pressed urgently for answers.

Under the circumstances, basic assumptions and inferences count for a lot in drawing initial conclusions. Because U.S. analysts had expected to see a ballistic missile, that is what they thought they saw. But the trajectory of a missile can differ significantly from the trajectory of a space-launch vehicle. A long-range ballistic missile climbs steeply and arcs high through space; a rocket attempting to place a satellite in low Earth orbit takes a shallower course and levels off before releasing its payload.

North Korea had hinted at its intention to try a satellite launch. Kim Il Sung's desire to do so, expressed as far back as the winter of 1993–94 at a meeting of the Korean Workers' Party, had led to an expansion of the country's nascent space program. But the notion that North Korea would skip from the single-stage Nodong to a triple-stage Taepodong I, and then use it to carry a satellite, had never been considered by U.S. experts. The conventional wisdom was that the Taepodong I would be a Scud mounted on a Nodong with a range of up to 2,000 kilometers.

After five days of listening to denunciations by U.S. and Japanese officials, who kept referring to the event as a missile test, North Korea issued a statement angrily asserting that its flight had been misportrayed. It was a satellite launch, North Korea said, meant to promote "scientific research for peaceful use of outer space." Washington officials reacted with deep skepticism, although several acknowledged to reporters some uncertainty and confusion about the North Korean claim. One U.S. government official quoted in the *Washington Post* said that the trajectory of the rocket and other characteristics of its flight had appeared "a bit odd" and lent some credibility to North Korea's claim of a satellite launch.

Indeed, the data did contain some anomalies that had begun to puzzle analysts. For one thing, the trajectory appeared flatter than it would be for a ballistic missile, rising to an apogee of only about 130 miles. For another, the second-stage burn was longer than the normal minute or so expected of a Scud in a two-stage-missile configuration. Moreover, there were indications that something had separated from the second stage, something with perhaps some thrust behind it. Although these oddities failed to fit the assumption of a missile trajectory, they did seem appropriate for a space-launch vehicle.

"Since you're expecting a missile launch, and the trajectory should have been one way but it was something else, you figure it just wasn't burning the way it should have," Walpole recounted. "Then you realize that it was burning exactly the way it should have been if they had wanted to depress this into an orbital trajectory. Then you start figuring that this isn't going to have enough velocity with those two stages to insert a satellite into orbit. And then it doesn't take long to figure that there's another stage. Ultimately, we were able to sort out the existence of another stage."

Faulty assumptions at the outset were not the only factor handicapping the initial U.S. assessments. Analysts lacked information about the last segment of the flight, when the third stage kicked in. That stage had not provided a large, notable thermal signature for military satellites to detect, and the main U.S. monitoring ship in the area, the *Cobra Judy*, had been poorly positioned to track the entire flight. Fortuitously, though, a Japanese destroyer with an Aegis radar system was able to catch the missile's final moments and provided the clinching piece of evidence about the third stage. Some U.S. officials found this particularly ironic, given the frustrations voiced by Japan at the extent of U.S. intelligence sharing.

Within a week or so after the Taepodong I had flown, American analysts had concluded that there had indeed been a third stage, which contained a solid-fuel rocket motor and a tiny satellite. It had broken up shortly after separating from the second stage, apparently because the release came at too low an altitude. Japan dispatched search ships to the area but found no parts to recover.

The realization that North Korea had managed to fly a three-stage missile as far as it did—and on its first try—shocked U.S. specialists. It also left them with a lot of explaining to do. North Korea had essentially leapfrogged several developmental levels. The surprising third stage gave the Taepodong I intercontinental-range capabilities that U.S. analysts had predicted would be seen only with the Taepodong II. The move reinforced North Korea's reputation as a resourceful, daring builder of missiles, a feat all the more impressive given the country's impoverished condition.

"When North Korea broadcast that it had launched a three-stage rocket, I called a friend at the CIA on a secure line and asked, 'What gives?'" recalled Bill Graham, who had served on the Rumsfeld commission. "By this time, they had been looking at it five days. It must have been enormously frustrating for the North Koreans. Here they had gone to all the trouble of launching a three-stage missile, with great belief in the technical collection powers of the United States, and we didn't give them credit for this feat. This was an example not just of our sensors being out of position or unavailable. But we weren't looking for a three-stage launch so we didn't see one at first."

In the days immediately following the launch, a flurry of Japanese and Russian reports spoke of a second North Korean missile being readied for an imminent flight test. U.S. officials also noted continued activity at the same military launch facility along North Korea's northeastern coast but played down the prospect of a follow-up shot. And none took place.

North Korean authorities must have known that their action would be seen as threatening to the security of their Asian neighbors and the United States, undercutting whatever hope they might have had of receiving additional foreign economic and humanitarian aid. So why would they do it? Why did they risk the ire of the United States and Japan and back themselves into a dangerous corner by showing off their ability to make an advanced missile?

Some analysts speculated that the launch was probably meant to stir up nationalistic pride in anticipation of the fiftieth anniversary of the country's founding on September 9 and the expected elevation of Kim Jong Il as North Korea's president, a post that had been vacant since the death of his father in 1994. In this view, any military value derived from the launch was not nearly as significant to the North Korean leadership as its political effect. The launch served as a domestic propaganda ploy to demonstrate North Korea's increasing technological prowess.

Other specialists saw the launch as a statement of political defiance linked to the negotiations in New York on the future of North Korea's nuclear program. Earlier in 1998, North Korea had signaled that it might halt testing and exporting of missiles in exchange for economic aid from the West, along the lines of the deal it had agreed to in 1994 to freeze nuclear weapons material production at its Yongbyon reprocessing facility. Its missile-firing may have been an effort to demonstrate the kind of weapons it could sell if no deal were reached.

Then too the launch may have been intended largely as an advertisement, a way to court purchase orders from such nations as Iraq, Libya, and Pakistan. Ballistic missiles have constituted North Korea's most lucrative means of obtaining foreign currency. As a weapon, the Taepodong I had little additional military importance for North Korea, which already could reach most regional targets with the Nodong. Striking the United States would require a substantially more powerful delivery vehicle, but other countries could conceivably have use for such an intermediate-range missile, and the North Koreans would be happy to sell it to them.

Whatever North Korea's motivation, the launch ended up having a profound effect on the Clinton administration's thinking about whether to step up development of an antimissile system that would guard against the possibility of rogue nation attack. Until then, the threat, while widely forecast, had been largely theoretical. Just one week before the test, General Shelton had written to Senator James Inhofe of Oklahoma, a vigorous Republican proponent of national missile defense, expressing confidence that U.S. intelligence agencies would be able to provide adequate warning of "indigenous development and deployment" of long-range missiles by a rogue state. Although the Rumsfeld commission had issued its report a month earlier stating that rogue nations might acquire an ICBM capability in a short time without the U.S. intelligence community knowing about it, Shelton had disagreed. "We view this as an unlikely development," he had written Inhofe.

Now the North Korean action had proven how possible it was to blindside the Pentagon and the CIA. It had pulled the rug right out from under the chairman of the Joint Chiefs. Coming on the heels of the Rumsfeld commission report, it was widely read as factual confirmation of the commission's central findings—not just the nearness of the threat, but the likelihood that U.S. intelligence would not be able to predict its arrival with much accuracy. In fact, the joke making the rounds in Washington at the time was that Rumsfeld had paid North Korea to conduct the launch. "I was over at the CIA in the days after the launch," said Richard Haver, who had served as the intelligence community's liaison officer for the Rumsfeld commission, "and I bumped into an analyst in the halls who said to me, 'You know that "Ode to

Kim Jong Il" that the North Korean satellite was supposed to be playing? It really was the "Ode to Rumsfeld" played backwards.'"

In fairness, the intelligence community's record on the Taepodong I was not a total flop. As early as 1994 U.S. analysts had reported the missile's existence and predicted that the first flight test could occur in 1996 or later. They also correctly predicted that North Korea, by the end of the 1990s, would demonstrate a missile capable of delivering payloads to parts of Alaska and Hawaii. At the time, however, the analysts were thinking not of the Taepodong I but its follow-on, the Taepodong II, which was envisioned as a cluster of three Nodong engines, or possibly a single giant engine, with a range of 4,000 to 6,000 kilometers, depending on the payload. Anchorage is about 6,000 kilometers from North Korea; Honolulu is about 7,400 kilometers away.

What caught U.S. intelligence officials completely by surprise was the configuration of the Taepodong I. They never anticipated the third stage—and for that matter, neither did the Rumsfeld commission. An eighty-six-page classified annex to the commission's report discussed different configurations that might be used. None envisioned a three-stage Taepodong I. Incredible as it may seem in retrospect, for the many years that analysts had brainstormed about how proliferation might occur, they had never imagined taking a Nodong, adding a Scud, then slapping on a kind of jury-rigged third stage. Instead, they had stuck with the rather narrow prediction of a two-stage rocket with a dummy warhead.

Part of the reason for the oversight, Defense Intelligence Agency (DIA) officials explained months later, was that intelligence experts had concentrated on militarily useful payloads. They had never figured that the North Koreans would choose to give up a normal-sized payload and substitute a tiny broadcasting satellite boosted by a small motor. "We assumed they were focused on achieving military utility for this system," one DIA official said. "Instead, they chose to go for political effect and replaced a military payload with one that was politically dramatic. This may reflect a kind of institutional bias we have, but if it's not militarily useful, we just kind of ignore it."

It was this kind of oversight that had drawn critical notices from the Rumsfeld commission. The episode appeared to confirm the commission's assertions that U.S. intelligence analysts tended to be confined to factual evidence and were reluctant to hypothesize beyond the known—to think outside the box.

"In my view, part of the problem is that our system has not supported speculative thinking or imagination or initiatives apart from the known facts," acknowledged Patrick Hughes, a retired three-star Army general who headed the DIA at the time. "Our decisionmakers and policymakers quite often lack the time or interest—nor is it their nature—to deal with what could happen, because there are so many possibilities. They want to know what will happen with as much clarity as possible. That drives the analysts back to a sometimes small knowledge base and causes them to send to the decisionmaker what they absolutely know and are sure of. They do not send speculation, and that's especially true when there are many possibilities. Deeply embedded in the intelligence system is a desire to be right, a fear of failure, and a resistance to change."

The intelligence lapse represented by the Taepodong I also drove home another institutional shortcoming: U.S. difficulty penetrating closed societies like North Korea. "We've built a magnificent technical intelligence-gathering capability, where we can gather anything from almost anywhere on Earth," Hughes observed. "But the critical issues of intent—and of what you might call the story behind the visual or electromagnetic data we collect—is something we quite often do not have good intelligence on. The reasons are obvious: this kind of information, generally speaking, comes from intercepts of communications when we're lucky enough to get them, and also has to come from human intelligence-gathering. However, that's something we aren't very good at, particularly in a restrictive environment like North Korea. It requires risk, which the United States has frankly avoided."

After the launch, the intelligence community issued new assessments of the ranges of the Taepodong I and Taepodong II. It estimated that a two-stage Taepodong I could be expected to fly about 2,000 kilometers with a payload of several hundred kilograms—enough for a small nuclear bomb. That was still within the medium-range category. But the three-stage configuration could achieve ICBM ranges, analysts concluded, delivering small payloads distances of more than 5,500 kilometers— bringing it within reach of Alaska and western parts of the Hawaiian Islands—although with "great inaccuracy." As for the Taepodong II, if it also had a third stage, it conceivably could reach anywhere in the United States, although again, only with a small payload and with "significant inaccuracy," U.S. intelligence officials said.

As a result of the Taepodong I launch, the likelihood that a Taepodong II would be successfully tested had gone up considerably. So had the likelihood that it would exceed 6,000 kilometers in range with a useful payload. And the chance that the United States would learn much about the capability of a Taepodong II before the event appeared lower than ever.

One of the most intriguing—and perplexing—questions for U.S. intelligence agencies was how the North Koreans had managed to make such a technological leap. Had they done it on their own, or had they received considerable help from Russia or China, or both? Experts on North Korea's missile program were puzzled. Although the North Koreans had proven themselves adept at copying production of the Soviet Scud in the 1980s, it was unlikely that on their own they could have made the advances in propulsion, staging, and guidance required by the Taepodong I. The very fact that the missile performed as well as it did on its first flight test strongly suggested outside assistance, since other countries, including the United States and the Soviet Union, had failed repeatedly in their initial ventures in this area decades before.

Much of the speculation centered on Chinese assistance with the satellite. Some analysts figured that Russia had a hand in helping with the booster and staging. One of the more radical theories along this line, put forward by Robert Schmucker, a German professor specializing in aeronautics who had served as a United Nations inspector in Iraq, held that the North Korean missiles were not really North Korean at all. They were Russian, secretly built with Russian components and the active and ongoing help of some errant Russian scientists inside North Korea.

Schmucker argued that a rogue team of Russian missile scientists—thrown out of work after the collapse of the Soviet Union—may have moved to North Korea. And there, for profit or glory or both, they directed the North Korean program, with the North Koreans themselves doing little more than putting the pieces together. Schmucker's evidence for this was that North Korea had performed few if any important missile tasks independently. Its missile assembly lines were built with Russian help, and designs for the Scud C and Nodong were derived from Soviet missile programs in the 1960s. In Schmucker's view, Russia was using North Korea to hide the origin of the seller and get new customers. This theory was regarded skeptically by U.S. government analysts ("Don't shortchange North Korea's indigenous capabilities," one senior CIA official said), but it did pique the interest of some American specialists outside the administration.

Whatever the source and motive of the North Korean launch in August 1998, the event set the stage for a new look at national missile defense by the Clinton administration. The launch was seized on by missile defense proponents as concrete evidence that the United States could come under threat of missile attack by rogue states much sooner than previously thought. What had been mere conjecture about the likely pace of foreign missile development suddenly became all too real.

Still, it is curious that a single missile launch—and one that failed in its stated mission of placing a satellite into orbit—could have had such a galvanizing effect on U.S. policy. Even government intelligence analysts differed over the military significance of the Taepodong I launch. The CIA's Walpole was ready to call the North Korean missile an intercontinental-range weapon because it went the distance—or appeared able to do so. And it fit the Pentagon's formal definition of a long-range ballistic missile. But DIA analysts disagreed, noting that the Taepodong I neither looked like a conventional ICBM nor carried a military payload. The CIA came up with a notional payload of biological warfare agents that was very small and that conceivably could be transported by a three-stage

Taepodong I to Alaska or Hawaii. But the DIA argued that even assuming North Korea would fire such a missile armed with germ agents, the chances of it striking U.S. territory with any accuracy were remote given the lack of any demonstrated guidance capability.

Then too it was only one launch, and the last time the North Koreans had tested a missile had been five years earlier. The country had yet to show its mastery of the other major technological challenges associated with an ICBM, including development of a warhead capable of withstanding the heat and stress of reentry.

The very primitive and protracted nature of North Korea's missile development effort was driven home months after the Taepodong I launch when the Federation of American Scientists (FAS) paid a commercial satellite firm to take overhead photos of the North Korean launch site. The pictures showed the site in a rural area not far from the town of Nodong and about six miles from the town of Taepodong. The complex was very modest in size. It lacked any railway connections or even paved roads connecting it with the outside world. Nor were there the kind of propellant storage facilities and permanent housing that would support a sustained testing schedule. "The antithesis of Cape Canaveral," the FAS called it in a statement, which went on to describe the test site as "a facility barely worthy of note, consisting of the most minimal imaginable test infrastructure." The place could hardly support the kind of extensive test program needed to fully develop a reliable missile, the statement said. Since the facility's construction in 1988, only two missile tests had been conducted there—in May 1993 and August 1998—indicating a prolonged development schedule.

Indeed, there were other valid reasons not to get carried away by the North Korean action. The Pyongyang leadership had shown a willingness four years earlier to suspend its nuclear activities at the Yongbyon facility in exchange for foreign economic assistance. Perhaps a deal could be struck that would compel the North Koreans to dismantle their missile-building and missile export business and spare the United States the expense of constructing an antimissile system.

More generally, the notion of a destitute, bankrupt North Korea posing a serious missile threat to the far more powerful, technologically advanced United States struck many missile defense critics as far-fetched. North Korea's economy was a shambles. Its agriculture had collapsed, resulting in widespread starvation and malnutrition. True, by devoting an enormous proportion of its gross national product to military purposes, North Korea had created large forces of tanks and artillery, many of them deployed within range of South Korea's capital of Seoul. But there seemed little logic to the notion that this tightly sealed nation—which had become a caricature of Stalinist tyranny— would use a few ballistic missiles to attack the United States and invite annihilation.

Nonetheless, the sudden evidence that North Korea might actually possess such a capability had a profoundly jarring impact in Washington once the extent of the technical accomplishment was fully realized days later. It came on top of not just the Rumsfeld commission report but other developments earlier in the year indicating startling advances in Third World missile and nuclear weapons technology. In April, Pakistan had conducted its first test of the Ghauri medium-range missile. Then in May, both Pakistan and India set off underground nuclear tests. And in July, Iran for the first time flew its Shahab 3 medium-range missile. These events made the missile threat appear more certain and much closer in time. They also highlighted the difficulty that the U.S. intelligence community was having in providing timely predictions of foreign weapons developments.

Against this background, the North Korean launch served as a kind of final straw. By the autumn of 1998, even top officials in the Clinton administration, once reluctant to acknowledge a near-term missile threat, were coming around to the recognition that the world had changed. And that however little enthusiasm they had for national missile defense personally, they were confronting a new strategic reality. The launch had altered the administration's ability to be fairly relaxed about timelines for developing an antimissile system. It also made

Democratic lawmakers anxious about their ability to continue voting against Republican efforts to legislate accelerated timetables or the development of more robust antimissile systems.

One other subtle but significant shift was also under way that would add to the pressure to respond quickly to the North Korean action. It had to do with the point at which a foreign missile under development was deemed an actual threat. Prior to the Rumsfeld commission report, the determinative event for the intelligence community and policymakers had been the moment a missile achieves "initial operating capability," or IOC—that is, when it gets deployed. The deployment date tended to be calculated by figuring when the first flight test would be, then adding a few years for more testing.

But this approach was based on U.S. and Soviet patterns of missile development. The point underscored by the Rumsfeld commission was that Third World countries were as likely as not to test just once or twice and then deploy.

"The reason we had used IOC was because it had meaning in the military context," explained Hughes, the former DIA chief. "It meant a country had gone beyond research and development and had achieved a workable system with a known, measurable capability, which could be produced in adequate numbers to represent a true military weapons system. And this used to take a fairly predictable period of time. It was dependent on the industrial base of a country, access to resources, and other features that took time to assemble. But technology had changed that, and the pace of change of development of something like a missile had become much faster than it used to be. We had seen that not just in North Korea, but in Pakistan, India, Iran, and Iraq."

Besides, the new thinking went, it might not be so important for these ICBM-seeking countries to have a missile that could actually be fired with any confidence of working. It might simply be enough to have one that could be brandished against the United States or regional adversaries, since just the threat of attack can provide valuable leverage. "For the most part, these are not military capabilities that they're seeking to acquire, but political capabilities," said Bill Schneider, a

Rumsfeld commission member. "They want to be able to deter intervention and gain some leverage for their position, either bilaterally or
in the region."

In any case, the Taepodong I launch had stunned the intelligence
community and the Clinton administration and shaken the conventional wisdom about foreign missile development. Suddenly, the skies over
the United States looked a lot more threatening.

II

Policy and Politics

Chapter 4

DISCONNECTED TIMELINES

As an Air Force lieutenant colonel assigned to the Pentagon's Joint Staff, Glenn Trimmer spent his time working issues related to the ABM Treaty. It was a backwater assignment, full of the arcane aspects of international arms control. Changes, if they came, tended to require endless negotiations with the Russians, lasting years.

But Trimmer enjoyed all the technicalities and subtleties. He took pride in his thorough knowledge of the treaty. And by the summer of 1998, he was convinced he had spotted a looming problem that seemed to have escaped the notice of the administration's senior policymakers.

The problem had to do with the administration's plan for developing a national missile defense. Several years earlier, President Clinton had set 2000 as the year in which he would decide whether research on an antimissile technology had progressed enough, and whether the threat was grave enough, to warrant building a defensive system. But a decision to proceed then would run the United States smack into violating the ABM Treaty. So if the president really intended to make a deployment decision in 2000, Trimmer figured that the United States had to start raising with Russia as soon as possible the prospect of revising the treaty to permit such a move—or else resign itself to dumping the treaty.

Given Clinton's abiding interest in preserving the accord, Trimmer expected that administration officials would want to go to considerable lengths to try to negotiate treaty changes. What concerned Trimmer was that senior officials were paying scant attention to the need for the United States to start talking to the Russians. The prevailing assumption appeared to be either that some way might still be found to deploy a system that would not violate the treaty or that the United States would have several years—after a deployment decision had been made and while the system was being built—to negotiate any necessary revisions.

Both assumptions suited the administration's deep reluctance to raise tensions with the Russians by pushing the idea of national missile defense and getting into sticky treaty talks over it. But Trimmer considered both assumptions dead wrong. In his view, the Pentagon was unlikely to devise a treaty-compliant antimissile system, and time was running out for negotiations. There was little advantage to waiting much longer, and a lot of risk in further delay. In just twelve to eighteen months, Washington would have to give Moscow notice of its intent to withdraw from the ABM Treaty if no deal could be reached between the parties by then.

Trimmer was convinced that he had identified a significant disconnect between the administration's existing national missile defense timetable and the actual requirements of the ABM Treaty. But his dilemma was this: How does a junior officer at the Pentagon get word up the ranks, politely, that the people running the government are asleep at the wheel?

From his modest perch well down the military totem pole, below all the generals and admirals who crowd the ranks in Washington, political prudence dictated that Trimmer remain seen but not heard. He had received a painful reminder of this lesson earlier in 1998 after he had sent an electronic memo to his boss and five other senior Joint Staff officers critical of Representative Curt Weldon, the volatile senior member of the House Armed Services Committee and a major proponent of building up U.S. defenses against missile attack. Trimmer had heard Weldon deliver a speech lambasting the administration for its missile defense policies. In the memo, he characterized Weldon's remarks as "more a prolonged shout than a speech" and described the congressman as "someone so convinced of his own ideas that logical discussion" with him would be "of little value." He cautioned against sharing information with Weldon, concluding: "His remarks led me to believe that little is privileged with him, and most is strongly filtered by his own views."

The memo found its way into the *Washington Times* and created some waves for Trimmer. So Trimmer was not exactly eager to stick his neck out again and sound critical of any senior political authorities. On the other hand, the treaty issue appeared to him quite real and potentially very serious.

To deliver his message up the chain of command, Trimmer finally decided to put together a briefing. He enlisted an Army captain, John Fitzgerald, who was a whiz at designing computer graphics—and doing so quickly. Good graphics had become essential to getting a point across in a Pentagon world oversaturated with PowerPoint charts and briefing slides.

In early August 1998, several weeks before the Taepodong I launch, Trimmer presented his case to his boss, Rear Admiral Chip Griffiths, deputy director of international relations on the Joint Staff. Trimmer argued that the timelines for possible deployment of a U.S. antimissile system and for negotiations with the Russians were out of sync. The official U.S. government position, he noted, was to wait to engage the Russians on the issue until after a deployment decision had been made. But waiting that long, Trimmer said, would leave little time to reach a deal. The United States would come up hard against a deadline either to violate the treaty or delay construction.

To illustrate his point, Trimmer presented a series of charts outlining the decisions about the architecture of the system and alterations in the treaty that would have to be made, and when, if the president wanted to preserve the option of a deployment decision in March 2000.

Trimmer's pitch quickly persuaded a previously skeptical Griffiths of the problem. And as word of the lieutenant colonel's arguments spread, he was invited to present his briefing to top Pentagon and White House officials. But the administration was not quite ready to jump into talks with the Russians or even conclude that changes in the treaty would be required.

One sign of the political sensitivities came during a briefing Trimmer delivered to Deputy Defense Secretary John Hamre, the second-ranking civilian at the Pentagon. The lieutenant colonel rolled through his arguments without incident, until he got to a chart underscoring the first article of the ABM Treaty, which explicitly prohibits the use of a missile defense system to defend all national territory. Trimmer said that such language would appear to rule out an antimissile system that was both national in scope and treaty-compliant, a situation that some in the administration still thought might be possible. This legal interpretation drew sharp objection from Judy Miller, the Pentagon's general counsel. The next day Trimmer was summoned to the office of the Joint Staff's lawyer, who said that Miller had called to complain that Trimmer's briefing had been all about the meaning of the treaty's first article. Trimmer disputed this, and gave the officer the full text of the briefing to see for himself.

Trimmer needed no reminders about the sensitivity of the whole issue. In previous months, while preparing talking points for U.S. negotiators attending meetings in Geneva of the joint U.S.-Russian Standing Consultative Commission, which reviews treaty matters, Trimmer and other Joint Staff officers had tried to insert some mention of the fact that the United States might want to seek amendments to permit a national antimissile system. But time after time, they had been rebuffed. Midlevel officials in the Pentagon's policy branch appeared reluctant to address the issue. And no one at the State Department was eager to push for negotiations either.

Then North Korea launched the Taepodong I, and suddenly missile defense acquired a sense of urgency. The possibility that the United

States might actually erect a system began to appear more real. This, in turn, reinforced Trimmer's message that the United States had about a year and a half at most to gain treaty changes and avoid having to walk away from the ABM Treaty.

"Here was an officer on the Joint Staff—a non-ideological, solid guy with no parochial or political angle—who, doing due diligence, just doing his job, had put together this briefing saying that if we were really serious about deploying, we'd better focus on the treaty timelines," said Robert Bell, the National Security Council's senior specialist on defense policy and arms control.

The briefing had the effect of alerting senior administration officials to the fact that they had not yet made basic programmatic decisions about how an antimissile system would be structured and how that structure would or would not fit with the ABM Treaty. In the absence of that kind of planning, it was hard to determine when the U.S. effort might press up against the treaty. But if those decisions were not made, the United States ran the risk of bumping into the treaty without having given the consequences enough thought. That recognition at various levels began to sink in during the autumn of 1998.

"Until Trimmer's briefing, nobody had talked about the timeline," Hamre recalled. "That was the first time that any of us, I think, got the gravity and complexity of the political situation that this initiative created."

Among those moved to action after hearing Trimmer's exposition was Air Force General Joseph Ralston, vice chairman of the Joint Chiefs. Politically astute and action-oriented, Ralston had gained a reputation in the administration and on Capitol Hill as an effective Washington operator. Convinced that the treaty issue needed attention, he raised it at one of his regular Tuesday "deputies' lunches" in September. The lunches brought together the second-ranking officials from the Pentagon, State Department, CIA, and National Security Council staff for informal chats with no preset agendas.

The site of the luncheons rotated week to week from department to department. The CIA served the best food, everyone agreed; the State Department got the award for the worst. The day Ralston brought up the ABM Treaty, lunch was at the CIA. He told his colleagues that he and Hamre were due to testify before the Senate Armed Services Committee during the first week in October and would surely be grilled about missile defense in the wake of the North Korean launch. He knew the Republicans would accuse the administration of being too constrained by treaty considerations and sacrificing U.S. national interests for the sake of not upsetting Russia and European allies. He also knew what his response would be.

He would insist, he told the others, that the ABM Treaty had inhibited nothing, that the administration was doing everything it could to rapidly develop and field a national missile defense system. But Ralston also said he would not be able to make that statement much longer. "We have to start talking to the Russians, because if we don't—and if we have to make treaty modifications and they refuse, and we have to give six months' notice to withdraw—then the United States, all of a sudden, will be up against a wall," the general said. This had been Trimmer's basic message. Now Ralston was telling the other deputies that in his public testimony to the Senate, he intended to put the administration on record for the first time saying it would have to confront the issue with the Russians.

It was then and there, Ralston would later recall, that some of the administration's top officials—Strobe Talbott, deputy secretary of State; Jim Steinberg, deputy national security adviser; Leon Fuerth, Vice President Gore's national security adviser; and John Gordon, deputy director of central intelligence—started focusing on the fact that the treaty issue was coming at them sooner than expected. No one at the lunch argued with Ralston about his intention to acknowledge the treaty issues publicly. "I didn't get a pushback," he recalled. "Rather, the reaction was, 'Yeah, okay'—kind of a sober realization that we have to do that."

In his prepared testimony for the October 2 hearing, Ralston included a sentence saying that negotiations on treaty changes needed to be completed by May 2000. But it was Hamre, speaking before Ralston, who actually broached the subject at the hearing. The deputy Defense secretary also had come around to the view that time was running out.

Senator Strom Thurmond, the Republican from South Carolina who at ninety-five years of age was still presiding over the Armed Services Committee, had opened with a fierce salvo against what he called the administration's "wait-and-see" approach to missile defense. He said developmental delays had put any potential deployment years behind where it should be. "Another result of the wait-and-see policy is the administration's failure to face up to the ABM Treaty dilemma that is incoming on the horizon," the senator said. "Any decision to deploy an NMD [national missile defense] system capable of defending the entire United States will require the United States to seek amendments to the ABM Treaty or to withdraw from the treaty altogether. The administration has not taken one step to lay the groundwork for such change. In my view, this delay only increases the likelihood that a few years hence, the United States will be faced with a difficult choice between not deploying the defenses we need and creating a diplomatic crisis with Russia. Although I personally believe that the ABM Treaty has outlived its usefulness and has no legal standing following the dissolution of the Soviet Union, I acknowledge the need to engage Russia as we transition from today's ABM Treaty regime. The administration should get on with this immediately."

Hamre answered by underscoring the administration's interest in preserving the treaty. He said the accord still represented the "kind of bedrock of confidence in working with the Russians every day" that enabled the two countries to pursue joint programs for eliminating nuclear weapons. But then he gingerly, almost offhandedly, raised the prospect of amending the treaty, indicating the subject had begun to receive serious administration attention. He said the administration was looking at the question of where to locate an interceptor site. "The treaty right now designates that it should be in North Dakota," Hamre said. "We have also done modeling that shows that there are very good reasons why you may want to put it in Alaska. I mean, it's going to take a fair amount more work—we're not talking about years, we're talking about months—to kind of home in on that and figure out what's the right answer. Now, if it goes to Alaska, that requires us to sit down and make a change in the treaty."

Senator Carl Levin, the Michigan Democrat who was the ranking minority member on the committee, had spoken privately with Hamre and Ralston before the hearing and knew what was coming. Behind the scenes, he also had been urging the administration to start talking to the Russians about the possible need for treaty revisions. While certainly no fan of national missile defense, Levin wanted at least to avoid a sudden treaty crisis. With Republican pressure mounting in Congress to pass some kind of legislation mandating deployment of a national antimissile system, Levin and other senior Democrats were eager to see the administration get something going with the Russians.

As far back as August 31, as President Clinton was preparing to leave for a meeting with Russian President Yeltsin, Levin had written Clinton's national security adviser, Samuel "Sandy" Berger, suggesting it would not be too early for the two leaders to announce "preliminary discussions" on amending the treaty. "Such action is not only right on the merits," Levin wrote. "It will help us prevail in the debate on this issue in the weeks ahead, since it would be so patently irresponsible and destructive to press forward with resolutions calling for a commitment to unilateral deployment of an NMD system when bilateral negotiations are under way."

But the White House had been reluctant to open talks. "We realized that you couldn't negotiate with the Russians until you had decided on an architecture," one senior administration official recounted. "And we still thought there might be a treaty-compliant option, in which case there wouldn't be any treaty notice or any negotiation with the Russians. But no one wanted to put that proposition to the test, because it would be difficult, very difficult, to pull the administration into a wrenching review that would finally settle all these questions about what article I really meant and what would be the range of engagement out of North

Dakota and so on. It would all have to be done in a very hypothetical way, because you weren't deciding to deploy this, you were just deciding on an architecture that would inform a negotiating strategy."

So Berger wrote back on September 28: "At this stage of our NMD program, it is too soon to say what modifications to the ABM Treaty, if any, would be required to proceed with NMD deployment. For this reason, we believe it is premature to initiate any negotiation on treaty amendments." He added, however, that the administration was keeping the Russians "informed of developments in our NMD program." At the summit, U.S. officials did at least advise the Russians that the Pentagon would be starting environmental impact surveys in both North Dakota and Alaska for the possible deployment in either place of an antimissile weapon.

At the hearing, Levin again encouraged the start of "preliminary discussions" with the Russians on amending the treaty, "since it may take some time for those discussions to bear fruit." Ten days later, on October 13, Levin wrote Berger once more, warning: "If we do not engage the Russians soon, it may lead to problems when the time comes to make an NMD deployment decision. In other words, if we wait until we make a deployment decision to engage the Russians, it may be too late."

Responding on November 24, Berger indicated a change of tone. He acknowledged "the need to closely integrate" the administration's approaches on national missile defense and the ABM Treaty and promised to give Levin's arguments "serious consideration." Although Berger stopped short of making a commitment to start talks with the Russians just yet, his more receptive attitude to the idea reflected the results of a formal policy review that was nearing completion.

The review had begun quietly a month before. After hearing Trimmer's briefing, the NSC's Bell had set up an interagency working group to look at where to go next with missile defense. Because the North Korean launch had made the missile threat more real, pressures to proceed with a national antimissile system were likely to mount. It was time to begin giving high-level consideration to types of system architectures and negotiating approaches with the Russians.

From mid-October through November, the working group drafted a discussion paper that would serve as the basis for a decision meeting in December attended by Clinton's top national security aides. In the working group, representatives of the Pentagon, State Department, and National Security Council had quickly reached consensus that the Taepodong I launch had changed things in a way that would affect U.S. national antimissile system development and the future of arms control. Recognizing that the Russians would have to be approached soon about amending the ABM Treaty, the group also considered how those talks would overlap with another set of negotiations on reducing offensive nuclear weapons under the START process.

Trimmer left the Joint Staff shortly after presenting his national missile defense briefing to the military chiefs on October 20. His next assignment took him to the State Department and another arms control desk job. Despite his part in mobilizing high-level officials to focus on disconnected policies and diverging timelines, it did little to advance Trimmer's military career.

The lieutenant colonel had worked fourteen-hour days since taking the initiative in mid-August. Sitting in the Senate hearing room in early October, listening to the Pentagon's top officials recite the points that he had brought to their attention, he was ecstatic. He had changed Washington, he thought; he had done something that had affected policy at the highest level. But as the months passed, Trimmer began to think that he had not really played much of a role at all. The considerations he had raised were rather obvious; senior administration officials would have awakened to them sooner or later; after all, the people running the government were very smart, he thought. It was just that sometimes, they got so busy that they failed to notice basic disconnects—until someone put it all together in a chart and drew a few red lines.

Chapter 5

SHOOTING FOR 2005

On December 7, 1998, the Pentagon's top acquisition and budget officials gathered with the military chiefs to ponder what to do about their
existing timetable for a possible national missile defense system. That
timetable, introduced in 1996 and known as the three-plus-three plan,
allowed for three years of development followed by a presidential decision to deploy and another three years of construction. But with the
first three-year period coming to a close and the first flight test still
months away, the plan clearly had lost whatever viability it had. Two
outside reviews—one by the Welch panel, the other by the General
Accounting Office—had declared the timetable rushed and unrealistic.

Pentagon officials, fearing Republican accusations of foot-dragging
and irresponsibility, were reluctant to extend the target date. Two years
earlier, forty-one members of the House had even filed a civil suit in U.S.
District Court against President Clinton and Defense Secretary William
Perry, seeking to compel compliance with deployment dates for the shorter-
range antimissile systems mandated in the 1996 defense authorization

act. The lawmakers alleged that the president had violated his constitutional duties by refusing to spend the appropriated funds. Judge Stanley Sporkin dismissed the complaint, saying the congressmen lacked standing because they had suffered no harm. He also called the issue "not yet ripe for resolution by the judicial branch." But the lawmakers had made their point, and the administration felt the jab.

Lieutenant General Lester Lyles, the director of the Ballistic Missile Defense Organization, had taken to speaking in a kind of code about the compressed timetable. He would not say publicly that the system could not be built by 2003. Instead, he characterized the schedule as "extremely aggressive" and fraught with a "high risk" of failure. Translated, that meant there was virtually no way the system could be up and running in the time specified. Other Pentagon officials likened the missile defense rush to earlier crash efforts to develop new weapons in times of great military crisis. "This is as close as we can get in the Department of Defense to a Manhattan Project," Deputy Defense Secretary John Hamre told a Senate panel in October 1998.

Not only did the Pentagon have to decide what to do about the program's unrealistic schedule, but it also had to figure out future funding. Money had been set aside only for the first three years of the three-plus-three program, but that phase was ending in 1999. Funds for the next three years—the actual deployment phase—would have to be included in the fiscal 2000 budget if the administration's plan were to maintain any semblance of credibility. At the same time, the money would have to be added in a way that did not appear to represent a presidential decision actually to deploy the system. Clinton wanted to wait until mid-2000 for that.

By December, decisions already had been taken on other major weapon systems in the 2000 budget plan, which the White House was readying for submission to Congress. Missile defense had been saved for last because it was among the most controversial items and because Pentagon officials had needed more time to consider ways of revamping the program. Seated around a large table in the third-floor Pentagon conference room adjoining Hamre's office, the group of high-ranking civilian officials and military officers—known formally as the Defense Resources Board—studied three options.

One was to hold to the three-plus-three plan for the time being.

A second proposal, drafted by the Ballistic Missile Defense Organization, involved letting the deployment date slip to 2005 to allow for more testing and design work. But the prospect of shifting beyond the 2003 deadline was so politically sensitive that defense officials avoided mention of 2005 in the charts showing this alternative. Instead, the option was referred to as the "modified '03" program.

For some Pentagon analysts, even shooting for 2005 seemed overly ambitious. So the Office of Program Analysis and Evaluation (PA&E), a kind of internal truth-squad agency for the secretary of Defense, put forward a third option with a deadline of 2007. If adopted, it would place national missile defense on a schedule more in line with the pace of other major military acquisition programs.

Whatever option was selected, the program would require an additional $5 billion to $10 billion over the next few years. The prospect of national missile defense siphoning billions of dollars away from other defense programs made the military chiefs grumpy. In 1998 they had begun voicing concerns about signs of sagging readiness, noting eroding conventional capabilities, shortfalls in recruiting, and declining retention rates. They had gone before Congress that autumn to complain of being shortchanged at least $17 billion a year to cover unmet procurement and refurbishment needs. The chiefs viewed every cent spent on national missile defense as money that could be better invested in modernizing and revitalizing the armed forces.

Normally, the chiefs would have relished playing up a new and growing military threat. Such developments help attract public support for a strong military and higher defense spending. Making the case for large defense budgets, however, had become more challenging since the Soviet Union's collapse. The prospect of a Third World missile attack looked like just the kind of new post–cold war menace that would reinforce the Pentagon's view of the world as still a dangerous place, filled with adversaries determined to exploit U.S. vulnerabilities. But the chiefs doubted the immediacy of the missile threat, even in the wake of North Korea's surprising missile launch. Besides, they had what they considered more pressing priorities.

Among the four-star service leaders there was no more vocal a critic of a national antimissile system than General Charles Krulak, commandant of the Marine Corps. Known for being outspoken and opinionated on many matters, Krulak had a particular interest in the missile defense controversy. His own exposure to the issue dated back to 1988, when he had spent a year working at the Pentagon for Don Latham, assistant secretary of Defense for command, control, and communications. Krulak had oversight responsibility for the network of battle management computers and communications facilities that was to serve as the electronic nervous system for Reagan's Strategic Defense Initiative. A Marine infantry officer with no background in the subject, Krulak wondered why the assignment had fallen to him, and he asked Latham about it. "Because I think you'll be able, more than anybody, to evaluate through a giggle test whether this makes any sense," Latham replied.

The experience thrust Krulak into frequent meetings with missile defense program officials. He observed their classified tests. He listened to their experts. And he eventually became convinced that an antimissile system was doable; the technology, he concluded, either had arrived or was about to.

So a decade later, as a member of the Joint Chiefs, Krulak had little doubt about America's ability to build a system if it wanted to. But he did dispute the need for it. The issue for him was priorities. He believed more pressing threats existed than the prospect of North Korea firing a missile at the United States.

At the Defense Resources Board meeting, Krulak sat silently through a presentation of charts showing various timelines and associated budget increases for missile defense. When time came for comments, the Marine commandant let loose, blasting the whole premise for a national antimissile system. He questioned the extent of the threat posed by North Korea. He also challenged the notion that an antimissile system was the only means of deterring such a rogue state from attacking the

United States, noting that any real threat would probably be obliterated by a preemptive American air strike.

"Do you really think North Korea would send a missile?" Krulak said, his voice rising. "We'd act first. We'd take them out. We'd vaporize them. They would never get a missile off."

Besides, Krulak remarked, U.S. troops in South Korea have had shorter-range North Korean missiles aimed at them for decades. For Krulak, the fear was not a rogue state armed with a long-range missile. Rather, it was a terrorist attack—an anonymous bomb in a skyscraper, for instance, or the silent release of deadly anthrax spores in a crowded big-city subway station. "This nation only has a finite amount of money that it's willing to spend on national defense," he observed. "Now where are we going to spend it? That's the issue."

Other members of the Joint Chiefs chimed in with similar concerns, questioning whether this was an appropriate expenditure, whether national missile defense should rank above modernization of the current force and whether rogue states really would use ICBMs. They also expressed doubts about the maturity of the technology and urged more testing ahead of any more decisions. "If I were going to attack the United States," said General Hugh Shelton, chairman of the Joint Chiefs of Staff, "I'd just get a mortar shell filled with germs or poison gas and drive down the Hudson River and lob it into New York City. I wouldn't shoot a long-range missile."

But for all their reservations about national missile defense, the chiefs were reluctant to openly oppose a weapons system that had strong political support. They would express their misgivings in closed-door sessions, but they would not protest publicly. Their position came down to this: they would dutifully—if unenthusiastically—support deployment of a national missile defense, but only if their individual service budgets were not drained to pay for it. They would insist that the cost of building any national antimissile system be covered by extra White House allotments.

● ● ●

The chiefs were not alone within the Pentagon in their lack of enthusiasm for national missile defense and their inclination to take a more gradual approach to development. Hamre, the deputy Defense secretary, was inclined that way, although he kept his views largely to himself. So were William Lynn, the Pentagon comptroller, and Robert Soule, who headed the influential Office of Program Analysis and Evaluation, which had proposed extending the deployment date to 2007.

It is PA&E's job to serve as a sort of in-house Pentagon critic, questioning the acquisition proposals pushed by one military branch or another. The office performs its own cost estimates and prepares decision briefs for the secretary and deputy secretary. Often it ends up challenging a program's projected costs and timetables as overly optimistic, as was the case with national missile defense. "They're perhaps the only organization that helps the secretary avoid spending money," one veteran Pentagon official said. "They are constantly searching for reasons not to spend money."

Soule, a seasoned analyst with nearly two decades of experience assessing Pentagon acquisition programs, was skeptical about the threat of ballistic missile attack. Like the chiefs, he figured that if an enemy wanted to deliver a nuclear bomb or germ weapon against the United States, a ballistic missile probably would be the least likely means chosen, since it would carry a definite postmark and leave the rogue state subject to obliterating retaliation. At the same time, he saw some merit in the argument that the mere possession of an ICBM by a rogue state could affect U.S. actions if America remained undefended. Most of all, Soule prided himself on his professionalism. He tried hard to keep any personal bias out of his analysis of weapons systems and to maintain an evenhanded approach in his briefings. It was his judgment that the development of a national antimissile system should be slowed to conform with the customary schedule of ten years or more for designing and testing complicated weapons.

Arguing the opposite case and pressing to retain the 2003 deadline was Walter Slocombe, top policy adviser to Defense Secretary William Cohen. A lawyer with years of experience in and out of government working on defense matters, Slocombe was widely regarded as one of the brightest minds in the department. He was probably the last hold-out in the Pentagon's upper ranks for sticking with the three-plus-three scheme. His main point was that in adopting the plan originally, the administration had indicated a desire to be ready to deploy an antimissile weapon if the threat worsened. Well, the threat had done just that, so to delay now would undercut the administration's credibility.

There was some irony to Slocombe's emergence as, in effect, the administration's most senior civilian hawk. He had worked on the ABM Treaty as an aide to Henry Kissinger, then national security adviser to President Nixon. He had opposed President Reagan's Strategic Defense Initiative. Yet here he was, arguing forcefully for a crash program to put up an antimissile system. When asked to explain his seeming conversion, he would talk about the "change in the strategic problem"—a reference to the missile-building rogue states—and note that he always had been "sympathetic to the idea" of a limited antimissile system, provided the threat was real and the system could be shown to work.

On the question of how fast to proceed, Slocombe drew a distinction between a new weapons system intended to replace an existing capability and the invention of a whole new capability. He could see good reason to be cautious and proceed gradually in the first case. There was no need to rush ahead with, for instance, a new fighter jet meant to provide an important but marginal improvement over an existing fighter jet. But in venturing after a capability that was lacking altogether—as was the case with national missile defense—Slocombe saw great advantage to getting at least something in place, however imperfect, and then improving it.

Reluctantly, Slocombe could rationalize a delay of a year or two in the deadline for a missile defense system on grounds that the project always had been considered very aggressive technologically. More development time, he figured, should boost confidence in the system's operational effectiveness. But Slocombe needed some convincing that this was the real reason for any administration move to delay. He was wary of those who would find whatever reason they could to put off ever constructing a national missile defense.

Lyles, the BMDO director, certainly did not fall in that camp, yet even he did not feel comfortable staying with the 2003 deadline. So 2005 became the obvious compromise, although many at the December meeting recognized that even 2005 was a stretch. "I think there was a widespread recognition that we would go for 2005 but that over time it would become 2007," Soule recalled months later. "I think virtually everyone thought that was where things likely would end up, but that didn't mean we should start there."

No decision was actually taken at the meeting. Such Defense Resources Board gatherings had become a venue more for laying out the issues and allowing people to vent than for reaching formal conclusions. Ultimately, the decision fell to Cohen to make.

A former legislator with a career interest in military and intelligence matters, Cohen was a longtime proponent of a national missile defense—not the expansive Star Wars vision promoted by President Reagan but a limited version to counter Third World adversaries and accidental or unauthorized launches by major powers. As a member of the Senate Armed Services Committee in the 1980s, Cohen actually had disparaged Reagan's notion of an all-encompassing shield against massive Soviet attack as simply beyond technological reach. He also had opposed moves by Republican colleagues to mandate deployment by a specific date, believing it was not possible to order technological progress. But Cohen felt a particular sense of paternalism toward the three-plus-three plan, whose origins could be traced in part to the legislative compromise that he had drafted with John Warner, Sam Nunn, and Carl Levin in 1995.

In joining the administration in 1997 as the lone Republican in the cabinet, Cohen knew he had been recruited in large part to ease strains between the administration and the GOP-led Congress, although when he had talked to Clinton about the job, the president never mentioned missile defense. "We really didn't talk about that, and I don't really know why," Clinton recalled years later. "I think I assumed he knew what my position was, which was that we ought to pursue the research. I had made no decision on any deployment because at the time we didn't have enough information to justify it or know where we were going with it."

Nonetheless, Cohen's presence in the cabinet would bolster the administration's credibility in claiming it was pursuing a viable missile defense plan. Outside, he would defend the plan to his former congressional colleagues, while inside, he would push to strengthen it. Soon after taking over as Defense secretary, he had increased funding for the program and now sought to transform it into a major acquisition effort with a real budget for deployment. He resolved to work quietly to bring the president's national security adviser, Sandy Berger, and Secretary of State Madeleine Albright—and ultimately Clinton—around to the idea of supporting a modest antimissile system.

For their part, Berger and Albright approached the whole notion of a national missile defense with skepticism. Both had opposed the idea during the Reagan years, as had their deputies, James Steinberg and Strobe Talbott. But being in power at a time when the Russian threat appeared to be waning and the Iraqs, Irans, and North Koreas of the world posed a rising menace, the Clinton team members found themselves rethinking their views of strategic deterrence and the relevance of antimissile weapons.

Just reading the daily intelligence reports about foreign missile developments—especially items about the transfer of missile parts and technology among the rogue states—had a jolting effect. "I got kind of a sense that there was a whole new element of danger to the United States," Albright recalled. "It's hard not to see it that way. I was fascinated by the networking among these countries."

In her previous job as U.S. ambassador to the United Nations, where she also frequently dealt with proliferation issues, Albright had taken to using the metaphor of Saturn and its rings to describe the problem. Saturn represented the United States and other nations with shared treaty obligations and a similar view of the rules that should govern the spread of dangerous technologies. The rings stood for the rogue states that were trying to circumvent the core group by constructing their own trading arrangements in pursuit of advanced missiles and weapons of mass destruction.

As they thought about what it would be like to confront a missile-wielding rogue state, the Clinton advisers began to doubt that the cold war notion of deterrence through threatened annihilation would work. "I wasn't fully satisfied that we knew how to think of deterrence in this changed environment," Steinberg recounted, "but my confidence that a Saddam [in Iraq] or Kim [in North Korea] wouldn't risk the annihilation of his country was much lower than my confidence that a Brezhnev or Andropov would," referring to the former Soviet leaders. To test his traditional views against recent real-world scenarios, Steinberg tried to imagine whether the United States would have reacted differently in the 1994–95 crisis over North Korea's nuclear processing facilities or the various standoffs with Iraq over U.N. weapons inspectors if those countries had possessed long-range missiles at the time. "I came to the conclusion that it might have made us more cautious and less willing to take on more challenges," he said.

But in the autumn of 1998, the administration's senior national security officials were just beginning to reason their way through such issues as a group. Cohen was more firmly decided about the need for an antimissile system than the rest. He would have to tread carefully. As a senator, he had gained a reputation as a maverick willing to break ranks with his party leadership. As Defense secretary, he had hoped to be seen as a team player and had taken pains to avoid appearing out of step with the White House. The missile defense issue would test his ability to stay true to his beliefs and credible to his former Republican colleagues in Congress while also remaining loyal to his cabinet associates.

As for the immediate decision he faced about adjusting the administration's missile defense plan, Cohen felt that he would lose all credibility on Capitol Hill if he went with the 2007 option proposed by PA&E. Besides, the advancing status of North Korea's missile program convinced Cohen that slipping the U.S. antimissile program by several years would be too risky. At the same time, he had to acknowledge that the existing timeline was unrealistic. He figured he could justify a delay of two years to 2005, particularly if he could persuade the White House to put deployment money in the defense budget.

On December 11, Cohen joined Berger, Albright, and Shelton, along with several aides, in the White House Situation Room to take stock of where the administration was headed on missile defense and arms control. The meeting would prove to be a pivotal moment, marking the administration's formal turn toward missile defense and setting the agenda of strategic military issues for the final two years of Clinton's term.

In a series of meetings in October and November, an interagency working group led by the NSC's Bob Bell had prepared a set of recommendations covering the range of matters up for consideration—future funding of national missile defense, preliminary talks with the Russians on amending the ABM Treaty, and renewed negotiations on deeper cuts in offensive nuclear arms. The recommendations were presented in a classified document that Bell titled the "Nexus Paper," because it dealt with the convergence of ABM, START, and National Missile Defense, or NMD, the official name of the Pentagon program. It proposed adding deployment dollars to the budget while putting off an actual decision on deployment until 2000. It also affirmed the need to begin talks with the Russians on revising the ABM Treaty in order to preserve the option of going forward with a national antimissile system. And it acknowledged the inevitability of linking the ABM discussions with future negotiations over arms reductions.

The official U.S. position had been to refrain from formal negotiations on START III until the Russian parliament, or Duma, had ratified START II. But to get a deal on national missile defense, it was becoming apparent that the United States would probably have to accede to Russia's interest in nuclear warhead reductions well below the START II level of 3,000 to 3,500—or even below the level of 2,000 to 2,500 agreed to in Helsinki in 1997. The Russians wanted a new START deal because they could not afford to maintain the launchers for as many as 3,000 single-warhead strategic missiles. At the time, administration officials still harbored hope that START II would be ratified by the Duma before the end of December, enabling the United States to begin START III negotiations without a shift in policy. But officials were resolved to proceed even if ratification was further delayed—which is what happened when U.S. warplanes launched four days of bombing runs against Iraq in late December.

The Situation Room was often the site of some of the administration's most sensitive national security deliberations. A compact space, especially compared to conference rooms at the Pentagon and the State Department, it encourages intimacy and close debate. Located in a basement corner of the West Wing, across from the entrance to the White House mess, the room contains a rectangular cherry-wood table that seats ten people in large black leather chairs. Another dozen or so chairs rim the room, up against dark cherry paneling that makes the setting feel even more confined. A presidential seal hangs on the wall at one end of the table. On a side wall runs a row of digital clocks showing various times: wherever the president is, wherever the latest crisis is, and Greenwich Mean Time.

Berger, who presided over the meeting, opened by saying that the decisions the group was confronting on the National Missile Defense program, the ABM Treaty, and START represented some of the most consequential in the administration's remaining time in office and would affect not only arms control but relations with the Russians and NATO governments. He also made clear that ultimately the president would be the one to decide which way to go. The message was that Clinton, not the Pentagon, would be the final arbiter on the shape of any national antimissile system.

Bell then briefed the group on the Nexus Paper. Talk quickly turned to the issue of funding. Cohen said an additional $7.8 billion was needed over the coming six years to cover deployment. Sylvia Matthews, the deputy director of the Office of Management and Budget (OMB), asked whether the money could be found in the existing Pentagon budget. Not all of it, Cohen replied. Matthews responded that OMB could not support the program until the financing was worked out. Although she voiced a willingness to work with the Pentagon to figure out some funding arrangement, she cautioned against a false sense of optimism that the White House could cover the gap.

Albright, worried about how the announcement of additional money for missile defense would be perceived, interjected that she wanted to be clear that the increase would in no way constitute a deployment decision; it was just meant to keep the Pentagon's program credible and allow for financing in the event of a deployment. Berger seconded her remarks. "We're not saying we'll deploy this system," he affirmed. "Our decision reflects the potential for deployment."

Berger also noted the recent efforts on Capitol Hill to pass legislation mandating deployment of a national missile defense system. Twice that year—in May and again in September—Senator Thad Cochran, a Mississippi Republican, had sponsored a bill requiring that a system be fielded "as soon as technologically possible." Both times, Democrats successfully filibustered the measure, defeating motions for cloture by one vote, fifty-nine to forty-one. (To invoke cloture and force a vote in the Senate requires sixty votes.) Berger stressed the importance of being careful now about how the administration's funding increase would be portrayed.

"What we're saying is, we're putting deployment dollars in the budget, not—repeat *not*—that we've made a decision on deployment," Berger remarked.

With the principals unable to decide on the source of the extra money, details were left to be worked out after the meeting by officials from the Pentagon, NSC, and OMB. Here the existence of a large projected federal surplus ended up making a big difference. Clinton had pledged an additional $112 billion to the Pentagon earlier that month in response to pleas from the chiefs; $6.5 billion of the increase had been reserved for "White House initiatives" covering counterdrug operations and other special programs. In a conference call after the Situation Room meeting, Bill Lynn, the Pentagon's comptroller, and Bell appealed to David Morrison, a White House budget official, to assign the full $6.5 billion to national missile defense. Morrison resisted at first, expressing concern about the fate of the other programs for which he had been holding the money. Lynn assured him that ways would be found to deal with them. Finally, Morrison relented.

The rest of the White House session focused on the timeline for negotiating possible changes in the ABM Treaty, given a presidential deployment decision in the summer of 2000, and on proceeding with START III talks in the absence of START II ratification. Relations with Russia were strained at the time as a result of NATO's decision to extend membership to some former Soviet-bloc nations and to intervene in Bosnia. Several participants at the meeting expressed concern that approaching the Russians about missile defense, without having even worked out the particulars of a system or made a deployment decision, could look altogether premature. But the consensus was that further delay served the interests of neither side. Part of the challenge, everyone seemed to agree, would be convincing the Russians of the need for early engagement—to make them understand that it was not in their interest to wait until a sense of crisis diplomacy developed in 2000.

"We had to impress on them that it's better for them to get out in front of this problem rather than wait until it lands on their doorstep with literally only days to deal with it," a participant recalled. With Albright scheduled to visit Moscow in January and meet Russian Prime Minister Yevgeny M. Primakov and Foreign Minister Igor Ivanov, she could begin to make this case in person.

Clinton himself did not attend the session or meet soon after with his national security team to discuss the new course. The team's own attention quickly ended up riveted on a week-long bombing operation against Iraq. At the time, too, Clinton was preoccupied by House impeachment proceedings against him and the prospect of a Senate trial early in the new year on various charges associated with his attempt to cover up his affair with a White House intern, Monica Lewinsky. Briefed by Berger on the missile defense recommendations, Clinton readily approved them. They would allow the administration to appear responsive while leaving the president flexibility in the months ahead. After all, he had made no deployment decision yet—that was still a year and a half away. In the interim, U.S. intelligence analysts could get a better idea of the extent of the missile threat; the first flight tests of the proposed system could be run; negotiations with the Russians could get under way; and U.S. allies in Europe and Asia could be consulted. So it was far from certain that the United States would end up fielding an antimissile system at all.

But Cohen and his aides considered that an important threshold had been crossed. The principals had agreed that the threat was or would be present and therefore the Pentagon could move ahead with development. A real countdown toward a deployment decision had begun, with significant money to construct the system now in the budget. Berger and Albright had acknowledged that the threat of missile attack was growing and that a national missile defense could well be an appropriate response. "It had been a groundbreaking meeting in the sense that people who had spent most of their professional lives not being favorably inclined toward missile defense made a turn," an aide close to Cohen asserted months later. "While there was no commitment to deploy, it was the meeting at which there was a general consensus on the need for a system and authorization to fund it. Both for substantive and political reasons, people were forced to reassess their views."

Of all the president's team members, Albright was perhaps the most self-conscious about the turn the group was making. "I remember that meeting very well," she said months later, "because I looked around the table, and here was the cast of characters who had all been the ones saying that Star Wars was crazy, that we didn't need anything like that, that basically not only did it not work but it could be dangerous because it would make people feel they were being protected when they were not."

But at the time, Albright also felt that the administration was doing the logical thing, the responsible thing—that she and the others had checked their ideology at the door and were responding as U.S. government leaders to a changed strategic environment. The one-two punch of the Rumsfeld report and the Taepodong launch had produced a consen-

sus that something more had to be done, that previous counterproliferation efforts had proven insufficient, that missile defense was perhaps
inevitable.

The December meeting could have played out differently; it could
have treated the funding issue as just a normal progression in the three-
plus-three plan and directed the Pentagon, for example, simply to place
greater emphasis on testing a variety of architectures to ensure that the
United States was doing everything reasonably possible to develop an
effective system. But it went further. It suggested a change in course and
conveyed a sense that the administration's view of the threat had darkened and its opposition to missile defense had eased.

The Clinton administration had started down what would prove a
slippery new policy slope, pushed partly by the Republican-controlled
Congress, but also pulled by its own alarmist reading of North Korea's
missile program.

Albright and Berger drew reassurance from the thought that the ABM
Treaty was not being abandoned. There had been no argument—no real
discussion actually—about the possibility of withdrawing from the
treaty, just as there had been little debate about alternatives to missile
defense in dealing with the ballistic missile danger. In fact, the whole
meeting had gone quite smoothly, with little apparent division.

But it did not take much poking below the surface to recognize that,
if the choice came down to one between the system and the accord, a
fault line would appear. The issue quickly broke into the open in
January 1999 at a Pentagon news conference when Cohen announced
the administration's decision to increase funding for national missile
defense and to approach the Russians about the possibility of revising
the treaty.

During a question-and-answer period, Cohen was asked his view of
the treaty. He noted its importance in limiting offensive weapons, suggesting that if defensive systems were allowed, then "Russia or other
countries would feel free to develop as many offensive weapons as they

wanted, which would then set in motion a comparable dynamic to off-set that with more missiles here." He said the treaty was "in our overall interests," but added that "it should be modified to allow for a deploy-ment" of a national antimissile system.

What if it could not be amended, he was asked. "Then we have the option of our national interest indicating we would simply pull out of the treaty," Cohen said.

The remark was factually correct. The treaty did allow either side to withdraw, with six months' notice, if it determines that "extraordinary events related to" the treaty "have jeopardized its supreme interests." But it was possible to interpret Cohen's comment as suggesting this was what he would recommend if the Russians refused to revise the accord, and the *New York Times* played the remark prominently in its story on the news conference, which led the front page.

Officials at the State Department and the White House were jittery. They wished Cohen had skirted the whole question without raising the prospect of withdrawal. They did not want to think about the option, since withdrawing from the treaty would have profound diplomatic repercussions for the United States. It would be interpreted as a lack of American commitment to other treaties. It would undercut efforts to get the Russian parliament to ratify START II. Privately, Albright asked Berger to speak with Cohen, and she spoke to the Defense secretary her-self, urging him to be more circumspect in his remarks on the subject. Publicly, Bell appeared in the White House press room the next day to carefully articulate the administration's position on the treaty.

All issues involving national missile defense, he stressed, "must, of course, be addressed within the context of the ABM Treaty. The ABM Treaty remains, in the view of this administration, a cornerstone of strategic stability, and the United States is committed to continued efforts to strengthen the treaty and enhance its viability and effective-ness. Secretary Cohen underscored yesterday that he believes it's in our overall interest to maintain the treaty and that the treaty is important to maintaining the limitations on offensive missiles that are contained in the START treaties."

"But," a reporter noted, "Cohen also pointed out that we would withdraw, probably would withdraw, if Russia doesn't go along."

"Well, I think it's important to be clear on exactly what the secretary said," Bell replied. "The secretary did not threaten to withdraw from the treaty, as has been reported. The secretary merely noted that the ABM Treaty, as in the case with every arms control treaty, contains a clause that gives that option."

As hard as administration officials sought to deny any internal rift, their nervous reaction even to the mention of the withdrawal option reflected a general ambivalence among most of them about the course on which the president had just embarked. Exactly how a national missile defense might be reconciled with the treaty was at that point quite vague in their own minds. But the hope was that the treaty's basic premise could be preserved while the text itself was stretched to allow for the sort of limited national missile defense being contemplated. A limited defense, it was argued, would not necessarily upset the strategic balance of power with the Russians, since the threats it aimed to counter lay elsewhere.

Still, the spectacle of the White House arms control specialist seeking to clarify the remarks of the secretary of Defense marked an inauspicious launch of Clinton's revised course, reinforcing skepticism among missile defense advocates about the administration's real commitment to its own program. It was just such skepticism that Cohen had sought to dispel by securing an increase in funding for national missile defense and ensuring dollars were in the budget to cover a possible deployment. The money, in Cohen's mind, would show his former colleagues on Capitol Hill that the administration's missile defense plan was not just a smoke-and-mirrors game. He also figured it would help bind the administration to its own program. The increased spending certainly succeeded in blunting the distress among missile defense hawks over word that the timetable for deployment had slipped from 2003 to 2005. But as the hand-wringing over the ABM Treaty indicated, a U.S. national missile defense remained far from a sure thing.

Chapter 6

A POLITICAL TIPPING POINT

As the administration began to move toward a tentative embrace of a national missile defense, Democratic lawmakers on Capitol Hill also were starting to rethink their resistance to the idea. Since the Republican takeover of both houses in the 1994 elections, Democratic leaders had opposed renewed GOP efforts to mandate deployment of a national antimissile system. In the previous year alone, they had successfully stalled action on Thad Cochran's bill calling for construction of a national missile defense "as soon as technologically possible." But their one-vote margins in sustaining filibusters against the bill were nail-bitingly narrow. The defection of a single Democrat would make it impossible to continue to hold the line. And by early 1999, it was evident that some Democratic senators had begun to waver.

For their part, the Republicans, after a string of legislative flops, had learned some lessons about how to draft a bill on missile defense that might entice enough Democrats to support its passage. The Cochran measure was notable for its simplicity. It consisted essentially of a single sentence: "It is the policy of the United States to deploy as soon as is technologically possible an effective national missile defense system capable of defending the territory of the United States against limited ballistic missile attack (whether accidental, unauthorized or deliberate)."

This unfettered approach had been the brainchild of Mitch Kugler, staff director of the Senate Governmental Affairs Committee's panel on international security, proliferation, and federal services, which Cochran chaired. A tenacious promoter of missile defense, Kugler had written Cochran in February 1998 reflecting on the repeated failures by the Republicans to gain passage of legislation ordering a national antimissile system. Since assuming control of Congress, the GOP had introduced legislation three times—in 1995 in a defense authorization bill, then in a stand-alone bill sponsored in 1996 by Bob Dole, followed by another stand-alone measure offered in 1997 by Majority Leader Trent Lott. But none of the language had made it into law.

Kugler concluded that the drafts had tried for too much by stipulating specific deployment dates, system architectures, and ABM Treaty considerations. They had been full of hooks for opponents to grab on to and use to pull down the legislation. Kugler had an idea for a spare, stripped-down approach. Stay away from specific dates, he advised, since history had shown how impossible it was to predict completion dates for defense programs anyway. And avoid micromanaging the Defense Department by telling it what system to field. The matter should come down simply to this: when the technology is ready, field it.

Kugler, who had served for a short time in the Army, invoked military terminology in his bottom-line assessment for Cochran. The previous legislation, he said, had "ignored the most basic of infantry lessons: fighting on ground chosen by your opposition makes it difficult to win." The bills had been "overly broad in scope, making it too easy for the administration and its Senate proxies to caricature." The new bill, Kugler argued, should be succinct.

The previous year, Cochran's subcommittee had held eleven hearings documenting the spread of missiles, nuclear weapons, and biological and chemical agents around the world. This too had been part of a revised strategy to build support for missile defense by elaborating on the rationale for it. "We had spent too much time talking about the solution and not enough time educating colleagues about the problem," Kugler said months later. "We were losing people before they even got to a discussion of the solution. We figured we had to go back and provide people with facts about the problem."

Based on evidence compiled in the hearings, the subcommittee had published a report in January 1998 entitled "The Proliferation Primer" that challenged the adequacy of the Clinton administration's responses to the emerging threat. It accused the administration of "speaking loudly but carrying a small stick" and declared, "Now is the time to take decisive action to protect ourselves from the proliferation of weapons of mass destruction and their delivery system." Kugler figured that the hearings and the "Primer" provided the best case to date for moving ahead with missile defense. "If we cannot make our case on the need to decide *now* to deploy NMD, after all the information garnered from these eleven hearings, it is unlikely we ever will be able to—short of an actual long-range ballistic missile launch by a rogue nation," Kugler wrote Cochran.

The one aspect that worried Kugler most about his own proposal was the lack of any firm deployment date in the new bill. Earlier bills had set 2003 as a target date, reasoning that a deadline would lend urgency that had been missing from the Pentagon's research program. Even the administration's three-plus-three plan had envisioned a possible deployment by that year. Kugler fretted that some missile defense advocates would decry the absence of a date in the new bill as a step backward.

But Kugler had concluded that legislation with a deadline could not get enacted. He also figured it could well prove technologically unachievable. And besides, the panel had heard testimony that a rogue-nation ICBM could materialize at any time, facilitated by the sale of technology—or even a whole missile—from Russia or China or both. So fixing on a specific date like 2003 made little sense.

There was a political advantage as well to omitting a date. It would deprive the Democrats of a favorite argument against past bills, which was that by mandating a deadline, the Republicans would be forcing the administration to make the oldest, least effective technology part of the national missile defense system. "Though a distortion and incorrect, this argument has been used effectively against us," Kugler wrote.

Most of all, Kugler liked the bill's simplicity because it provided an unambiguous policy statement. The battle needed to be fought one step at a time, Kugler argued. Once the bill passed, advocates could press for other details they had sought to pour into previous legislation.

Cochran approved of the approach. He was ready to lead the effort.

Cochran was an uncustomary champion of national missile defense. A courtly, white-haired figure, the Mississippi senator had not been among the most ardent Republican promoters of missile defense on Capitol Hill, who included Jon Kyl of Arizona, James Inhofe of Oklahoma, and Bob Smith of New Hampshire. But his conservative voting record had been impeccable during more than a quarter-century in Congress—as a congressmen from 1972 to 1978 and then as a senator. He had a reputation for toiling quietly but effectively, and he was not without ambition. He had vied with his fellow Mississippian, Trent Lott, for the majority leader's job.

Cochran's leadership of the bill would provide an advantage in wooing wavering Democrats, who might be more willing to go along with the mild-mannered southern senator than with one of the GOP firebrands normally associated with the missile defense issue. Cochran was even able to enlist a Democratic cosponsor, Daniel Inouye of Hawaii, with whom he served on the Appropriations Committee. The Hawaiian senator had a parochial stake in seeing a national antimissile system built as soon as possible since his state, along with Alaska, provided a much closer target for North Korea than did the continental forty-eight states.

Cochran's bill was first introduced in the Senate on March 27, 1998, after Lott agreed that its simple statement stood a better chance of passage than the more complicated measure he had authored the year before. A month later, the Armed Services Committee approved the bill for Senate consideration in a ten-to-seven vote that split along party lines, with Senator Joseph Lieberman abstaining. The Connecticut Democrat tended to be hawkish on many military matters and was coming around to the idea of supporting a limited national missile defense. But the Senate's Democratic leadership, along with the White House, strongly opposed any move to mandate deployment of a system. So when the Cochran bill came to the Senate floor in the middle of May, the Democrats successfully blocked consideration by filibustering. A cloture motion fell just one vote shy of the sixty needed to end the filibuster as only four Democrats—Inouye, Lieberman, Daniel Akaka of Hawaii, and Ernest Hollings of South Carolina—joined all fifty-five Republicans in voting to proceed with the legislation.

During the summer, missile defense advocates made clear they were spoiling for a fight over the issue. Republican National Committee Chairman Jim Nicholson identified missile defense as "the most important" security issue of the 2000 election. Additionally, Jack Kemp's Empower America organization was trying to win over some members of the Senate to the cause by running pro–missile defense radio ads in such targeted states as Nevada. Listeners were told: "We are only one vote shy of ensuring the safety of you and your family. But the people standing in the way are Nevada's own senators."

Moreover, it seemed that nearly every month another development abroad or at home underscored the proliferation of ballistic missile technology, including medium-range missile tests by Pakistan and Iran and underground nuclear tests by Pakistan and India. Following the North Korean launch at the end of August, Cochran and Lott were ready to try again with the missile defense bill, moving it to the floor on September 9. Once more, the Democrats mounted a filibuster. And when the vote for cloture came, the result was the same as the previous spring: fifty-nine to forty-one. The Republicans remained one vote short.

But through the autumn and into the winter, as the Clinton administration itself began to look more seriously at missile defense as a necessary response, so did some leading Democrats. The administration's announcement in early January 1999 of a boost in funding for missile defense and plans to engage the Russians in ABM Treaty talks signaled a swelling political sea change.

When Cohen visited Capitol Hill in mid-January to brief members on the administration's revised program, Cochran announced his intention to reintroduce his bill that day. As the senators filed out of the room, Lieberman approached Cochran and offered to become a cosponsor. The involvement of the Connecticut senator sent a strong signal to other Democrats.

Still, the administration appeared determined to oppose the legislation. In a letter to Warner and Levin on February 3, Sandy Berger, the national security adviser, threatened a presidential veto. Administration officials insisted they were moving as fast as possible to develop a missile defense. But they wanted to defer a decision on whether to deploy the system until mid-2000, partly to try to reconcile deployment with the ABM Treaty. They argued that because the Cochran legislation would peg a deployment decision solely to the defense system's technical feasibility, it would amount to a unilateral declaration of U.S. intent to disregard the treaty in order to deploy missile defenses. Thus, the argument went, Cochran's measure would undermine administration efforts to secure Russia's agreement to treaty amendments that would allow a limited antimissile defense while preserving the basic structure of the pact.

In his letter, Berger said basing a deployment decision solely on when the system became technologically possible constituted an "unacceptably narrow definition." Technology would undoubtedly be a factor in the ultimate presidential decision on whether to proceed with the system, but that determination, Berger said, would also take into account three other factors: assessments of the threat, deployment costs, and the impact on arms control negotiations.

It was the first time that a senior administration official had spelled out the criteria on which Clinton intended to base a deployment decision in 2000, and from then on, these four points served as the administration's decisionmaking framework. Drafted by Bob Bell and his deputy Steve Andreasen, they had antecedents as far back as 1995, when similar language could be found in the Senate compromise worked out by Warner, Cohen, Nunn, and Levin in an attempt to win passage of the 1996 defense authorization bill. That compromise eventually failed, but the idea of rooting a deployment decision in these four criteria survived and was seized on by the White House in early 1999 as a way of broadening the context for weighing the net costs and benefits of a national missile defense.

But even as administration officials were arguing for broader context and more deliberate consideration, momentum toward deployment was building among Democratic lawmakers. When the Armed Services Committee again took up Cochran's bill on February 9, not only did Lieberman join the panel's Republicans in voting the bill out of committee, but Senator Mary Landrieu of Louisiana, the ranking Democrat on the strategic forces subcommittee, abstained. With the bill headed to the floor, anticipation grew on the Hill over which of the Democrats would be the first to break ranks and provide the sixtieth vote needed to overcome another filibuster.

"We kept getting asked by Lott's office, 'What's your vote count?'" Kugler recalled. "I kept saying, 'Bring the bill to the floor. We have more than sixty. We just don't know who the sixtieth is.'"

To plot strategy, Minority Leader Tom Daschle huddled with Carl Levin, Joseph Biden, and Robert Kerrey, the ranking Democrats on the three committees that dealt with one aspect or another of national security policy—armed services, foreign relations, and intelligence. Their inclination was to attempt another filibuster, but they worried that party unity might not hold, particularly among the freshmen elected to the Senate the previous November—notably, Evan Bayh of Indiana, Blanche Lincoln of Arkansas, and John Edwards of North Carolina, all of whom came from states where there might be pressure to support missile defense.

The tactic used by the Democratic leadership to stall the bill—that is, blocking a motion to proceed with it—also unsettled some party members, who favored confronting the issue head-on in open argument rather than invoking a parliamentary procedure to forestall debate. Nonetheless, Daschle and the other party leaders resolved in a critical strategy meeting in March to try to work with the new members and sustain a united Democratic front against the Cochran bill.

But the next day, when the same senior senators met in Daschle's office to confer with some of the freshmen members, Kerrey suddenly indicated that he had changed his mind and would be supporting the missile defense legislation after all. The announcement stunned Daschle, Levin, and Biden—and left them angry. Prior to the meeting, their aides had conferred with the staffs of the new members to brief them on past confrontations over the Cochran bill and seek their cooperation in holding the line against it.

Then again, it hardly was out of character for Kerrey to go his own way. During his two Senate terms, the Nebraska senator and onetime presidential candidate had gained a reputation for independent-mindedness. Discussing his about-face later, Kerrey gave two explanations to associates. One was that he had been trying to balance his advocacy of the idea of further substantial reductions in U.S. nuclear weapons. The second was that he had been persuaded by the Rumsfeld commission report and the North Korean launch to regard the threat of possible missile attack on the United States as more imminent than he had thought.

One Senate aide involved in the missile defense discussions offered still a third explanation. He had the impression that Kerrey simply "was tired of carrying the administration's water." A senior administration official who conferred with Kerrey at the time added: "He was making a facts-of-life argument that we were just getting killed, that the Taepodong flight had knocked the props out from under the administration's credibility on this, and that he just couldn't stay with us."

Kerrey's shift clearly left the other Democratic leaders in the lurch, for it was evident they could no longer hope to sustain a filibuster. Even had Kerrey not broken ranks, it was questionable whether party unity could have been sustained in the face of doubts by other members. In any case, Daschle had never been defeated on cloture before—and he was not about to risk a defeat over missile defense. The Democrats would need a new strategy if they were to salvage at least a compromise.

If they could no longer block the bill, the Democrats hoped at least to have it revised in a way that might constrain any future deployment move. Two amendments ended up attached to the measure. One provision declared that any money spent on an antimissile system would be subject to the regular process of annual defense appropriations. The other asserted that the United States would continue to seek negotiated cuts in nuclear weapons with Russia.

Republicans willingly allowed the amendments, regarding them as too obvious or too general to be meaningful. GOP leaders figured the changes were hardly much of a price to pay if the Democrats would abandon their long-standing opposition to a bill favoring deployment of a national antimissile system. Cochran even took it upon himself to sponsor the provision on funding. The other amendment was put forward by Landrieu and Senator Olympia Snowe, a Maine Republican.

But to the Democrats, the changes amounted to considerably more than face-saving. After joining with Republicans on March 17 to pass the bill in an overwhelming ninety-seven-to-three vote, Democrats gleefully claimed to have inserted the legislative equivalent of a Trojan horse, substantively disarming the measure with amendments whose real significance had not been apparent to GOP strategists. "Kugler's strategizing had been brilliant," conceded the White House's Bob Bell, who had worked closely with congressional Democrats to blunt the initiative. "He really did get everyone in line to avoid the shoot-yourself-in-the-foot problem the Republicans had before. But then we set out to outmaneuver them, and I don't think they saw it coming. We didn't deliver the punch line until after the vote."

It was delivered by Levin, a Harvard-educated lawyer and onetime civil rights attorney from Michigan who had served in the Senate for two decades. An able debater with a deep appreciation for the complexities of missile defense, Levin had proven adept at using legislation to constrain pursuit of a national antimissile system. With his small reading glasses perched far down his nose, and a patient but firm manner of questioning and speaking, he would probe a witness at a Senate hearing or engage an opponent at a Council on Foreign Relations debate. His tireless focus on legal details sometimes bordered on nit-picking, but his grasp of the facts and political shrewdness made him a respected adversary among missile defense proponents. "He's great to watch," Kugler said admiringly. "He's always the first guy we think of whenever we're going to do anything on the floor, we're always trying to figure out what he'll do, because he's the brains of the operation over there."

Within minutes of passage of the Cochran bill, Levin appeared before reporters to explain what he said was the real significance of the amendments. The provision that made funds for a national antimissile system dependent on the annual budget process, he argued, meant that the legislation did not constitute a deployment decision. And the language on nuclear weapons cuts, he said, amounted to implicit support for the ABM pact, since in the past Russia had agreed to large cuts in its nuclear arsenal only because of the treaty. By endorsing the goal of further reductions, he said, the Senate was endorsing the means of achieving it. In this way, Levin sought to claim that any future decision on deploying a national antimissile system would have to be taken in the context of arms control. His basic point was that the legislation contained not one policy but two, and if they came into conflict it was not clear how they would be resolved. The Senate had not spoken to which was more important than the other, or so Levin asserted.

Such wrangling over the real significance of the amendments left both the Republicans and Democrats claiming victory, although clearly the GOP had the bigger win. Even with the strings attached to the bill, Cochran's measure had advanced the cause of missile defense. The lopsided vote represented a watershed event, reflecting bipartisan support for fielding a national antimissile system. Not since the Missile Defense Act of 1991 had Republicans and Democrats been able to find common ground on such a measure. The new legislation stepped up pressure on the administration, which was widely perceived as having been forced to give some ground at the risk of being swamped not only by political critics on the right but by centrist Democrats. The congressional action reinforced impressions abroad as well that the United States this time might really be serious about building a national missile defense.

At the same time, the measure set no details about the type of system and provided no specific money for it. It also allowed the administration to avoid any real commitment to deploying a system. In a cable to U.S. embassies—sent by the State Department but carefully reviewed by the White House—the administration sought to minimize the significance of the legislation and reassure nervous European governments. Highlighting the four criteria for proceeding, the cable said that no antimissile system would be built unless the missile threat materialized, the technology was shown to work, the cost was deemed affordable, and such a defense was determined not to conflict with arms control considerations.

A day after the Senate acted, the House also decisively approved a bill to field a national antimissile system by a 317–105 vote. Its measure was worded even more simply than the Senate version. It said nothing about technical feasibility, stating only, "It is the policy of the United States to deploy a national missile defense."

The House effort was led by Representative Curt Weldon, one of the Republican Party's most enthusiastic missile defense experts. Weldon had sought to coordinate with Cochran, but the senator worried that the chances of his own measure being accepted by his Senate colleagues might be jeopardized if it were associated with the hard-edged House member. "We wanted to build a coalition from the center out rather than from the right in," Kugler explained. "We didn't want people to think this was just a bunch of right-wingers trying to push the same old thing through again."

While the Senate had debated its version of the bill for three days, adopting two amendments, House Republican leaders had barred all amendments and completed the debate in one afternoon. The vote came after a rare ninety-minute closed briefing on the floor attended by about 250 lawmakers, during which they heard from Rumsfeld about his commission's findings on the missile threat.

With convincing approval of the missile defense measures in both houses in one week, the Republicans had scored big. But their victory soon risked being undercut by a fit of intraparty bickering as Senate and House staff members differed over whose version would be adopted as final by Congress.

Weldon was widely regarded as one of the hardest-working and best-prepared legislators on Capitol Hill, but also as a polarizing figure. A behind-the-scenes pragmatist, he could be an impassioned partisan fighter and downright bully in debate, especially on subjects dear to him like missile defense and U.S.-Russian relations generally. Weldon prided himself on having made his fifteen-word missile defense bill even sparer and more to the point than the Senate version. He referred cuttingly to the Senate's addition of two amendments as an act of cowardice and stressed that his bill had faced a straight up-or-down vote. Cochran, in turn, expressed some willingness to recast the amendments but stressed the need to include language that would ensure the support of enough Senate Democrats for passage. He also defended his version as containing not just a policy statement but a trigger for deployment.

Weldon's adamant posturing placed the House and Senate leaderships in an awkward position. No formal conference session was ever convened to resolve the differences. Instead, Republican staff members sought to work out a deal. Ultimately, a senior aide to House Speaker Dennis Hastert phoned a top aide to Lott and offered to accept the Senate's language in return for naming the bill HR4 as a gesture to Weldon. Although Weldon was not pleased with the outcome—and said so publicly—he had little alternative at that point.

The substantial bipartisan support for the Cochran and Weldon bills constituted a significant political consensus for the idea of a national antimissile system. But like earlier moments of political convergence on missile defense in the late 1960s and the early 1990s, this one was narrow—it supported only a limited architecture and assumed the international arms control framework would remain intact—and it was fragile, as was made plain by the simplicity and vagueness of the legislation. Few details had been included because sponsors understood that the consensus could easily break down over specifics, as it had done before.

Among the unaddressed issues, the most critical was the fate of the ABM Treaty. No mention of the treaty appeared in either the Cochran or Weldon bills, indicating that there was no agreement over the accord's relevance.

Clinton viewed the treaty—and the principle of mutual vulnerability embodied in it—as the foundation for past and future arms control deals with the Russians. It was the first treaty made with the Soviet Union that had established strategic weapons limitations during the cold war, and its continued existence provided a sense of stability and reassurance in East-West relations.

By contrast, missile defense advocates regarded the accord as a barrier to sensible technological development. Without the treaty, they figured, the United States would be able to build not only a missile defense system, but the right kind of system. In their view, the treaty had frustrated development of the most effective missile defenses and would continue to do so. Its ban on the testing and deployment of sea-based or space-based interceptors and other restrictions on radars and satellite sensors had led the Pentagon to pursue a limited land-based system that arguably was among the least effective and most costly alternatives.

There is no way of knowing whether alternative courses would in fact have proven more promising or economical. As things were, the Pentagon encountered substantial technical difficulty and cost overruns trying to develop just the scaled-down, ground-based version. But those who had overseen missile defense efforts under the first Bush administration insisted that their plans for a global defensive system, which included space-based interceptors, held considerable promise and would have stood a good chance of being fielded had the Clinton administration stuck with them. Henry Cooper, who headed the

Pentagon's missile defense office in 1991 and 1992, has characterized the Brilliant Pebbles system of space-based interceptors, which was under development at the time, as his "soundest program from management and technical perspectives." Cooper maintains that the technology was mature enough that it could have led to a first-generation space-based defense later in the decade. He also has been a leading advocate for a sea-based antimissile system that would use the Navy's existing fleet of Aegis-equipped ships.

Testifying before the Senate Foreign Relations Committee in 1996, Cooper detailed the argument that the ABM Treaty had cost the Pentagon's missile defense efforts time, money, and operational effectiveness. He said strict adherence to the treaty had contorted testing programs and constrained work on boost-phase systems that targeted missiles early in their climb rather than in space during the midcourse phase. It had limited the design of the Navy's Aegis ship radars, denying them the ability to track long-range ballistic missiles, and it had restricted the velocities of interceptors for the Army's and Navy's shorter-range antimissile systems, preventing use of these weapons against ICBMs.

"Such policy constraints—I call them a deliberate 'dumbing down' of our fighting capabilities—are a direct consequence of an excessive commitment to the ABM Treaty and, by extension, to its underlying idea that intended victims of ballistic missiles are most safe when vulnerable to attack," Cooper declared. He blamed the ABM Treaty for frustrating development of the "most effective" antimissile systems. "It has accomplished its purpose—no effective defense of the United States of America can be built consistent with its terms."

Republicans faulted Clinton not just for neglecting the most promising antimissile technologies but also for choking off a diplomatic opportunity to alter or dissolve the ABM Treaty. In the final months of the first Bush administration, U.S. and Russian negotiators had sought to capitalize on Russian President Yeltsin's offer to work jointly on a global defense system. Just how close the two sides were to a deal when Bush left office, and what U.S.-Russian cooperation would have meant for the ABM Treaty, remains debatable. But Clinton chose not to continue the talks as part of a general shift away from national missile defense. Instead, the administration reaffirmed the ABM Treaty's importance and embarked on what became four years of negotiations on treaty changes intended to breathe new life into the accord and ensure its viability for years.

The changes addressed two main areas of concern. One was determining which, if any, of the newly independent former Soviet republics would succeed the Soviet Union as signatories to the treaty. The other was deciding where to draw the line between the long-range antimissile systems prohibited by the accord and permissible shorter-range systems. Clinton and Yeltsin reached agreements in principle on these succession and demarcation questions in early 1997, and U.S. and Russian representatives settled on details in September. Under the succession agreement, the list of signatories would expand to include Belarus, Kazakhstan, and Ukraine, three of the fifteen independent states born out of the Soviet Union's demise. Under the demarcation terms, all low-speed interceptors (that is, under 3 kilometers per second) would be considered outside the treaty if not tested against target missiles with speeds faster than 5 kilometers per second or with ranges greater than 3,500 kilometers. The conditions for high-speed interceptors were a bit more involved. Not only would they have to adhere to the same testing limits as low-speed interceptors, but any concerns that parties to the treaty might want to raise about these more powerful antimissile weapons would be subject to consultations.

Missile defense advocates did not like the provisions. They objected to expanding the number of signatories out of concern that doing so would complicate the resolution of conflicts. And they criticized the demarcation accord for increasing the reach of treaty restrictions on missile defenses. In fact, Senator Jesse Helms and his fellow hawks appeared eager to receive an administration request for ratification of the proposed amendments in order to use the proceedings to attack the treaty's legal standing. So the administration stalled, declining to send the amendments to the Senate for approval.

Initially, the administration took the position that it did not have to submit the succession agreement in particular to the Senate. Administration lawyers argued that the president had exclusive responsibility for recognizing the dissolution of a country and establishing successor states, citing as precedent a host of other treaties with the Soviet Union that had not gone back to Congress for re-ratification. But the Republicans asserted otherwise. Eventually, the administration indicated it would submit the succession agreement in return for Senate approval in May 1997 of alterations to a 1990 treaty governing conventional forces in Europe. Those treaty changes, which U.S. officials hoped would soothe Russian fears over NATO's planned expansion, allowed Russia to adjust troop concentrations on its northern and southern flanks, enabling it to keep more heavy weapons for a longer time around troublesome republics like Chechnya. Republicans joined Democrats to give unanimous approval to the ratification resolution, which contained a stipulation obliging the White House to submit the proposed ABM Treaty amendments for ratification. Some Republicans even declared they were going against their better judgment in voting yes in order to lock Clinton in to his pledge to submit the treaty changes. One such member was Senator Kay Bailey Hutchison of Texas, who in a floor speech expressed "great reluctance" about supporting the measure but said she was putting aside her reservations because of the ABM Treaty provision.

Still, the administration continued to sit on the ABM Treaty amendments. A year later, in May 1998, nuclear weapons tests by India and Pakistan triggered another Republican assault on the general relevance of international arms control agreements, including not only the ABM Treaty but also the Comprehensive Test Ban Treaty, which prohibited nuclear weapons tests and was awaiting Senate ratification. In an op-ed piece in the *Washington Post* following the tests, Henry Kissinger, who had been President Nixon's national security adviser when the ABM Treaty was written, declared it no longer relevant. "The nuclear explosions by India and Pakistan have knocked the last prop out from under the administration's doctrinaire opposition to ballistic missile defense," he wrote. "It is reckless to stake the survival of a society on its vulnerability or on genocidal retaliation—even against an accidental launch. National and theater missile defense must become a higher national priority."

The renewed criticism prompted a passionate response from Madeleine Albright in defense of the administration's arms control policy. She described suggestions that nuclear nonproliferation was doomed as "dangerous nonsense," saying the Indian and Pakistani tests actually illustrated the logic and necessity of the existing arms control structure. "Efforts to halt the spread of nuclear weapons do not come with a guarantee," she said in a speech at an arms control forum. "But to abandon them because they have been dealt a setback would be a felony against the future."

Some missile defense proponents refused even to recognize that the treaty was still in force, contending that it had expired when the Soviet Union collapsed in 1991. According to international law and a long history of legal precedents, they argued, the United States was no longer bound by the treaty. In October 1998, eight Republican leaders, including Lott and Helms, sent Clinton a letter declaring the ABM Treaty null and void because of the dissolution of the Soviet Union. They were angry, they said, at the president's continued refusal to submit the treaty amendments and his reneging on a commitment to do so the previous year. "It is our position that the ABM Treaty has lapsed and is of no force and effect unless the Senate approves" the proposed succession amendment to revive the treaty, they asserted.

The fact that anyone was even raising the issue of whether the treaty had remained in force surprised some scholars, who figured it had been settled under the first Bush administration. A month after the dissolution of the Soviet Union, Secretary of State James Baker had affirmed that the United States accepted the ABM Treaty and regarded Russia as a successor to the Soviet Union's rights and obligations under it. At a joint press conference on January 29, 1992, following a meeting with President Yeltsin, Baker stated: "I made the point to President Yeltsin that the United States remains committed to the ABM Treaty." Since both the Bush and Clinton administrations had thus assured the

Russians that the ABM Treaty was still in force, government legal experts maintained that the United States had established by "state's practice" that Russia was the legal successor to the treaty. And state's practice, it was said, was more compelling than legal precedent for determining the status of the treaty.

In any case, not all Republican lawmakers agreed with calls to trash the ABM Treaty. As united as the party was in support of missile defense, several of the GOP's most prominent national security experts—including Thad Cochran, John Warner, and Richard Lugar in the Senate and Curt Weldon in the House—worried that dumping the treaty would rupture relations with the Russians and upset the fragile balance between the world's largest nuclear powers. Their preferred approach was to make room for missile defense while still keeping at least the essence of the treaty and possibly even pursuing some kind of cooperative missile defense arrangement with the Russians. Clinton administration officials saw in this group the potential core of Republican support for a revised treaty. "We ought not to be cavalier about this," Cochran told *Congressional Quarterly* in December 1999. "We have to recognize . . . the nuclear threat that Russia poses."

The administration's own reluctant shift toward accepting the possible deployment of a national missile defense system appeared to occur without much questioning, if any, of the value of the ABM Treaty. All of the president's senior advisers remained committed to the notion of preserving the treaty, although in amended form. Bill Cohen said as much when he announced the administration's plan for increased funding in January 1999, and Sandy Berger reaffirmed the administration's view of the ABM Treaty as "a cornerstone of U.S. security." This statement drew renewed huffing from Republican conservatives about the impracticality of trying to pursue missile defense while abiding by the treaty. "The ABM Treaty is the root of our problems," Helms intoned in the *Wall Street Journal.* "So long as it is a 'cornerstone' of U.S. security policy, as Mr. Berger says, we will never be able to deploy a nationwide missile defense that will provide real security for the American people."

With Albright about to go to Moscow in January 1999 to outline

U.S. interest in treaty changes that would permit a limited antimissile system, Helms appeared ready to force the issue of whether to keep the treaty at all. He demanded that the administration submit the three previously proposed amendments for ratification by June 1, saying that otherwise he would block review of other agreements that the administration wanted approved, most notably the Comprehensive Test Ban Treaty. "The single most important item on the Senate's foreign policy agenda this year must be clearing the way for a national missile defense to protect the American people," Helms declared. "And there is one thing standing in our way: the Clinton administration's stubborn adherence to the antiquated and defunct 1972 ABM Treaty."

The administration refused to comply with the deadline, taking the position that it would send along the changes only after the Russians ratified START II. Clinton appeared as determined as ever to fight a holding action on the ABM Treaty in hopes of keeping Moscow at the negotiations. If there were any doubts about the president's own commitment to the treaty, he dispelled them in March. At a joint press conference with Italian Prime Minister Massimo D'Alema on March 5, Clinton was asked by UPI's Helen Thomas how the United States could justify "chipping away at the ABM Treaty." He gave a convoluted answer that reflected his own discomfort over where his administration appeared headed, but he also made clear the limits of how far he was willing to go:

> Doing the research on a missile defense system, which is not a violation of the ABM Treaty—it is theoretically possible that we could develop a missile defense system that, either by its nature or by where it was deployed, would be a violation of the ABM Treaty. I, personally, have told the Russians over and over again I have no intention of abrogating the ABM Treaty. Anything we do, we will do together.
>
> But the only threat we have—excuse me—the threat that the United States is likely to face ten or twenty years from now from missiles coming in is by no means—not just from North Korea. It is a fact that many countries with whom we have serious differences now are making vig-

orous efforts either to build or to buy missiles with increasing ranges, that go distances far beyond anything that would be necessary to protect their own territory.

General Shelton has said that this missile defense is tough, it's like trying to hit a bullet with a bullet. That's what missile defense is. I think if we believe that the technology might be there, we owe it to ourselves and to all of our allies—not just our old allies, but some of our post–cold war allies—to try to develop that, along with an adequate warning system, to try to prevent countries that are desperately trying to get missiles—that they could not possibly need to defend their own territory—from ever taking offensive action against us or anyone else.

But I have no intention of supporting or initiating a unilateral abrogation of the ABM Treaty. I will not do that. We have been very candid with the Russians. We have talked to them about what we are doing. We have talked about what kinds of information we might share in the future. But I have never advocated, initiated, encouraged, sanctioned, or blinked at the possibility that we could unilaterally abrogate the ABM Treaty. I personally would be very opposed to that.

It would be eighteen months before Clinton confronted his self-imposed deadline for deciding whether to initiate steps toward deployment. During that time, in hopes of maintaining some pressure on Moscow to strike a deal, the president's aides would attempt to keep the Russians and the rest of the world guessing over Clinton's actual willingness to withdraw the United States from the treaty. But at the news conference with the Italian leader, Clinton effectively undercut any threat of a U.S. withdrawal. If push came to shove, he had little intention of walking away from the treaty.

When it comes to signing acts of Congress, a president can choose to hold a public ceremony or do the deed in private. Staged fanfares tend to be reserved for legislation that reinforces major aspects of the president's own agenda. Most signings are handled quietly, which was the way Clinton and his aides preferred to deal with the Missile Defense Act of 1999. As much as White House officials sought to claim victory for defanging the measure with amendments, they hardly wished to feature Clinton putting his name to a law that called for deployment of a national missile defense as soon as technologically possible. So the president signed the document out of public view, issuing only a four-paragraph statement on July 22 to mark the occasion.

The statement itself was a display of verbal gymnastics. Clinton insisted that the legislation was compatible with his plans to base a deployment decision on considerations of cost, missile threat, operational effectiveness, and impact on arms control—the same four criteria spelled out in the February letter to Senators Warner and Levin. However much Republicans were hailing the new law as advancing missile defense from a question of "whether" to "when," Clinton maintained that fielding a system was still not a foregone conclusion.

He said the legislation, by making missile defense funds contingent on annual appropriations, meant that no decision on deployment had been taken. He also cited the section stipulating that the United States would seek continued reductions in Russian nuclear forces; that section, he said, reaffirmed his administration's position that missile defense policy had to take arms control objectives into account.

"Next year, we will, for the first time, determine whether to deploy a limited national missile defense," Clinton said.

At the time his statement was issued, he and top aides were nearing completion of an intense internal policy review on just how to get to that moment of decision. They had spent weeks arguing over what kind of system to pursue and how to structure negotiations with Russia. The new law, even if largely symbolic, reinforced their inclination to take the next step toward a deployment decision. But it would be a step hardly taken enthusiastically.

Chapter 7

GETTING THE RED OUT

As the senior NSC staff member for defense policy and arms control, Bob Bell was the official responsible for coordinating interagency deliberations on missile defense and teeing up options for the president. Bell also happened to be the one person in the administration most convinced that a way could be found to deploy a national antimissile system without violating the ABM Treaty.

What Bell had in mind was actually a two-step approach. Step one would be to field a limited system that required no major treaty change. The system could then be expanded later into something more capable that would necessitate major revision of the treaty.

Bell had believed in this possibility for a decade, ever since his days working as a top aide to Senator Sam Nunn, the Georgia Democrat who had chaired the Armed Services Committee in the 1980s and early 1990s. It certainly would simplify things if it could be done. It would allow the United States to move ahead while avoiding a confrontation with the Russians and Europeans, who would then have time to get accustomed to the idea of missile defense.

The idea of a two-phased approach had appeared in the Missile Defense Act of 1991, which Nunn had co-sponsored and Bell had played a central role in drafting. It surfaced briefly again in 1995 when the Senate tried to break a political stalemate over a national antimissile system. The compromise plan negotiated that year by Senators Nunn, Levin, Warner, and Cohen would have made it U.S. policy to develop a treaty-compliant system by 2003, followed by a second phase that would have more sites and interceptors and require alterations in the treaty. That plan was rejected by House Republicans.

When the issue of national missile defense came surging back in 1998, Bell assumed that a way could be found to proceed with a similar two-phase plan. Informal discussions with a senior Russian diplomat had even given him some reason to believe the Russians might be persuaded to accept such an approach. But Bell had two substantial hurdles to jump before he could gain acceptance of the two-phase idea within the administration. One was technological, the other legal. Technologically, he had to demonstrate that a system based in North Dakota—the one place that the United States was allowed to operate an antimissile weapon under the treaty—could defend all fifty states. Legally, he had to make a case for why a system purporting to provide national coverage—even one based in North Dakota—did not on its face violate Article I of the treaty, which explicitly banned national systems. The North Dakota site had been intended to protect U.S. missile silos, not the whole country.

Through the winter and spring of 1999, Bell combed through the historical record of the ABM Treaty negotiations and consulted experts in an effort to build his case.

To the engineers in the Pentagon's Ballistic Missile Defense Organization, the North Dakota option was a nonstarter. All the computer models showed that interceptors based in North Dakota simply would not have enough time, after learning of a missile launch from North Korea—not to mention Russia or China—to reach Alaska and

Hawaii, the most distant states. By contrast, positioning the interceptors in Alaska would allow coverage of these two states in addition to the U.S. mainland, although defending the southeastern United States might then become a stretch against missiles from the Middle East.

Bell kept pressing the Pentagon to consider various options that might allow the interceptor site to remain in North Dakota. What would happen, he asked, if just the main radar used for tracking enemy missiles—known as an X-band radar—were placed, not in North Dakota near the interceptors, but in Alaska, closer to North Korean missile sites? Wouldn't that provide earlier warning of an enemy missile launch and thus allow U.S. interceptors to be fired sooner? And with earlier warning, couldn't an interceptor from North Dakota beat a North Korean missile to Alaska or Hawaii?

Pentagon officials responded that moving the radar would make some difference, but not enough to ensure total fifty-state coverage. In fact, even with the radar site positioned at the western tip of the Aleutian Islands, the North Dakota interceptor option would still leave the western half of Alaska and the western Hawaiian Islands unprotected.

What, then, if the radar were pushed even farther west and placed in South Korea? In such a case, BMDO replied, coverage would expand to all of mainland Alaska. But part of the Aleutian Islands would still be vulnerable, as would the westernmost Hawaiian Islands.

The population of these islands totaled only a few thousand people. Should the United States risk a major confrontation with the Russians and a split with NATO allies essentially just for the sake of a relatively few isolated communities in the Aleutian and Hawaiian Islands? Bell wondered. He also knew that the chance of the notoriously inaccurate North Korean missile actually hitting any of these islands was next to zero.

But the notion of settling for a system that provided anything less than full coverage never received serious consideration. For one thing, the military chiefs insisted that any national antimissile system had to leave no state undefended. It was a matter of principle for them. And in any case, doing otherwise would have resulted in political suicide. Both Alaska and Hawaii had powerful representatives in Congress. Senator Ted Stevens, a

Republican from Alaska, chaired the Senate Appropriations Committee, and Senator Daniel Inouye, a Democrat from Hawaii, was the ranking minority member. "Cohen in particular felt that he simply was not going to go up and try to tell Stevens and Inouye to support something that wouldn't defend all of their states, even if the islands were uninhabited or involved only several thousand people at risk on islands that the North Koreans couldn't hit anyway," Bell recalled.

The Alaska legislature had adopted a resolution in May 1997 demanding that the federal government deploy a missile defense system capable of defending all Americans on an equal basis. Alaska's representative in the House, Don Young, introduced a resolution making the same demand in 1998. Language from this resolution was incorporated into the House version of the 1999 defense authorization bill.

Nonetheless, Bell pressed BMDO for months for ways to close the gap between North Dakota–based interceptors and the westernmost U.S. states. He sought alternatives for speeding up warning times or reducing the amount of early-warning data needed before interceptor launch. He explored the possibilities of building faster interceptors. But always he got back the same bottom-line answer: nothing short of moving the interceptor site to Alaska would ensure coverage of all fifty states from a single site.

BMDO produced color-coded charts that showed the areas that would be protected by various site locations. The fully protected areas appeared in blue, the unprotected areas in red, and various other colors denoted partial protection. For Bell, the challenge boiled down to getting all the red out. "Every time they would come back, there would always be this red, so their summary chart of my option would say: 'Fails to meet JCS [Joint Chiefs of Staff] requirement to defend all fifty states,'" Bell said.

At the NSC, Sandy Berger supported Bell's efforts, as did the top officials at the State Department, although they remained skeptical that Bell could make a North Dakota solution work. Even at the Pentagon, the notion of a treaty-compliant option drew some sympathy, particularly from Deputy Defense Secretary John Hamre.

In February 1999, Hamre visited Alaska and North Dakota for briefings on the siting options. He came away with the troubling impression that BMDO had slanted the briefings in favor of an Alaska site. On the return flight to Washington, Hamre confronted Major General Willie Nance, Jr., the National Missile Defense program director, accusing the organization of bias. "You're presenting effectiveness charts that give Alaska preferential analytic grounding compared to North Dakota," he told Nance.

Hamre could not dispute that under phase one an Alaska site would provide more comprehensive protection than a North Dakota–based facility. His principal concern focused on phase two, when the threat would expand from North Korea to Iran and other Middle Eastern countries and protection of the East Coast would loom as a larger consideration. The Alaska site might be optimal against a North Korean scenario, Hamre noted, but it did not offer as quick a response time for defending the eastern United States against Middle Eastern launches. Of course, this disadvantage could be remedied by the construction of additional U.S. interceptor sites. But Hamre's question for Nance was: What if the United States ended up confined by treaty considerations and other factors to a single site?

"You can't do it from Alaska alone," he told Nance. "You've given up the East Coast in exchange for getting the Aleutian Islands."

Ultimately, the physics of siting became a moot point. Legal considerations prevailed. By June, the administration's lawyers had concluded that any national antimissile system—whether in North Dakota or Alaska—would require relief from the treaty's first article.

Bell knew the history of the treaty as well as anyone in the administration. As a senior aide to Senator Nunn in 1987, he had spent months (nine hundred hours, he once calculated) poring over the treaty's negotiating record in a windowless room in the Capitol reserved for classified material. At the time, Reagan administration officials had insisted that the treaty permitted not only research but the testing and development of defensive systems based on such exotic technologies as lasers and particle beams. This so-called "broad interpretation" of the treaty, formulated by Abraham Sofaer, a legal adviser to the State Department and a former U.S. District Court judge, marked a sharp departure from the way the American government had read the treaty for the previous thirteen years. Bell's exhaustive research resulted in Nunn delivering an authoritative and devastating critique of this view on the Senate floor.

Now, twelve years later, Bell found the tables turned. This time he was the one searching desperately for a broad interpretation that would permit deployment of a limited antimissile system in North Dakota. The irony was not lost on him—or on others in the administration. And yet Bell believed a credible argument existed for reconciling a limited national missile defense with the ABM Treaty's blanket prohibition against a national defense. After all, Sidney Graybeal, who had negotiated the treaty, had told Bell many times that U.S. negotiators intended the "no national defense" clause to mean no defense of all the territory at the same time against one attack—in other words, no "thick" defense. The treaty did not prohibit, Graybeal said, the "thin" defense of any spot against a limited attack.

Much to Bell's own amazement, he found that Tom Graham, general counsel for the Arms Control and Disarmament Agency, had presented a similar argument in internal Reagan administration deliberations. And Max Kampelman, Reagan's chief negotiator at the Defense and Space Talks in Geneva, had laid out this case in a plenary session with the Soviets.

Bell was convinced that if only the right lawyers could be found to look critically at the negotiating record, they could affirm a credible legal argument for reading the treaty differently than Clinton administration attorneys were now doing. Besides, Bell argued, the Russians probably would welcome such a flexible interpretation, since it would avoid a confrontation with the United States. It also would spare the White House from going to a critical Congress—and the Kremlin from having to petition a resistant Duma—for ratification of any treaty changes.

But Bell was not a lawyer and had no seat at the table of attorneys from the National Security Council and the Departments of State, Defense, and Justice conducting an interagency review. State's representative, Mary Lib Hoinkes, the department's chief arms control lawyer, was adamant that no national system, even one based in North Dakota, would comply with the treaty's ban on territorial defense. "Where national missile defense ran into trouble with the treaty," she was fond of saying, "was at the letter 'N'."

Hoinkes argued forcefully that if the United States wanted to build a national shield, it would have to get Russian agreement to modify the treaty—or withdraw from the accord altogether. And Article I was not the only spoiler. Other treaty provisions affecting the location of tracking radars, which had to be within 150 kilometers of the interceptor, and the use of early-warning radars, which were not supposed to perform tracking or other missile defense functions, also would be violated even if the interceptors stayed in North Dakota.

Ultimately, Bell found that technological and legal considerations left no easy way out. If the system were designed to try to comply with the ABM Treaty, it would fall short of the military and political requirement of fifty-state coverage. If it satisfied the requirement, it would bust the treaty.

In his spacious third-floor Pentagon office, Edward "Ted" Warner kept a large whiteboard on a wall that he used to jot down words, phrases, and symbols, often arranged as complex diagrams. That was the way he tackled complicated subjects, posting elements on a wall and looking for interconnections. It was an approach he had learned at RAND, the defense think tank he had joined after leaving the Air Force in 1982. Since the start of the Clinton administration, he had served as assistant secretary of Defense, first for strategy and requirements and subsequently for strategy and threat reduction.

Much of the working-level brainstorming on national missile defense within the Clinton administration during the winter and sum-

mer of 1999 took place in Warner's office. It was there that Pentagon
and State Department planners gathered to review options for configur-
ing a system—the possible numbers of interceptors and radars, where
they might be located, the relative effectiveness of each configuration,
and the degree to which each would depart from the ABM Treaty. They
also looked at associated timelines for construction and for negotiations
with the Russians.

Although it was clear to the group that the deployment of any effective system would require amendments to the ABM Treaty, it also was
clear that they were not to stray too far from the treaty's constraints in
reviewing alternative architectures. So no consideration was given to
incorporating sea-, air-, or space-based interceptors into the plan. Some
ardent missile defense advocates had regarded these technologies as
more promising and effective over the long term, since they offered the
possibility of intercepting enemy missiles in their boost phase soon after
launch rather than waiting for the midcourse phase, as in the case of the
land-based system. But treaty considerations, along with the fact that
ground-based interceptor systems were the most technologically mature,
had compelled the Clinton administration during much of the 1990s to
concentrate research and development on the land-based approach, thus
ensuring that this option would continue to lead the others.

A package of designs had emerged from the Ballistic Missile Defense
Organization, based on a projected evolution of the missile threat and a
determination of when the technologies necessary for various parts of
the national missile defense concept would reach maturity. In its basic
form, the system would work like this: Satellites would detect the hot
plume of an enemy missile soon after launch and send an alert to the
U.S. monitoring center in Colorado's Cheyenne Mountain. One or
more of a total of five upgraded early-warning radars based in Alaska,
California, Massachusetts, Greenland, and Britain would start tracking
the missile. A high-frequency X-band radar, with greater tracking and
discrimination capabilities than the early-warning radars, also would
key on the target from its planned location on the island of Shemya in
Alaska. All this data would then feed into a computerized battle man-

agement network that would prepare a target map and relay it electronically to the interceptor in its silo in central Alaska. Once the interceptor was launched, it would receive target updates. Minutes later, the kill vehicle would separate from its booster and home in on the target, obliterating it with the kinetic force of a high-speed collision.

This initial architecture could then be expanded by adding interceptor sites and radars, enabling the system to combat larger numbers of warheads and more sophisticated decoys. Pentagon planners had divided the expansion program into three phases: C-1 (for Capability-1) would be designed to cope with no more than five "simple" warheads—that is, warheads with decoys no more advanced than balloons and chaff. C-2 would deal with a maximum of five "complex" warheads—targets accompanied by more complicated decoys, radar jammers, and other deception measures. The C-3 system would be required to defeat up to twenty warheads loaded with complex deception devices.

Warner and the rest of his white-board group had identified a gap between C-1 and C-2. They concluded it was unrealistic to think that a rogue state would stop at a handful of missiles while trying to develop more complex warheads. It was more likely, they figured, to continue to produce simple missiles. So the group conceived of a fourth category dubbed "enhanced C-1," or "C-1-prime." It consisted of twenty "simple" targets and became the basis for phase one of the system ultimately proposed by the administration.

To combat twenty targets, defense officials calculated that at least eighty interceptors would be needed. This figure was derived from a firing formula that envisioned shooting four interceptors at every target to raise the likelihood of intercept to a very high level. But since the ABM Treaty allowed for a maximum of one hundred interceptors at a given site, the Pentagon's plan for phase one became one hundred interceptors in Alaska (twenty by 2005, then eighty more by 2007). This would be more than enough to cover the C-1-prime threat, officials figured.

To get to C-2, plans called for three more X-band radars, including two potentially controversial ones overseas—in Fylingdales, England, and Thule, Greenland. Authorities in Britain and Denmark (which governs Greenland's foreign relations and defense policies) already would have been asked as part of the C-1 deployment to allow for upgrades of existing early-warning radars at Fylingdales and Thule. But U.S. officials anticipated that gaining approval for entirely new radar facilities at these sites would be more difficult given European concerns about the U.S. missile defense effort.

The full C-3 threat would require a second interceptor site and a total of 250 interceptors, one more early-warning radar, and several more X-band radars. In addition, the four-satellite Defense Support Program (DSP) system on which the United States had depended for thirty years to detect missile launches around the world was scheduled to be replaced by the six-satellite Space-Based Infrared System-High (SBIRS-High), which promised faster scanning rates and a new "stare" capability, and thus more accurate data for launching antimissile interceptors. Still greater tracking features, plus the ability to distinguish between warheads and decoys, were said to be coming with SBIRS-Low, an elaborate constellation of two dozen or more satellites that was planned for 2010.

So deployment of the National Missile Defense system was going to have to occur in phases, just as the threat was projected to develop in stages. The question then arose: Didn't it make sense for the U.S.-Russian treaty talks to go in phases as well and to focus first on just what would be needed to field the initial C-1 architecture? Or should the United States, in a single negotiation, seek all the treaty amendments it would need to build the entire two-site system, with the capability to counter the C-3 threat?

This became a hotly contested point in midlevel interagency discussions. The Pentagon favored a single negotiation with the Russians, preferring to nail down early everything that would be required for the complete C-3 system. But officials at State and the White House argued for splitting the talks into two phases and pursuing a modest one-site deployment in the opening stage. The rationale was that the administration lacked enough time still in office to do a bigger deal, and besides, this approach would be much easier for the Russians to accept. It also would give the next U.S. administration more time to research alternative second-phase architectures, including possibly some kind of cooperative arrangement with the Russians.

It was not until mid-June, six months after their decision to step up work on national missile defense, that Clinton's most senior national security aides returned to the issue at a Principals' Committee meeting. They had been preoccupied with NATO's air war against Yugoslav troops in Kosovo. The allied victory had fortified U.S. resolve to stand tough against dictatorial regimes abroad, but it also strained relations with Russia, which had strong cultural and political ties to Serbia, thereby complicating the outlook for any deal on national missile defense and nuclear arms cuts.

With Clinton scheduled to see Yeltsin later in June in Cologne, the president's top advisers convened in the Situation Room to refocus attention on moving forward with the ABM Treaty and START talks. Hard-liners at the Pentagon had argued against linking START III and the ABM Treaty, or even beginning discussions on START III, as long as the second START accord was not ratified. But it seemed inevitable that the two tracks would be joined.

As the White House session got under way, Cohen urged that in proceeding with START III discussions the United States not concede a defeat of START II and instead continue to insist that the Yeltsin government focus on getting it ratified. This led to a lengthy exchange over the term that would be used to describe the upcoming START III conversations between U.S. and Russian negotiators. Would they be "negotiations," "discussions," or "talks"?

The term "negotiations" was sensitive for both the United States and Russia. U.S. officials did not want to be negotiating START III while START II was unratified, and the Russians did not want to acknowledge that they were negotiating actual changes to the ABM Treaty. It seemed better to employ a less formal term. For this reason, Albright, who was about to leave for a meeting in Helsinki with

Russian Foreign Minister Igor Ivanov to plan the summit communiqué, was authorized to use the term "talks" or "discussions"—whichever the Russians would prefer—but not "negotiations."

The principals then turned to a review of national missile defense. A freewheeling discussion ensued. How capable a missile defense system did the United States ultimately need—C-1, C-2, or C-3? What was the best way to sell it to the Russians? How would the Chinese react? The European allies? Would they support the positioning of any radars on their soil?

Berger reaffirmed that a presidential decision on deployment would be made in June 2000, so that talks with the Russians were to get under way by the coming September. This left only a month or two to make some tough decisions about the system architecture and negotiating approach. John Podesta, the president's chief of staff, noted that the Russian election cycle would be a factor as well, with a presidential vote there due in the summer of 2000. Berger ended with instructions to keep working the issue over the next three to four weeks, and he noted that the president would have to be brought into the process.

The president's top aides remained divided over the question of whether to proceed in one phase or two. Cohen favored going for a single, all-inclusive negotiation with the Russians; his worry was that, if the talks were phased, the United States might concede too much in the beginning and have little to bargain with later. He also was concerned that a phased approach would turn the missile defense issue into a continual source of friction in U.S.-Russian relations for years to come. Each time the antimissile system needed to be modified to respond to a changing threat situation, the treaty would have to be renegotiated.

But Albright, Talbott, and Berger preferred phasing in order to avoid overburdening an already fragile U.S.-Russian relationship by trying for too big a deal too quickly. They argued that recent tensions over NATO expansion and the air war in Kosovo had prompted Moscow to regard the United States more critically and soured prospects for cooperative ties. They said that by asking the Russians initially to accept only a limited modification of the ABM Treaty, an important precedent would be set and some confidence gained that would then facilitate a follow-up negotiation. In this view, the first and most significant step was to achieve a conceptual breakthrough—namely, Russian acceptance of the idea of a limited national antimissile system.

Berger was determined not to go to the president for a decision until all the senior advisers were in agreement. This would mean bringing Cohen around to accepting a two-phased negotiation, and bringing Albright and Talbott around to accepting not just one interceptor site but a second one for an eventual C-3 capability. But Berger also wanted to avoid trying to hammer out the differences in more group meetings, which might only lead to a hardening of positions. So he engaged in a series of one-on-one sessions with Cohen and some private conversations as well with Albright and Talbott.

"Sandy's brain was the equivalent of the Situation Room during all this. He did a lot of what I call 'virtual interagency,'" a senior State Department official said. "What he wanted to avoid was a classic schism, with the president having to play King Solomon with a baby—that is, having to choose between a secretary of Defense saying we absolutely have to have national missile defense and a secretary of State saying we can't do this."

At least twice, Berger traveled across the river to visit with Cohen in the Defense secretary's office. "You know Sandy is really working a problem when he comes over here," said a Pentagon official who sat in on the sessions.

The conversations were always amicable but frank. "You're going to get it shoved down your throat, Sandy," Cohen warned at one point, meaning that unless the administration acted forcefully, the Republican-led Congress would mandate a deployment. And Republican lawmakers, everyone knew, were wary of any two-step approach with the Russians, as were Cohen's top lieutenants, John Hamre and Walt Slocombe. Somewhat ironically, given their ambivalence about spending significant sums on national missile defense, the military chiefs also favored a one-shot approach.

Berger argued that only so much was achievable in the relatively little time the administration had left in office. He told Cohen that there was no way to modify the treaty substantially and get it approved by Congress by November 2000. "We're going to have to find a two-step process so that you can get an achievable first step that buys us time for a second," he said.

Berger also warned of the consequences of going for broke and failing. Such an outcome, he said, was likely to alienate the NATO allies and jeopardize U.S. ability to put the X-band radars in England and Greenland, as envisioned for phase two.

Cohen pressed his concern about U.S. negotiators retaining bargaining leverage in a second phase. Berger said the threat of withdrawing from the ABM Treaty would remain. It would apply to a second phase as much as it would to a first one.

By mid-July, Cohen, though he still regarded the one-shot approach as the most logical, was willing to accept a two-step plan as the most pragmatic. But he insisted that the Russians be told that any first-phase deal would be followed quickly by negotiations for a second phase, with the ultimate U.S. aim being deployment of a C-3 system.

For their part, Albright and Talbott were willing to commit to a second phase, particularly in view of a new National Intelligence Estimate that they knew would soon be issued predicting that a more sophisticated rogue-state missile threat could materialize sooner than previously anticipated. They were relieved to have gained Pentagon acceptance of the two-phase approach and to have limited the initial phase to a single site of one hundred interceptors.

Albright figured it was only natural, given the built-in institutional biases of Defense and State, for officials at the two departments to hold contrasting perspectives. But she worried about the Pentagon getting ahead of the story—about being so focused on building a robust antimissile system that it would lose sight of the effect that deployment of such a system would have on the Russians, the Chinese, and the NATO allies. Diplomacy, of course, was her primary job, and she sometimes wished Cohen and his aides were more sensitive to that aspect of the issue.

But she was determined to stay on friendly terms with the Defense secretary. One incident in the spring of 1999 showed the care she took to preserve the friendship. It involved a bit of history that the *Washington Post* reporter Michael Dobbs had come across while researching a biography of Albright. In President Jimmy Carter's archives in Atlanta, Dobbs discovered an assessment that Albright had written of Cohen in the 1970s. At the time, Albright was working in the Carter administration as a legislative assistant to the national security adviser, Zbigniew Brzezinski. Her job was to assess the positions of lawmakers and advise Brzezinski on how to deal with Congress on major issues. One politician she watched was Cohen, then a freshman Republican senator from Maine. In one memo, Albright noted that Cohen "is on [the] Armed Services [Committee] and has already begun arguing about SALT," the Carter administration's big strategic arms limitation foreign policy issue, which it then was trying to sell to Congress. "He has an overestimation of his intellectual capabilities, but is a clever self-promoter and will, therefore, be talking more than most," she wrote. "It would be useful to try to influence him."

Tipped off that word of Dobbs's discovery was going to appear in a *Washington Post* Sunday magazine column by Al Kamen on April 11, Albright went to a bookstore and bought two books on forgiveness. She handed them to Cohen. "Why are you giving them to me?" he asked. "Because," she replied, "you're going to have to learn to forgive me after you see what they're saying I said about you in 1978."

Albright remained ambivalent and very self-conscious about the administration's steady movement toward a possible deployment decision. Although she had gone along with the decision in December 1998 to increase spending and begin talks with the Russians, she continued to feel odd about finding herself part of a team that would support even a limited antimissile system. She was acutely aware of the irony of her role in supporting a missile defense development program now, years after her opposition in the 1980s. Before the Principals' Committee meeting in June, she had asked two aides who had been staunch, card-carrying members of the arms control community—James Rubin, her press secretary, and Mort Halperin, her senior policy adviser—whether the administration should be taking the path it was.

"You know, it's funny," she remarked to them. "I remember in all my political campaigns and all the work I've done, I was always fighting for these treaties. I feel funny about this and how strange it is for someone who was battling Reagan on these subjects to now be on a different side than the arms control community."

Rubin and Halperin assured Albright that the proliferation problem required new solutions.

"I think that helped convince her that it wasn't some right-wing, radical, Republican, Defense Department thing," one close associate said. "I mean, Halperin is one of the intellectual fathers of arms control, and Rubin was a young apparatchik of the arms control business. And if they were both telling her that this was a responsible course, I think that helped her."

To some senior Pentagon officials, Albright's lingering doubts represented what one top Cohen aide called a "lapsing back" from the administration's decision in December, which the Pentagon had viewed as a fundamental policy shift. At the June meeting, Albright had expressed concern about the effect that a U.S. deployment might have on China. What if it triggered a Chinese buildup, which in turn could upset India and Pakistan? She suggested that John Holum, the State Department's senior arms control official, brief a subsequent meeting on such potential ripple effects. "There was nothing wrong about her raising such issues," said one Pentagon official who was present. "But it clearly reflected a sense of her searching for reasons not to go forward."

Indeed, if the Pentagon was wary of State getting weak knees, the State crowd viewed the Pentagon suspiciously as trying to rush the issue. Through the winter and spring of 1999, Halperin had closely monitored Cohen's public pronouncements on missile defense, looking for deviations from the administration's stated policy. Several times, Halperin had negotiated precise wording with defense officials for a Cohen statement, only to see the Pentagon leader take a harder line in his actual remarks—whether on his inclination to advocate withdrawal from the ABM Treaty or on the extent to which the administration already had decided to deploy a national antimissile system.

"Mort was very conscious of the Pentagon moving ahead of the interagency consensus," a senior Albright aide recalled, "and he was constantly offering up a quote, where Cohen or somebody went one word beyond the interagency consensus. He would urge Albright to call Berger or call Cohen directly to get everyone back to what had been agreed."

Halperin and other State officials also remained reluctant to support the idea of an Alaska site. Meeting with Albright in early July, Halperin raised the prospect of getting the Pentagon to look once again at the North Dakota option. To protect the few thousand people in the Aleutian Islands and Hawaiian Islands who would not be covered from North Dakota, he suggested using ship-based interceptors for that contingency alone. Or maybe the Pentagon could come up with some other creative approach, he remarked. "He wanted to go back one more time to the Pentagon and say, 'We think in terms of the negotiability with the Russians and acceptance by our allies, North Dakota is much better than Alaska, and we'd like you to take another run in a creative way,'" one participant in the meeting recalled.

Several days later, however, an item about the discussion in Albright's office appeared in the *Washington Times*. It gave a somewhat distorted account, saying that Albright and her aides had considered leaving parts of Alaska and Hawaii uncovered. The State Department issued a statement denying that the secretary would ever consider such a move, but the story also killed any further effort by Albright's team to explore some kind of alternative protection for the western islands.

On July 16, the president's advisers again convened in the Situation Room to consider national missile defense. Berger had met several days earlier on Capitol Hill with Daschle, Levin, and Biden, the three key senators who served as a kind of political barometer for the administration on missile defense. "Their message to Sandy was: 'There's a train wreck coming. The Republicans will be unstoppable on this issue unless the administration comes up with a credible and proactive position soon, preferably before the end of July,'" said a senior administration official familiar with the conversation.

By that time, Berger had managed to achieve a consensus between the Pentagon and the State Department on proceeding in two phases toward the ultimate objective of a C-3 system in 2010. Now the issue was deciding just what to include in the first phase of negotiations. The main sticking point came over the issue of the X-band radars that would be needed abroad—at Fylingdales in England and Thule in Greenland. They were not required for the C-1 system, which was focused on the North Korean threat, but they would become critical later in combating any missiles launched from Iran or other Middle Eastern states.

Whether to include them in the phase-one negotiations with the Russians hinged on a complex of military and diplomatic considerations. The Pentagon wanted the X-band radars in the first round because of concerns about the pace at which Iran might develop long-range missiles. But the State Department worried that early construction of the X-bands in Europe would be hard to sell to the Russians— and to the Europeans as well, since it would deepen their own role in the controversial weapon system.

"The radars became the central point of contention," a senior defense official said. "How soon we built them affected how capable the system would be. If we waited, it meant the system remained quite a bit less capable against Iran than North Korea. The argument for waiting was that they weren't needed yet and they presented a heavier load to carry both to the Russians and the Europeans."

The issue was not resolved at the July meeting. Berger was reluctant to have the president's advisers argue out their differences around a table, even in the secretive setting of the Situation Room. He would resort again to his personal diplomacy and lean especially on Cohen to defer the X-band radars to the second phase. For the moment, however, Berger wanted to impress on the others the significance of their deliberations. He concluded the meeting by observing that what the administration decided on missile defense would constitute the most important strategic decision in the eighteen months it had left in office.

Noted almost in passing at the July meeting, as it had been in June as well, was the issue of the system's cost. Both times it was someone from the White House Office of Management and Budget who broached the matter.

In June, OMB Director Jack Lew had briefed the meeting on the potential costs, emphasizing that over the next ten years the system's multibillion-dollar price tag would converge with the cost of other big-ticket weapons programs, including new generations of jet fighters and combat ships. The Defense Department, he stressed, would be facing an enormous bow wave of expenses just when the bill for missile defense was coming due as well.

In July, Sylvia Matthews, the deputy OMB director, again raised the bow wave problem. She said the added cost of missile defense could profoundly affect not only overall defense procurement but the whole domestic federal budget. Looking directly at Cohen and Shelton, she asserted that much of the price would have to be borne by the Pentagon out of its existing budget; it was unlikely that any new money would be forthcoming. Shelton, always sensitive to the cost issue, noted his concern. He said the cost did not make the project undoable; it just compounded the Pentagon's funding problems.

Berger sidestepped the issue. He responded that the question of how to pay for missile defense would require a separate conversation. But budget officials continued to worry about it. One senior OMB aide who worked on the issue considered the costs of missile defense outlandish. "The whole GDP of North Korea is $21 billion," he observed. "We could buy the whole country, carpet it, use it as a Club Med or something, for what we plan to spend on national missile defense. Of course, I'm being a tad facetious—but not much. There's a cost-effectiveness issue here that we may never get our arms around because of the politics and the threat and everything else."

• • •

Within a few weeks, Cohen had agreed in conversations with Berger to exclude the additional X-band radars from the first phase but to make clear to the Russians that negotiations on them would have to start in the spring of 2001, immediately after the planned conclusion of the first phase. "The last piece to fall in place was this notion that we would push the Russians sort of right after phase one to accept new radars at Thule and Fylingdales so that the system could grow fairly quickly to handle a bigger and more complicated attack from the Middle East," a senior State Department official recalled. "So to that extent there was a sort of blur in the distinction between phases one and two."

Cohen still worried about how he would justify the two-step approach to missile defense advocates in Congress. To help, the NSC prepared answers to an extensive list of possible questions that conservative critics might raise about the plan. For instance: Doesn't this deployment option ignore the extent of the rogue-state threat in order to preserve the ABM Treaty? Why should we not just deploy everything we think we need right now with or without Russian support? Doesn't the deployment provide uneven coverage of the fifty states? Is it true that the deployment option you're pursuing will not be able to address any threat that develops in the Middle East between 2006 and 2010? If we simply deployed ABM radars in the United Kingdom and Greenland, couldn't we have a greater capability against the Middle Eastern threat?

On August 16, Berger convened another Principals' Committee meeting to affirm the consensus and prepare to go to the president with a unanimous recommendation. Although the senior advisers had resolved their differences over the planned structure of a national missile defense system, they still faced the question of what they would be willing to trade for Russian acquiescence in terms of reductions in offensive arms. Two years earlier, at a meeting between Clinton and Yeltsin in Helsinki, agreement had been reached in principle to lower the maximum number of strategic nuclear weapons to a range of 2,000 to 2,500 under a START III accord. Now the Russians were pressing to go even lower, down to 1,500.

But the Pentagon was reluctant to take the offer without a thorough review of U.S. nuclear doctrine, which would take months to complete. At the August meeting, Admiral Richard Mies, the head of U.S. Strategic Command, which oversees the nation's nuclear arsenal, briefed the principals on the risks of dropping precipitously below the 2,500 level. Cohen and General Joseph Ralston, the vice chairman of the Joint Chiefs, also expressed a tactical negotiating concern: If the United States gave away the basic trade-off in phase one by agreeing to cut its nuclear arsenal, what would be left to bargain away in phase two?

State Department officials contended that just getting the Russians to change their position and acquiesce to missile defense was a big step—arguably the biggest—so that taking it would warrant the biggest U.S. concession. Anything after that was bound to be more modest and thus worth less significant concessions.

But the principals clearly were not ready to set a bottom line. That issue would require more study. Given the link between offensive and defensive weapons, the president doubtless would want to know what flexibility his advisers were recommending on the offensive side in negotiating for a national missile defense system. But the most they could tell him at this point was that the talks with the Russians should proceed; a decision on the U.S. position on reductions in the offensive force would have to come later.

The whole question of whether to proceed in phases and what to include in phase one was of particular interest to Vice President Al Gore, who had his eye on the presidential race in 2000 and what he would inherit if he won. Gore did not attend the Principals' Committee meetings, but his national security adviser, Leon Fuerth, was at the table. Fuerth was not happy at the realization that much of what could be bargained with would be offered in phase one, while major items like the X-band radars would be left for phase two. He worried about how it would be possible to negotiate phase two, which also would involve changes in the ABM Treaty. The hill to climb would be steeper, and there would be less in reserve from which to draw.

Nonetheless, Fuerth, who had favored a one-shot approach, eventually concluded that phasing was the only way to move forward and that it was a reasonable course to take given the limited time remaining in the Clinton administration. At the same time, he wondered whether the whole plan might not eventually collapse, spilling the pieces into the lap of Clinton's successor. In that case, Fuerth figured, the new administration would have to start from scratch.

On August 18, Clinton met with his national security team in the Cabinet Room to review their recommendation on national missile defense. He sat on one side of the long table with Steve Richetti, his deputy chief of staff. Facing him on the other side were Berger, Cohen, Albright, Talbott, Ralston, Steinberg, Bell, and Fuerth. Only two aides were present in a back row of chairs: the NSC's Steve Andreasen and the Pentagon's James Bodner.

Entering the room, Clinton started things off with a joke at Talbott's expense.

"Well, Strobe, what have you done today to sell out U.S. national security to the Russians?" the president quipped, glancing at the deputy secretary of State, the administration's top Russian expert and an old friend of Clinton's from their days as Rhodes scholars at Oxford University in England. Congressional Republicans had been accusing Talbott for years of cozying up to the Russians. A White House photograph taken just at that moment shows everyone around the table laughing and looking over at Talbott, who was leaning forward, uttering some retort now lost to history.

Berger then proceeded with a five-minute introduction framing the issues. He characterized the decisions that the president had before him as far-reaching ones that were likely to have an impact on the rest of his term. Then he turned to Bell to detail the package that the president's advisers were recommending.

It was a particularly poignant—and difficult—occasion for Bell. The meeting marked the culmination of months of interagency staff work largely coordinated by him. It also was Bell's last day on the NSC after seven years; he had landed a senior management post at NATO head-quarters in Brussels. And he was very sick: he was awaiting surgery for prostate cancer diagnosed only several weeks earlier. He had run a temperature of 104 degrees for a week—a reaction to a biopsy-induced infection—and had dragged himself out of bed to attend the White House meeting.

Bell described for Clinton the proposed system architecture. It envisioned deployment of twenty interceptors at Fort Greely, Alaska, by 2005, and a total of one hundred by 2007. The first-phase system also would include an X-band radar on Shemya Island and the upgrade of five early-warning radars in Clear (Alaska), Beale Air Force Base (California), Cape Cod (Massachusetts), Thule (Greenland), and Fylingdales (England), plus the new six-satellite SBIRS-High network. A second phase, due by 2010, would involve a second site—most likely in North Dakota but also possibly in Maine—with up to 125 interceptors at each location. The number of X-band radars would grow to nine, and the number of early-warning radars to six. And the SBIRS-Low constellation of twenty-four to thirty satellites would start operating to provide added capability not only in detecting and tracking missile launches but in discriminating between warheads and decoys in space.

No final decision on deployment would come until June 2000. But because no system ensuring fifty-state coverage could be built consistent with the ABM Treaty, amendments would have to be negotiated with the Russians within the next six to nine months. As a result, decisions needed to be made on the system architecture and the negotiating approach. The president's advisers were unanimous in recommending a two-phase approach that followed the pattern of the deployment plan. That is, the Russians would be asked to approve just enough changes in the treaty initially to permit the phase-one architecture, but they also would be advised of U.S. plans to begin a second phase of talks almost immediately after conclusion of the first phase. On the question of reductions in nuclear weapons, the president was told that his advisers had yet to reach a consensus and would have to come back with a proposal on how to manage an offense-defense trade.

Clinton was full of questions—one aide counted at least nineteen—which he proceeded to fire at his aides over the course of an hour-long discussion. His first query picked up on the fact that the military chiefs had expressed a reluctance to drop below 2,500 warheads—even just to 2,250—while the Russians were ready to slash their arsenal to 1,500.

"If 2,250 warheads will not deter the Russians, will their 1,500 deter us?" the president asked, clearly trying to understand the chiefs' resistance. The question struck at least several of those in the room as boring into the heart of the argument over how many nuclear weapons were enough. It did seem, Clinton was suggesting, that the Russians figured they could deter the United States with many fewer warheads than the U.S. military thought were necessary to deter Russia.

Without waiting for an answer, Clinton rattled off several more questions. Were he and his administration pursuing missile defense because they did not think U.S. forces could deter Iran and North Korea or as insurance against an accidental or unauthorized launch? Were his aides recommending that if the flight tests worked out, the United States should build the system? What was the real probability that the system would work?

Cohen made clear that if the tests succeeded, he thought the system should be built. But neither he nor Ralston could offer a guarantee of success. "Their basic message was, 'Let's wait and see,'" one participant said. "We all understood this was a high-risk enterprise in terms of just meeting the testing schedule and being in a position to make a deployment decision by the middle of 2000."

On the question of nuclear force levels, Ralston presented a sort of tutorial about U.S. nuclear targeting plans, noting some of the asymmetries with Russia's approach. He suggested the United States needed a larger arsenal in part because it had a larger concentration of its nuclear force in bombers, which were more vulnerable to a first strike. Cohen added that Russia also had a much larger number of tactical nuclear weapons. And he said that U.S. targeters also had to take into account China's growing nuclear force.

With the military chiefs still not in support of deeper nuclear cuts,

Berger said the thinking was not to put offensive weapons into play with the Russians at first. "The president said, 'Yes, but we have to understand where we're going,'" one participant recalled. "'You may not be putting it on the table, but we have to understand where we're going.'"

Much of the meeting focused on the administration's negotiating approach with Russia. Albright and Talbott stressed the difficulty of negotiating with the Russians given recent tensions over the Kosovo war. "Everyone agreed the Russian situation was delicate—not impossible, but delicate," another participant said.

It already had been decided that Talbott would take the lead in working out any deal with Russia on national missile defense. Talbott was well versed in the history of arms control. In his previous career as a journalist for *Time* magazine, he had written several books detailing the behind-the-scenes deliberations that led to the SALT and START I accords as well as the ABM Treaty. Although the Pentagon had pushed the hardest for moving ahead with missile defense, it was Talbott, a State Department representative, who ended up in this meeting leading the president through the reasoning that had brought the national security team around to supporting the plan.

Talbott sought to reassure Clinton that the approach being recommended did not represent a fundamental departure from the spirit of the ABM Treaty. It was wrong, he argued, to buy into the absolutist line pedaled by the Russians—and by many die-hard opponents of missile defense in the arms control community—that the treaty was meant to ban missile defense. In fact, Talbott noted, the treaty permitted certain limited antimissile systems and could be adjusted to permit the kind the administration had in mind. So it was possible, he concluded, to have a limited missile defense to protect against one kind of threat while retaining mutual deterrence to deal with the larger Russian threat.

As for how to negotiate the desired treaty amendments with the Russians, Talbott and Albright advocated putting the bottom-line U.S. position on the table at the outset of the talks—that is, making clear what changes in the ABM Treaty were being sought and what the United States would trade in exchange. And then not budging from that position. The State Department officials had a term for such a tactic: "table and stick." It was the approach that the United States had taken with the Russians over NATO enlargement and again over Kosovo. And it became a standard refrain in interagency discussions on how to handle the upcoming talks on missile defense.

Fuerth then spoke up and made clear that while Gore supported the interagency recommendations, the vice president was concerned that the administration might be front-loading so many of the available U.S. "offsets" to get phase one that not enough would be left to bargain with in phase two. "This was Leon saying, 'Okay, we'll support this, but we're not particularly happy about it,'" one official recalled.

The president concluded the meeting by saying: "Okay, I think you've done about the best you can. I've read all the memos, and I don't think we have any choice. If we don't bankrupt the nation, this is the right thing to do—politically and substantively. Is the military on board?"

"Yes," Ralston replied. "The chiefs support the recommendation provided it takes us in time to C-3-level protection for all fifty states—and it works."

Clinton, still troubled by the issue of workability, remarked that he was about to make a "consequential decision" about national missile defense "without knowing if the system will work."

The approach that Clinton took toward missile defense amounted to a search for a middle way through a thicket of concerns. Both in its limited scope and in its effort to keep missile defense embedded in a long-standing international arms control framework, Clinton's scheme had much more in common with Lyndon Johnson's approach in the 1960s than with Ronald Reagan's in the 1980s. Like Johnson, Clinton was acting reluctantly under mounting Republican pressure, proposing a very modest system aimed at a rogue threat while also hoping to draw Moscow into an arms control deal.

Still, it promised to be a tough balancing act. By deciding to sell the Russians and Europeans on a defensive system before demonstrating that technically it would even work, Clinton was letting the diplomacy get well out in front of the technology. A series of initial flight tests, scheduled for late 1999 and early 2000, could—if they were successful—provide important leverage with the Russians. But this was a risky bet, particularly given the trail of failed tests that had marked the history of hit-to-kill systems in the United States—fifteen missed intercepts out of twenty-four attempts since the 1980s, using various experimental system designs.

Clinton and his team felt they had little choice. They had not wanted to think about missile defense during much of their time in office. Other seemingly more immediate East-West concerns had preoccupied them, including conflict in the Balkans, expansion of NATO, and the security of Russia's nuclear arsenal. Only after the Rumsfeld commission report and the Taepodong I launch had the issue of missile defense come crashing down on them with apparent urgency. By the summer of 1999, congressional politics already had jumped ahead of U.S. policy. Having underestimated the issue's political momentum and the extent of planning by the Pentagon bureaucracy for a national antimissile system, the president's advisers raced to catch up and impose some decision criteria on the process. "It was like trying to lasso a train to get this thing to slow down," Berger said months later. Added Talbott: "We let ourselves be blind-sided. We had not realized the extent to which this train was coming at us and would drive us."

At the same time, the administration had arrived at a consensus view that the world had indeed changed, that the threat of attack from a rogue state was real, and that this required some kind of strategic policy adjustment by the United States. Underscoring a perceived need for action were reports in the spring and summer of 1999 that the North Koreans might be preparing for another launch, this time of a Taepodong II missile. Since May, U.S. reconnaissance satellites had monitored construction of a new 100-foot tall launch scaffolding, about 50 percent higher than the old one. By early August, it appeared that the Taepodong II vehicle was already complete. It could be moved into place and fueled for launch in a matter of a few days.

But no launch came. Instead, a month after Clinton approved the missile defense recommendations of his advisers, North Korea agreed to forego new missile tests, at least for the time being. The move was quickly followed by an easing of some U.S. economic sanctions that had barred commerce with North Korea for nearly half a century. The reciprocal actions had emerged from a quiet diplomatic initiative launched a year before when Clinton asked William Perry, the former Defense secretary, to oversee a full review of policy toward North Korea. The move had pre-empted a plan by House Republican leaders to order an outside assessment of their own. Traveling to Pyongyang in May 1999, Perry had told the North Koreans that they could either remain on a collision course with the United States and its Asian allies, or cooperate, which would mean giving up their missile program.

Although the North Korean gesture was welcomed in Washington as an encouraging first step, it did nothing to alter the administration's missile defense plan. Some Republican critics maintained that the North Koreans were being wrongly rewarded for easing a crisis of their own creation. And Perry himself stressed that North Korea had far to go before it no longer posed a threat to the United States and U.S. allies in the region.

Perry's own dealings with the North Koreans had reinforced the notion that they would remain reluctant to part with their missiles, not so much out of any perceived need for military protection but because of the political clout the weapons provided. This point was driven home to Perry and his delegation during their May visit. A top North Korean military officer, noting the bombing then under way by U.S. and other NATO warplanes against Yugoslav forces in Kosovo, told them: "Look at what you Americans are doing there. If you don't like someone, you just bomb them. You're not going to do that to us. We're not going to be another Yugoslavia."

III

Diplomacy and Technology

Chapter 8

AGITATION ABROAD

Seated around a giant circular table in the Council Room at NATO headquarters in Brussels were the permanent representatives to the alliance. They had gathered on this mid-November day in 1999 to hear Strobe Talbott, the State Department's second-ranking official, provide the first extended briefing to the allies about the U.S. missile defense plan and the reasoning behind it.

The administration's announcement two months earlier had alarmed the Europeans. Although there had been talk in Washington for months about a new U.S. initiative on missile defense, the NATO allies had not really expected Clinton to act—and then to sound so urgent about getting a deal with the Russians and authorizing the start of construction within a year. They were taken aback by how serious the American administration suddenly seemed about moving ahead, upset at not having been consulted, suspicious of U.S. motives, and fretful about the future of arms control.

The missile defense move had been followed by the U.S. Senate's refusal to ratify the Comprehensive Test Ban Treaty. Taken together, these developments reinforced a European perception of the United States as sliding into a new isolationism and abandoning the international strategic order that had bound it to Europe and prevented a nuclear war with Moscow. They were a double blow to an arms control status quo with which the allies had long since grown comfortable.

Talbott's aim in appearing before the North Atlantic Council—NATO's main political body—was to reassure the anxious Europeans that the U.S. administration was not intending to do anything rash. President Clinton, he said in opening remarks, would consider the impact on the alliance before deciding whether to proceed with any deployment. In the meantime, he pledged, the United States would make every effort to strike a deal with Russia to preserve the ABM Treaty and perhaps enter into some joint U.S.-Russian missile defense projects.

But no sooner had Talbott finished than the French representative, Ambassador Philippe Guelluy, let loose with a rush of objections to the U.S. plan. His diatribe contained more than a half-dozen criticisms, touching on nearly every European concern that would be voiced about the U.S. plan in the months ahead. While thanking Talbott for initiating an alliance debate on the American project, Guelluy pointedly added that such discussion should have begun months earlier. He said that a U.S. antimissile system would undermine deterrence. It would weaken the credibility of NATO's nonproliferation policies by appearing to presuppose a failure of such policies. And it would set off a new arms race by prompting the Russians, Chinese, and others to build more and increasingly sophisticated missiles.

And another thing, Guelluy said: if the United States erected a shield that would protect it alone from missile attack, it ran the risk of "decoupling" its security from the rest of NATO and becoming less inclined to come to Europe's defense in times of crisis. Then too the notion that the United States could end up deploying a system by making only modest modifications in the ABM Treaty was a pipe dream. The textual changes that the United States had in mind might appear minor, he said, but they challenged the whole spirit of the treaty.

What the United States was undertaking, Guelluy declared, was nothing less than a Copernican Revolution—that is, the injection of national missile defense threatened to upset the universe of international security. And besides, he concluded, the whole notion of "rogue" nations defying the logic of traditional deterrence doctrine was questionable. It was dangerous to assume that negotiation and engagement with certain states were not possible, especially since no nation had ever proved indifferent to the threat of nuclear annihilation.

With that, the French diplomat stopped. Other European ambassadors then leaned forward one by one to speak into tabletop microphones, echoing this or that part of Guelluy's litany. Although in the past Guelluy's perennial grumpiness about various U.S. actions had made him the odd man out, Talbott quickly realized that this time the trim, white-haired Frenchman was more like the pitch pipe of an eighteen-member chorus of dismay, suspicion, grievance, and warning.

The Spanish representative, who spoke immediately after Guelluy, was even more apocalyptic, lamenting national missile defense as an "existential threat to the very future of the alliance." The German ambassador concurred. The Belgian and Italian envoys underlined the risks of decoupling. The Danish diplomat warned of the prospect of an arms race. The Dutch and Canadian representatives expressed skepticism about the North Korean threat and the notion of undeterrable states. The new "best friends" of the United States in NATO—the Polish, Hungarian, and Czech envoys whose countries had just joined the alliance—had nothing to say. Only the British ambassador, John Goulden, avoided criticism of the American approach. He tried to be helpful but came across as only tepidly so, underscoring the importance for the alliance's future of holding a debate on the implications of any U.S. missile defense action. Afterward, Goulden took Talbott aside to confide that London had sent instructions to escalate British concerns about where the United States was headed.

Talbott was shaken. Normally he enjoyed visiting the NATO council and engaging with the Europeans, who in turn tended to view the journalist-turned-diplomat as an energetic and skillful briefer of U.S.

foreign policy. But caught off-guard this time by the attacks, Talbott found the experience emotionally bruising. If the ambassadors had not been so polite, he thought, the whole encounter would have amounted to a mugging. It had driven home to him just how alone the United States was on the issue and how strong the consensus was among the allies that national missile defense was either premature and ill-considered or an outright and dangerous mistake.

The NATO episode marked a pivotal moment, prompting the Clinton administration to deal more intently with European concerns. U.S. officials thought they had been responsive to European views by trying, in crafting their missile defense plan, to do as little harm as possible. They had emphasized preservation of the ABM Treaty and decided to negotiate treaty revisions in small steps rather than through a single, all-or-we-walk approach. The administration's decision not to press early on for construction of new X-band radars in Britain and Greenland also had reflected a consideration for European sensitivities.

But for all their months of deliberations, Washington officials had spent little time planning a response to a potential European uproar. They had badly miscalculated the extent to which the Europeans simply would not buy the premises of the U.S. approach—that a rogue-state missile threat existed, that traditional deterrence would no longer suffice, and that the United States would be able to deploy a national system without rupturing the ABM Treaty. "There was an assumption," a ranking U.S. diplomat said, "that by doing all we did to minimize the changes, the allies would be reassured that this was going to be so limited and evolutionary and modest in its overall impact that they would not get overly upset about it."

The weeks preceding Clinton's decision had carried several early-warning signs of mounting European angst over where the president and his aides were headed. French Foreign Minister Hubert Vedrine had sent a letter on July 13 to Madeleine Albright arguing that proceeding with a national antimissile system would undermine nuclear deterrence and generate a new arms race. Then, on the heels of the U.S. decision, French President Jacques Chirac fired off a letter to Clinton, calling the American plan "detrimental to international stability as a whole and to certain objectives we have in common." He said that "opening a breach" in the ABM Treaty could "pave the way" for later development of "even more ambitious systems" involving space-based weapons. It also would provide a pretext for proliferating states to avoid constraints. A "wiser" way to prevent proliferation, Chirac advised, would be to strengthen international disarmament agreements. "In the long run," the French leader wrote, "this is the only way to respond effectively and globally to the threat of weapons of mass destruction."

But these early French objections tended to be dismissed in Washington; they were perceived not as a signal of broader unease in the alliance but as a bilateral spat reflective only of a certain French pique. U.S. officials were offended by them, snapping back at the French for being too critical, too harsh. "The message we got was, 'That's not a way to talk among allies, and if you want a debate, that wasn't the way to go about it,'" a senior French diplomat said. U.S.-French strains were compounded by France's decision to cosponsor a United Nations resolution with Russia and China condemning national missile defense as destabilizing; the U.N. General Assembly voted eighty to four, with sixty-eight abstentions, to approve the resolution on December 1.

Earlier that autumn, U.S. officials had heard numerous objections to the missile defense initiative in contacts with European representatives. But Talbott's bracing encounter in Brussels in November seemed finally to register the extent of the European opposition and compel the administration into devising some kind of damage control mode. Convincing the Europeans to support the American plan seemed out of the question, at least in the short term. So the administration's goal became to keep the debate behind the scenes and the allies firmly on the fence. The idea was to avoid any move by the French or other allies to craft a NATO communiqué that would hamper U.S. aims.

• • •

Although there were differences in the degree and content of European concerns, they shared several striking common denominators. First, the Europeans did not accept the immediate trigger for the program—that is, they did not believe that the North Korean threat was either as real or as imminent as revised U.S. intelligence estimates were projecting. Political opinion in Europe was still oriented more toward realizing a post–cold war peace dividend than toward underwriting a new and more strenuous defense effort. People had little interest in hearing talk of new threats.

Their calmer attitude was not simply a matter of indifference. Europeans took sharp issue with their American counterparts over how to interpret evidence of rogue-state missile developments. While U.S. threat assessments focused on capabilities—that is, figuring out which missiles were likely to be acquired when and by which rogue states—the Europeans defined missile threats in broader terms. They looked at a combination of technical capability and hostile purpose, placing greater emphasis than the Americans on estimating what a country might do with a long-range missile once it got one.

Additionally, European analysts were uneasy with the way the Americans, after the Rumsfeld commission report, began linking the emergence of a particular rogue-nation threat with the first time that nation flight-tested the missile in question. The Europeans argued that a country's technical capability should be proven and fully tested before being declared menacing. A single test did not create an immediate threat, they said.

Determined to do a better job of selling its own threat assessment, the Clinton administration decided to send a veteran Pentagon official who was well known to the Europeans—Frank Miller, principal deputy assistant secretary for strategy—to brief NATO defense and foreign ministers on the missile threat at meetings in December. In Washington's view, part of the problem in winning over the Europeans was that they simply had stopped thinking about national missile defense since the cold war's end a decade earlier. Even as the debate had picked up again in the United States in the mid-1990s, the Europeans generally had ignored the issue. Now they needed time to catch up on the revised arguments for it.

"I think we appreciated the threat sooner than our allies," said one senior Pentagon official involved in easing European jitters. Indeed, as one European expert himself acknowledged, the attitude on his side of the Atlantic toward the changing strategic situation in the 1990s was, "Don't wake us up."

Miller went to Europe armed with photos of North Korean and Iranian missiles and charts marked with range arcs depicting where in Europe different missiles under development by rogue states in the Middle East and Asia would be able to strike. In his briefings, Miller emphasized that the threat of missile attack to the allies as well as the United States was real and growing. Much of Europe would fall within range of the North Korean Taepodong II missile that was under development, Miller said, particularly if the missile were loaded with relatively light chemical or germ agents. Iran's aggressive pursuit of intermediate-range missiles also posed a direct danger to Europe. Compared with traditional U.S. or Russian missile development, the time between initial flight testing and production in these rogue states was considerably less, he said. This would limit substantially the time that NATO governments had to react to the threat.

Additionally, Miller said, the rogue countries were very likely to try to disguise their ballistic missile prototypes as space-launch vehicles to deflect international public opinion and complicate the West's ability to respond. Finally, he said, North Korea was likely to sell Taepodong technology, components, and possibly entire missiles to countries closer to Europe, as it had done with its Nodong missiles.

The initial reaction from the European ministers was mixed. Clearly, much more than a briefing or two would be necessary to change minds. Although the European defense ministers appeared attentive to the U.S. rationale, the foreign ministers as a group continued to sound skeptical. British Foreign Minister Robin Cook, the first envoy to comment after Miller delivered his briefing, noted that the root of the threat identified by U.S. officials was North Korea. Describing himself as no great military strategist, Cook said that he nonetheless could not help but wonder whether there were simpler ways to solve the North Korean problem than national missile defense. He likened it to using a sledgehammer to crack a nut. If everyone worked together to solve the North Korean problem, he added, perhaps there would be no need for the United States to develop a national antimissile system.

Several other ministers and ambassadors made the same point. As for Guelluy, he did not attend the threat briefing. He left the room as it began, citing other demands on his time. A deliberate snub, U.S. officials thought.

To the Americans, it was evident that the Europeans simply did not trust where the United States was going with missile defense. The allies worried about whether Clinton would make his decision in 2000 on sound strategic, military, and technological grounds or opt for deployment because of a combination of congressional pressure and presidential campaign considerations. "We were under the impression that the United States was rushing and that Clinton would have to take decisions before the election and would be influenced by domestic political considerations," said Gebhardt von Moltke, Germany's ambassador to NATO.

John Goulden, Britain's ambassador to NATO, told Talbott flatly that he believed Clinton already had made the decision to begin construction and that the promised U.S. consultations were a sham. Another senior British official had been even blunter at a meeting in London with Talbott. He accused the administration of "cooking two sets of books"—that is, "creating a threat" to justify the deployment decision that already had been taken but not yet announced, and undertaking a phony, or dumbed-down, set of tests to "prove" that the United States had a system capable of dealing with the threat.

The allies also doubted that the United States would make a dent in Russian opposition to national missile defense. They figured that Clinton would face precisely the sort of either-or, lose-lose choice that he wanted to avoid—and they were deeply worried that he would choose missile defense over the ABM Treaty. This would drive another nail into the coffin of arms control, which was already filled with the ruins of the Comprehensive Test Ban Treaty.

Reflected in the U.S.-European split over missile defense was a fundamental divergence in approaches to security. Europeans had tended to see the U.S. fascination with missile defense since the late 1950s as evidence of a belief in the possibility of achieving absolute security, an idea that they had long since abandoned. Over the centuries, the Europeans had learned to live with a certain inescapable vulnerability, given the proximity and number of their neighbors and a long history of invasions. They favored "soft" approaches to security, which in the case of missiles and weapons of mass destruction meant relying on international technology control arrangements and such arms limitation agreements as the Comprehensive Test Ban Treaty, the Chemical and Biological Weapons Convention, and the Conference on Disarmament.

By contrast, the United States was seen as preoccupied with developing "hard" military and technological means of protection. To many Europeans, the U.S. program was a strategically and financially disproportionate response to an admittedly changing strategic situation. It was a technological approach to an essentially political and diplomatic problem. And it would simply encourage countries to develop a new generation of offensive weapons that could prove even more dangerous to strategic stability.

Some Europeans also thought that if they resisted, perhaps they could stop Clinton. After all, some U.S. officials had indicated privately that the administration itself was uncomfortable with the path down which it had embarked, portraying the move as a domestic political reaction to Republican pressure as much as anything else. "I had somebody at the NSC—not one of those directly involved in the decision but a senior official nonetheless—tell me in the autumn of 1999 that the administration was doing this because it was in a bind with Congress," said a European diplomat who was working the issue at the time. "He said that even if they had wanted to take European concerns into account, they couldn't have done much differently, because they couldn't even take their own concerns into account."

European officials also worried that if the United States proceeded, it would not stop at the limited phase-two architecture. Washington, they believed, was headed down a slippery slope of ever-expanding missile defense systems that would lead ultimately to a comprehensive, Star Wars–like shield against any and all missile threats. Future administrations, the allies feared, could not be counted on to remain as committed as Clinton said he was to striking a balance between defending America and maintaining arms control.

One joke going around Europe held that NMD really stood for No More Disarmament. Another version had it meaning No More Deterrence. Europeans simply had trouble accepting the argument that the threat of massive nuclear retaliation, which had been at the core of traditional nuclear deterrence doctrine, would not hold against rogue states. Where was the evidence that rogue-state leaders would behave any less rationally than had Soviet leaders? Why would a rogue nation fire off a nuclear attack on the United States, knowing that such a rash act would bring a devastating counterstrike? The Europeans still were inclined to argue that the classical deterrent threat of massive retaliation, if not preemption, would suffice to prevent a small country from actually implementing a suicidal threat—thereby making national missile defense unnecessary.

In fact, the whole field of deterrence theory had been neglected since the end of the cold war. Few military strategists had seriously examined the psychological and operational dimensions of deterrence in a world of multiple states with arsenals of long-range missiles of varying sizes and political grievances of different kinds. Some of the most ardent advocates of missile defense had been arguing the unpredictability of deterrence even during the cold war, but the advent of rogue states armed with weapons of mass destruction had given them a fresh breath. In speeches on deterrence, Keith Payne, one of the Republican Party's more prominent military strategists, made the case that it had been risky enough betting on rational behavior by the old Soviet Union, which was a familiar and reasonably informed adversary; Payne said that to apply a similar logic now to rogue states, about which even less was known, would be downright irresponsible.

But Clinton administration officials were not inclined to go so far as to declare that traditional deterrence had been voided by the modern-day rogue threat. Instead, they struggled to craft a modified theory somewhere between the deterrence-is-dead school and the deterrence-has-worked-up-to-now-so-why-change-it camp, which was the dominant European view. Through 1999 and into 2000, memos circulated around the Pentagon, the State Department, and the NSC staff, trying to devise a formulation that would portray national missile defense not as an alternative to deterrence but as a supplement to it.

To some extent, the administration was handicapped by its own terminology. The label "rogue state," by definition, suggested a country refusing to abide by international rules or norms of behavior. For years, U.S. officials had portrayed North Korean leaders as missile-obsessed and potentially reckless. At the same time, North Korea had agreed quite rationally in 1994 to freeze nuclear weapons material production in return for foreign assistance in meeting oil and nuclear energy needs. So what was North Korea: moody and mad or a nation receptive to dealing?

"At one point, the Europeans finally said to us: 'We have two groups of visiting Americans, one of which, explaining the missile defense decision, tells us that the North Koreans are totally crazy, that you can't negotiate with them, that deterrence doesn't work, and that we have to renegotiate the ABM Treaty,'" a senior State Department official said. "And another group shows up and says: 'We're negotiating with the North Koreans to get them to give up their nuclear program; we created this framework agreement, and we need some money from you to help pay for this heavy fuel.' Understandably, the Europeans asked: 'Are there two North Koreas or two American governments?'"

In time, senior administration officials refined their own thinking about what it was about rogue nations possessing long-range missiles that worried them most. The basic U.S. argument became not that rogue leaders might be so crazed or delusional as to mount a missile attack on the United States. It was rather that they might be so calculating as to think that by threatening an attack, they could prevent the United States from intervening in a regional conflict or taking some other action. The problem was not determining how to keep rogue states "from one morning launching a massive assault on the United States," explained Walt Slocombe, the Pentagon's policy chief, in a November 5 speech at the Center for Strategic and International Studies, a Washington think tank. The problem, he said, was finding a way "to deter them from attacking or otherwise coercing neighbors whose security is important to the United States."

This notion was elaborated in a speech at Stanford University in March 2000 by John Holum, the State Department's senior adviser for arms control and international security. Holum made clear that the main U.S. worry about North Korea, Iran, and other rogue states was not that they would use their missiles "as operational weapons of war." Rather, he said, "we have come to the view that such states seek missile and WMD [weapons of mass destruction] programs primarily as weapons of coercive diplomacy, to complicate U.S. decisionmaking or limit our freedom to act in a crisis by, say, coming to the defense of our South Korean ally." By reducing the chance that a missile attack might succeed, Holum said, "ballistic missile defense does not undermine but complements and reinforces deterrence." He said that development of an antimissile system "should not be taken as a signal that we have abandoned our efforts to prevent proliferation or our ability to deter. But to the extent prevention does not entirely succeed, and to the extent we cannot be confident deterrence will work, defense also has a role."

The United States thus would continue to threaten massive retaliation in the event of an attack. But missile defense, the new argument went, would enhance this traditional "deterrence by punishment" with a new "deterrence by denial." In a phrase coined by Slocombe, the idea was to make a rogue-state attack "not only fatal but futile."

Eventually State Department officials sought to ban the use of the term "rogue states" altogether, crossing it out in whatever government statements and speeches they were asked to review. In June 2000,

Albright made the shift official during an interview with the public radio talk-show host Diane Rehm. Asked about U.S. policy toward rogue nations, Albright mentioned that the State Department had stopped using the phrase, preferring instead "states of concern." Albright said afterward that she had not planned to make any announcement, that it had just popped out. In any case, not everyone in the administration adhered to the new terminology. At the Pentagon in particular, some officials very consciously continued to refer to "rogue nations," or jokingly, "states formerly known as rogues."

Probably the most emotionally charged concern that Europeans leveled at the Clinton administration's missile defense plan was "decoupling." The last time that potent word had been uttered in connection with a transatlantic nuclear issue was in the late 1970s and early 1980s, when the Soviet Union deployed intermediate-range SS-20 missiles that could hit Western Europe. The Americans had no equivalent—that is, no land-based intermediate-range missiles in Western Europe. This imbalance had led the Europeans to fret about whether Washington would be willing to counter the new Russian "Euro-missiles" with nuclear missiles based in the United States and on U.S. submarines—and risk Soviet retaliation against the U.S. homeland. The fix was the deployment of U.S. Pershing and cruise missiles in Europe, which eventually led to the elimination of all such weapons on both sides under the 1987 Intermediate-Range Nuclear Forces Agreement, but not before a grueling and divisive transatlantic wrangle had transpired.

Now European officials were warning that the United States was about to plunge the alliance into a similar strategic dilemma. By pursuing a national missile defense, they said, the United States seemed to be looking to its own interests and offering no such protection to its allies. The fear was that a U.S. antimissile system would establish zones of differing security within an alliance that had been founded on the basis of "one for all, all for one" in dealing with adversaries. A shielded United States, the argument went, would be less likely to intervene in a regional crisis overseas, since its security would be less intertwined with Europe's. The allies worried that they would end up more exposed to their enemies, less defended by their biggest, strongest friend, and under pressure from their frightened publics to undertake a hugely expensive European version of missile defense.

Tackling this line of attack, Talbott and other U.S. officials argued that the real lesson of the Euro-missile debate of the 1980s was that the United States, as a good ally, took very seriously a missile threat against Europe precisely because the missiles in question did not threaten the continental United States. The lesson of that experience, they said, was that the Europeans needed to regard the threat posed by the Taepodong as seriously as the United States had taken the SS-20s two decades earlier. Coupling, after all, was about shared risk and shared determination to act.

In the U.S. view, there was something downright perverse in the notion that a self-shielded United States would be less likely to protect an unshielded Europe. In fact, the opposite might be the case, U.S. officials said. If the United States was confident that it could defeat a rogue missile attack, it would be more likely to project power in crises in and around Europe. A national missile defense might therefore actually enhance U.S. ability to fulfill its commitments to NATO—and to allies in Asia like Japan and South Korea—since it would render less credible any possible attempts by an adversary to threaten or coerce the United States with ballistic missiles.

But as much as American officials argued that a U.S. missile defense would help bind the United States to Europe, the Clinton administration appeared reluctant to consider extending the shield over Europe. At the November NATO meeting, the Turkish representative had asked about just that possibility. Talbott sidestepped the question. He responded by urging an acceleration of discussions, already under way within the NATO alliance for several years, to develop other missile defenses—shorter-range systems meant to guard troops in the field or ships at sea, not European territory and civilian populations.

The issue of whether to extend the U.S. shield over Europe soon became a point of contention within the Clinton administration. The administration's approach had contained no provision for protecting the Europeans. And yet, at least two European governments—Britain and Denmark—were being asked to accept upgrades of early-warning radars on their territory, and probably the installation of new X-band radars later, to enable the American system. Then too, if the United States succeeded in persuading the Europeans that there really was a new ballistic missile threat against them, wouldn't it be logical for the Europeans to seek protection also?

Quietly, the Pentagon had attempted to sound out allies about whether they wanted to be included in the plan. In September 1999, Frank Miller had broached the issue while hosting a NATO group of his assistant Defense secretary counterparts in Omaha. He noted that the administration's missile defense plan would defend only the United States and possibly Canada, but not Europe. "If you're content, fine, then I've done my job," Miller told them. "But if not, you need to say something, because the current approach would preclude us from help-ing." Given the political sensitivity of missile defense in their own coun-tries, however, the Europeans were not eager to be seen as seeking to come under a U.S. defense umbrella. They also had real concerns about where they would find the money to finance their share of a missile defense network.

Nonetheless, sharp differences had begun to emerge within the U.S. administration over the advisability of excluding the Europeans from the initial phases. Miller and his boss, Walt Slocombe, found this approach unsettling. They considered it inconsistent or unjust to be ask-ing the Europeans to participate in the system without extending pro-tection to them. But the State Department and the White House argued for keeping the proposed system as limited as possible, saying that talk of a NATO-wide antimissile umbrella would further spook the Russians and complicate the ABM Treaty talks.

Lieutenant General Ronald Kadish, director of the Ballistic Missile Defense Organization, was instructed to begin quietly looking at designs for extensions of coverage to Europe. One tentative notion involved installing interceptors in England and an X-band radar in Turkey. But the effort never amounted to much, and the Clinton administration's official stance on using the planned system to guard Europe remained ambiguous. The administration neither explicitly offered help nor closed the door on the idea. Essentially, it left the ball in Europe's court. As one defense official said: "Our line was: 'It is the policy of the U.S. government that we have a program calling eventually for two sites in the United States. If allies are concerned about being defended, they need to raise that with us. If the issue is raised with us, we will engage.'"

By early 2000, administration officials had concluded that they had generally succeeded in keeping the lid on European opposition from boiling over into outright obstructionist moves. Some of the passion had already drained from the debate. Through repeated visits by senior U.S. officials to NATO committee meetings, along with extensive briefings on how the planned system would work and how the talks with the Russians were going, the administration had begun to debunk some of the European arguments and to chip away at some of the worst-case interpretations of what they were doing. "We did a full-court press, starting in late 1999 and into mid-2000," a senior U.S. diplomat said. "We made up for lost time, refuting some of the arguments and defusing some of the anxieties. At the end of the day, though, the allies still were hoping this would just go away."

The ultimate key to winning over the Europeans, U.S. officials figured, was not hammering away at them so much as persuading the Russians to accept changes in the ABM Treaty. In such a case, the Americans assumed, the allies would give their grudging support, however deep their apprehensions. The problem was that so long as the Europeans appeared at odds with the United States, the less compelled the Russians would be to negotiate and the more they would focus on exacerbating the division within NATO.

Russia's initial reaction to the U.S. plan had been sharply negative. Its denunciations of the American proposal to alter the ABM Treaty were punctuated by vows to overcome any U.S. defense system and military exercises showcasing Russia's long-range nuclear weapons and its own antimissile defense system.

If a U.S.-Russian deal on missile defense were to take shape, it would emerge most likely from talks between Talbott and Georgi Mamedov, a deputy Russian foreign minister. The two senior diplomats knew each other well. As the Clinton administration's longtime point man for Russian affairs, Talbott had worked through many tough issues with Mamedov. In April 1993, they had set up a formal contact channel, called the Strategic Stability Group, whose purpose was partly to anticipate problems and partly to control damage. Both parts seemed to apply to the missile defense controversy.

Talbott was not the only American diplomat with a long history with Mamedov. The Russian had served as a main point of contact for U.S. diplomats for more than a decade. Confident and talkative, speaking colloquial if accented English, Mamedov had acquired a reputation as something of an operator. He seemed to have good contacts with other Russian government agencies, including the Defense Ministry, and a good understanding of how to work issues through his government's interagency process. He could not always guarantee success, but the Americans viewed him as a straight-shooter.

Part of Mamedov's strength was his skill at diplomatic choreography. He was adept at directing complex talks and managing a sequence of events aimed at reaching a finale. Early on, he and Talbott established a set of parallel discussions between their governments. A working-level group, led on the American side by the State Department's John Holum, would focus on the technical details of possible missile defense and arms reduction agreements. At the same time, Talbott would pursue a higher-level channel with Mamedov, exploring ways around the central impasse.

They also settled quickly on the idea of drafting what Mamedov called a "legacy document," a statement of general principles of strategic defense. The thought was that Clinton and Yeltsin could issue this statement early in 2000 as a stepping-stone toward a possible final agreement later in the year. With Russian elections scheduled for the summer of 2000, the Russians were reluctant to complete any arms control deal beforehand. This would leave a narrow bargaining window of only a few months after the Russian vote and before the end of Clinton's term.

It was during one of Talbott's meetings with Mamedov—in Helsinki in October 1999—that the Senate rejected the Comprehensive Test Ban Treaty. "Black Wednesday," Talbott called it. He worried that the Senate action would complicate the missile defense talks by deepening Russian doubts that any agreement worked out with the Clinton administration could win congressional ratification. Still, after three meetings with Mamedov in September and October, Talbott was beginning to feel a tinge of optimism. He wrote Berger and Albright that it was "conceivable, although just barely," that the United States and Russia might be able to work out some deal. The key to doing so, he said, would be persuading the Russians that their strategic deterrent would remain intact, even after the full two-site, 250-interceptor system envisioned by the Clinton administration was in place. Also key to concluding a deal, Talbott advised, would be a major concession in the START talks—namely, agreement to drop to a range of 1,500 to 2,000 warheads from the 2,000 to 2,500 negotiated at the 1997 Clinton-Yeltsin meeting in Helsinki.

Although the Russians had remained firm in their opposition to missile defense, the history of negotiating with Moscow was one factor in Talbott's cautious optimism. In recent years, the Russians had initially said no to a number of proposals, including the enlargement of NATO and the intervention of NATO-led forces in Bosnia's civil war, only to come around eventually to accepting them. In Talbott's mind, therefore, a Russian no was not necessarily the last word on a subject.

Then Talbott's assessment darkened. In a surprise move on the last day of 1999, Yeltsin announced that he would be stepping down from the Russian presidency and that Vladimir Putin would take his place, followed by an election in March instead of June, as previously planned. Although acceleration of the Russian election opened a wider window for negotiations between the time a new president would be confirmed and Clinton's planned deployment decision in the summer, the change from Yeltsin to Putin was not encouraging. Talbott had met with Putin in late December and detected a hardening of the Russian position, a closing-off of options. Whatever potential for a deal that Talbott had sensed just three months before already appeared to have slipped away.

"No one ever quite said to me, 'Yeltsin would have done this, Putin won't,'" Talbott said. "It's just that things changed in a really elemental way. I think Putin had a strategy of defining himself as the un-Yeltsin, sort of like the un-Cola. Putin was clearly determined to find an issue fairly quickly where he could say, 'This is a Russia that can say no, and here's the issue on which we'll say it.' At the same time, what we got during all those contacts was that they didn't quite want to say totally no forever. I think they hadn't figured out what their end-game was yet."

Despite fresh pessimism about the potential for a deal, the administration pressed on. In January, the Americans presented the Russians with a draft text for amending the ABM Treaty. Rather than propose actually rewriting the treaty line by line, the United States put forward for adoption a protocol that would cover all the changes.

This approach had been somewhat controversial within the administration and on Capitol Hill. The Pentagon's Walt Slocombe had argued that a cleaner approach would be to rewrite directly those treaty articles that needed to be changed to permit the planned system. In Congress, Senator Thad Cochran also had pushed for a redrafting of the entire treaty—especially Article I, the section that explicitly banned deployment of national antimissile systems.

But State Department officials had wanted to keep the changes as minimal as possible. The proposed protocol would amend only those few parts of the treaty necessary to permit phase one of the planned antimissile system, with the rest of the treaty remaining unchanged. U.S. officials noted that amending the treaty by adding a protocol had ample precedent. A 1974 protocol, for instance, had reduced the number of permitted antimissile sites from two to one. Hoping to foreclose other approaches, State officials began passing a draft protocol around to the Pentagon and White House for comment.

Essentially, the proposed protocol exempted "limited" missile defense systems from the general ban on national missile defense articulated in Article I. At the same time, it affirmed that the limited system "will neither threaten nor allow a threat to the strategic deterrent forces of either Party." Other changes in the draft would allow the United States to relocate the planned interceptor site from North Dakota to Alaska (although the number of allowable interceptors would remain capped at one hundred) and permit construction of an ABM radar anywhere in the country (instead of within 150 kilometers of the interceptor site, as the treaty stipulated). The protocol also made clear that as soon as agreement was reached on phase one, a second phase of negotiations would begin after March 1, 2001, "to bring the treaty into agreement with future changes in the strategic situation."

But the Russians refused to discuss any of these proposed treaty text changes. They also declined to talk about several forms of potential cooperation that U.S. officials had offered to sweeten the prospects of a deal and assuage Russian fears about the proposed defensive system. These included helping Russia complete a partially constructed radar at Mishelevka in Siberia, expanding an agreement to share U.S. radar data, collaborating on two missile observation satellites, and participating in joint exercises related to battlefield missile defense.

"We had no intention of negotiating, and we always were clear on that," said a senior Russian diplomat involved in the talks. "But we kept meeting, telling them we wanted at least to understand what was behind this initiative. One thing we said was, we'd work with them in other ways to solve the issue of proliferation."

As challenging as it was to try to persuade the Russians, the Clinton administration at least had a plan and a process. But it never quite figured out how to address the concerns voiced by the other major nuclear player in the international drama over missile defense: China.

Like the Russians, the Chinese warned that a U.S. missile defense system would undercut their own nuclear force and compel them to modernize and expand their arsenal beyond what they might otherwise do. In fact, the Chinese tended to be much more passionate on the subject than the Russians, their heightened concern being a reflection of the fact that they had many fewer nuclear missiles.

While the Russians had more than enough missiles to overwhelm a limited American defense, the Chinese had followed a policy of "minimal nuclear deterrence": fielding only about twenty CSS-4 long-range intercontinental missiles capable of reaching the United States, China had just enough to strike back should it be attacked. Consequently, the United States could not claim—as it did with the Russians—that its planned antimissile system would have little effect on China. In fact, an interagency report drafted in early 2000 concluded that the first phase of the Clinton system might blunt China's current missile force, while the second phase might have some capacity against an updated Chinese deterrent. Further, notwithstanding the administration's own insistence that the planned system stay focused on combating just rogue nations, some of the more ardent missile defense proponents made no secret of their desire to see U.S. capabilities expanded eventually to counter the Chinese missile threat. "Everyone wants to be politically correct, but the truth is, China may be one of the countries that we will have to protect against some day," said Senator Jon Kyl, an Arizona Republican.

The Chinese had long planned to upgrade their nuclear force by introducing the DF-31, a solid-fueled mobile missile with the range to strike Alaska and the northwestern United States. A longer-range successor also was expected, along with the JL-2 submarine-launched ballistic missile. China also would likely have the potential, U.S. intelligence analysts said, to put multiple warheads on its existing CSS-4 missiles. Some U.S. officials worried that an American shield might goad the Chinese to expand this buildup. Others argued that China would hedge against the potential for an American antimissile system no matter what Washington actually did. In any case, a Chinese buildup, it was feared, could set off a chain reaction in South Asia as

India responded to China by increasing its arsenal and Pakistan responded to an Indian surge.

The whole issue of Taiwan further complicated the picture with China. Amid talk of a possible U.S. sale to Taiwan of destroyers equipped with the Aegis radar system, China feared that the ships could become a platform for a regional missile shield for Taiwan. For the Chinese, this raised the specter of a new level of American military cooperation with the island that could encourage Taiwan to resist Chinese pressure to accept Beijing's sovereignty.

Upping the ante, the Chinese had placed about two hundred short-range missiles near the island and were expanding that force at a rate of fifty missiles a year. Such regional tensions underscored another important difference between the Chinese and Russian cases—namely, that while Russia was a power in decline struggling to rebuild a devastated economy and establish new cooperative relationships in Europe, China was a growing regional power destined to compete with the United States for dominance of the western Pacific. In general, the Chinese saw deployment of an American missile shield as likely to embolden Washington to intervene militarily not just in the Taiwan Strait but elsewhere in Asia and around the world.

But the issue of how to deal with Chinese concerns had received only passing attention during the Clinton administration's high-level deliberations in the summer of 1999, preceding Clinton's decision on the architecture and negotiating approach with the Russians. At one Principals' Committee meeting in June, Albright did wonder out loud about the potential impact of a U.S. deployment on China and the possible ripple effect on India and Pakistan and suggested that John Holum, the State Department's arms control chief, brief the principals on the subject at a later meeting. But the briefing never took place. At a subsequent session, James Steinberg, Berger's deputy, also raised the issue. He found the Chinese case particularly vexing and intriguing, much more so than the Russian one, which he regarded as solvable over time given the diminishing importance that nuclear weapons were likely to play in the U.S.-Russian relationship generally.

Beginning in the autumn of 1999, the Deputies' Committee, which included the second-ranking officials at the Departments of State and Defense and the NSC, convened several sessions to consider the questions they needed to be asking about China and missile defense and how to answer them. One fundamental question was raised early on: Should the United States even offer China assurances that it had a right to a survivable nuclear deterrent? In other words, were the Chinese entitled to the same condition as the Russians?

"It had never been U.S. policy to acknowledge or grant the Chinese the survivability of their deterrent," said Talbott, who had posed the question of whether it was time for a change. A shift carried the risk of raising a furor among conservative Republican lawmakers, who were already critical of the administration's handling of China and alleged leaks of U.S. nuclear secrets to the Chinese. "We weren't about to make new policy on this point," Talbott recalled. Instead, administration officials settled for a vaguer declaration that its missile defense plan was not directed at China. "China posed a different problem than Russia in the sense that we have no treaty with them," Steinberg noted. "They are in effect a third-party beneficiary of the ABM Treaty, but we have no legal obligations to them."

The possibility of setting up an arms control regime with the Chinese that could coordinate U.S. missile defense with Chinese nuclear force modernization was briefly considered but dismissed. Additionally, the administration was in no position to offer the Chinese military the kind of incentives it had extended to the Russians, such as the sharing of missile launch data, because the relationship between the American and Chinese militaries was not nearly as developed.

The U.S. deliberations culminated in a visit to Beijing in mid-February 2000 by a delegation unusual for its large number of high-level officials, including Talbott, Steinberg, Slocombe, and General Ralston. It marked the first real opportunity to engage the Chinese on ballistic missile issues and provide threat and architecture briefings to a wide Chinese audience. But it was inconclusive, with the Chinese wanting to talk mostly about Taiwan. In fact, no sooner had the Americans left than China issued a "White Paper" broadening the reasons it would consider sufficient for using force against the self-governing island.

"We have a lot to overcome," Holum told a group at Stanford University a month after the visit. "The Chinese information level on national missile defense is very low, and the perceptual gaps are wide. The Chinese, like the Russians, connect dots and see national missile defense as part of a grand design aimed at China."

For the most part, the administration's focus remained on Russia. "They're the guys in the treaty and the ones our allies in Europe are concerned about," a senior defense official said. "To the extent China is an issue, we've laid out the basic issues, but nothing has changed."

Chapter 9

UP, UP, AND AWAY

Normally, when the Pentagon wants to invent a weapon—a new jet fighter, for instance, or a combat ship—it spends several years drawing up requirements for the weapon, then establishes a budget, hires a prime contractor, and proceeds with an extensive testing and evaluation effort. But the way the Clinton administration's National Missile Defense program was conceived and structured defied conventional approaches. It did not start from scratch with an optimal design and pursue a carefully sequenced course of development from drawing board through flight testing to production and fielding. Instead, it inherited work already done on key components for a land-based system. Its design was constrained by international treaty considerations that ruled out sea- or space-based interceptors. And its development was rushed as a result of a perceived need for a defensive system in as little as a few years.

These departures from the norm took their toll on the program, handicapping it with quality control problems, scheduling delays, and cost overruns. The Pentagon never tired of warning that the program was at "high risk" of further setbacks. Tensions would flare among some of the major contracting firms involved, and the Clinton administration would suffer the embarrassment of two missed intercepts in three attempts.

Still, in the mid-1990s, when the administration began to pick up the pace of development, there had been some reason to hope that the process would go more smoothly. For one thing, many in the missile defense business had offered assurances that the hit-to-kill concept needed no new scientific breakthrough to work; developing a successful system was just a matter of perfecting the engineering, they said. For another thing, work on major components of the system had been under way for several years, kept alive even in the first half of the Clinton administration as "technology readiness" initiatives.

Already in 1995, the kill vehicle was the object of a fierce competition between two major defense contractors—Hughes and Rockwell—and Raytheon was under contract to perfect the X-band radar. TRW was at work on figuring out a battle management, command, control, and communications network—known as BMC^3 and pronounced "BMC-cubed" in defense parlance—to relay target data to an interceptor. These programs fell far short of a full-fledged development effort. There was no corporate team leader and no firm, overarching concept for fitting the individual components together into a coherent whole—that is, for determining how many of what to produce, where to put them, and what trade-offs in capabilities to make among them to improve efficiency and save costs. Each subsystem was being designed according to its own list of requirements. Nonetheless, these projects gave the Pentagon something to build on when the administration ultimately got more serious about deploying a system.

With the adoption of the three-plus-three plan in 1996, which committed the Pentagon to being ready to deploy a national missile defense system as early as 2003, Lieutenant General Malcolm O'Neill, director of the Ballistic Missile Defense Organization, recommended forming an extraordinary consortium of the nation's leading defense contractors to build the system in the record time envisioned. Given the national importance and urgency of the system, O'Neill figured that bidding a contract for managing the system would be too time-consuming. And the government was hardly in a position to try to do the coordinating itself. But the consortium idea never got off the ground, and under O'Neill's successor, Lieutenant General Lester Lyles, bids were solicited in 1997 for selection of what the Pentagon termed a Lead System Integrator (LSI)—a corporate entity to oversee construction of the proposed system. Defense officials were relieved to hand off responsibility to a private firm, but the definition of just what the LSI was supposed to do—and how it differed from the more customary practice of a prime contractor—would dog the program for the next few years.

The leading contender to take on the role of LSI was a joint entity formed by the three defense firms that already had stakes in major elements of the system: Raytheon, TRW, and Lockheed Martin. Boeing, which had just acquired Rockwell and McDonnell Douglas, also decided to enter the competition. It argued against the idea of using a joint entity to manage the program, calling that approach risky and potentially fraught with internal differences. Boeing stressed its own skill at assembling complex systems and its reputation for quality control—in pointed contrast with Lockheed, which had suffered a string of quality problems managing THAAD, the Army's intermediate-range missile defense program. Boeing's proposed design also included some novel, cost-effective elements, especially for the system's network of radars. Instead of sticking to the government's concept of building a handful of X-band radars, Boeing proposed saving millions of dollars by simply upgrading the computers and software in existing early-warning radars to enable them to track targets with sufficient precision for a phase one deployment.

Coming from what had been very much an underdog position, Boeing won the LSI contract in April 1998. To head its missile defense management group, the company picked John Peller, an aeronautical engineer with experience on the Minuteman missile and space shuttle programs. Peller had led Boeing's bid team, and his own vision and enthusiasm for the project had helped propel Boeing to victory. Tall and lanky with a casual manner but intense mind, Peller brought to the effort a career's worth of knowledge of the missile and space business. He was confident about making the proposed system work—perhaps a little too confident. He told a congressional panel in February 1999 that while the task at hand was "indeed difficult," it was "far from the formidable challenge that a lot of our critics maintain. Frankly, this project is far simpler than the challenges we had on space shuttle." The remark troubled the Pentagon's chief acquisition official, Jacques Gansler; worried that Peller might be underestimating the job ahead, he phoned Boeing's president, Harry Stonecipher, to register his concern.

But Peller worked tirelessly to get the program under way, with no illusions about the constraints he faced. Handicapped at the outset by the vagueness attached to the LSI role, he labored to establish Boeing's authority and new contractual relationships with Raytheon, TRW, and the other subcontractors that had been accustomed to dealing directly with the Pentagon over their pieces of the national missile defense system. He also struggled to put together an adequate LSI team. Boeing had expected to be able to draw staff from its space station program, which was due to wind down. But that program ended up receiving an extension, prompting Boeing, in Peller's words, "to get out and really scavenge" for staff. "Many good people came in, although they weren't necessarily educated in national missile defense," Peller said. "In general, it took us an extra six months to get the staff up to where we wanted it. And in some areas, we had some quality problems."

One of the early tasks confronting Boeing was to select the design for the exoatmospheric kill vehicle (EKV), the 120-pound package of sensors, computers, and thrusters intended to home in on the enemy warhead and pulverize it with the sheer force of a high-speed collision. Both firms that had competed for the job had been swept up recently in a wave of defense industry mergers and acquisitions. Hughes had become part of Raytheon, and Rockwell had been swallowed by Boeing, which in turn had gone to some lengths to erect firewalls between its LSI team and the kill vehicle team.

The Pentagon had poured several hundred million dollars into the design competition to come up with the best technology that American industry could devise for one of the most challenging aspects of the missile defense effort. Both versions of the kill vehicle depended on infrared sensors to spot targets and computerized algorithms to process the data. But each offered certain advantages over the other. Boeing's sensors could detect targets at great distances but required considerable cooling; Raytheon's sensors required somewhat less cooling but could not see as far. Boeing's algorithms relied on a rather standard form of statistical analysis; Raytheon was trying to adapt a more novel analytical approach. Boeing's kill vehicle was equipped with a broad collar designed to increase the likelihood of hitting the intended spot on a target warhead; Raytheon's version had no such collar.

But the U.S. government never got a chance to choose between the two models. The kill vehicle went to Raytheon by default following a case—never publicly reported—in which several Boeing employees were found to have committed a grave ethical violation. The employees, members of Boeing's own kill vehicle team, had exploited a confidential Raytheon document that apparently fell into their laps. The episode not only compromised the high-stakes competition over the kill vehicle but badly roiled relations between the two defense giants, generating lingering corporate tensions within the National Missile Defense program.

The trouble began in July 1998 when an Army group from Huntsville, Alabama, involved in the missile defense project visited a Boeing facility in Downey, California, for a briefing on Boeing's kill vehicle. During the visit, a seventy-two-page document outlining a software test plan for the Raytheon kill vehicle was left behind—whether by accident or purposefully with instructions either to discard or mail back to Huntsville remains unclear. The paper revealed little about the specific design of Raytheon's kill vehicle, but it did provide insight into Raytheon's technical approach—and Raytheon considered it highly confidential. Dated July 8, 1998, the document bore the warning "unclassified competition sensitive" on a cover sheet and at the top and bottom of each page. Also inscribed on the cover was language making clear the document had been "prepared by Raytheon Company" and "prepared for U.S. Army Strategic Defense Command."

A software engineer on Boeing's kill vehicle team, assigned to collect and dispose of documents reviewed by the Army team during its visit, discovered the Raytheon document mixed in among other papers. She gave it to her supervisor, who not only passed it to another team manager but spent a weekend preparing a six-page analysis of it. The document also was slid under the door of a Boeing "blue team" that had been charged with learning all it could about Raytheon's EKV development effort. It was a member of this group who alerted Boeing's ethics office to the possession of the Raytheon document, even as some members of the blue team started making use of the information. A series of inquiries followed, conducted separately by Defense Department criminal attorneys, Boeing personnel, and an outside law firm hired by Boeing.

Several of the employees involved in the incident sought to defend their actions by noting that they had joined Boeing only relatively recently following the merger with Rockwell and were unaware of company policy on the handling of sensitive procurement documents. But Boeing executives, eager to distance the corporation from such errant behavior, dismissed that rationale, saying that all employees had received extensive ethics training. Ultimately, Boeing fired three employees in connection with the case, suspended two others for a month without pay, and reprimanded two more. "They thought they were doing the best thing for the company," said Stonecipher, Boeing's president, stressing disapproval of their action. "We work awful hard at it, but as hard as we work at it, we still can't stamp out all the wrong."

But that was not the end of it for Boeing. Raytheon, which had been ready to pull out of the competition in despair over what it had perceived as a competition biased in Boeing's favor, suddenly found new reason to stay in the running. It considered suing over the evident compromise of proprietary data. Meanwhile, senior Pentagon officials were deeply troubled that Boeing's conduct had jeopardized the fairness of the kill vehicle competition. By pitting the two defense contractors against each other, the Defense Department had hoped that the best solution to one of the most challenging aspects of the antimissile system would emerge. But now the contest's integrity had come under a cloud, and it was up to Peller to decide what to do next.

"I had a choice," Peller said. "I could have tried re-leveling the competition. But that would have delayed the process another year. So I just disqualified Boeing. That was my call in the end."

Even then, Boeing was not out of trouble. Both the Army and BMDO recommended that Boeing be compelled to recoup the cost to the government of the lost competition—or face debarment. Debarment makes a firm ineligible for government contract awards for a period usually lasting several years. Although there was no evidence that Boeing's senior management had encouraged or directed employees to obtain Raytheon's documents, federal regulations do provide for debarment if a contractor fails to exercise adequate training and oversight of employees.

"Clearly, the damage to the government was considerable," said a senior contracts manager for the Army's Space and Missile Defense Command. "I mean, we lost competition, and we had already funded hundreds of millions of dollars to maintain competition, and then Boeing had to drop out of the competition. The gross misuse of the document by Boeing's employees was inexcusable."

But the Army and BMDO differed over the extent of the damage. The Army argued that much of the $400 million invested in Boeing's EKV development effort had been lost. By contrast, BMDO contended that important technical data had been gathered from the Boeing project before it was aborted and estimated the loss at only $7 million to $13 million, based on the higher price that BMDO officials figured was paid for the contract with Raytheon in the absence of competition. There were differences too over how much of Boeing should be threatened with debarment. The Army recommended going after Boeing's Electronic Systems and Missile Defense Group, which, in addition to managing the National Missile Defense project, is involved in missile, satellite, and surveillance programs. BMDO preferred to aim lower, targeting simply the EKV unit.

For its part, Boeing insisted that it should not have to pay anything. The final decision rested with the Air Force, which was assigned jurisdiction because it did most of the Pentagon's business with the Boeing division in question. It is not uncommon for Pentagon officials to mull over a debarment recommendation and eventually reach an administrative agreement with a contractor in return for pledges by the contractor to strengthen internal training programs and abide strictly by all rules. In the summer of 2002, the case was quietly dropped.

The kill vehicle episode badly strained relations between Boeing and Raytheon, which already were bruised as a result of Boeing's victory over Raytheon and its partners in the competition for the LSI contract. Raytheon officials continued to suspect Boeing of maneuvering to salvage at least part of its kill vehicle effort and bump Raytheon aside in some kind of new contest for a follow-on model. Feeding these suspicions was the fact that the government had continued to pay Boeing to maintain its kill vehicle program, although at a reduced level, as a "hot backup" in case something went awry with Raytheon's approach.

This put Raytheon in what it felt was an awkward position—in effect, competing with its customer. Contract negotiations between the two firms over terms for Raytheon's delivery of kill vehicles for the missile defense program remained difficult throughout 1999 and into 2000. To resolve matters, Lieutenant General Ron Kadish, Lyles's successor as BMDO director, began to hold periodic meetings of the chief executive officers of Boeing, Raytheon, and the other major defense contractors shortly after assuming command in the summer of 1999. At the first meeting in November, the three-star general appealed directly to Boeing's Stonecipher and Raytheon's Dan Burnham to settle whatever was keeping their firms from signing a contract.

"So Dan Burnham and I had this conversation right there in the CEO meeting," Stonecipher recounted, "and Dan said, 'Well, it's a little hard to get down to the bottom line when our guys think that your guys, Harry, are still competing with us for the kill vehicle because you're building a backup design.' I said, 'The only reason we're doing a backup design is we're not sure you're going to get there at a price that we can afford, that the program can afford.' But then I said, 'Let's settle this right now. We want you to be the kill vehicle provider. So please get a contract negotiated, and let's get on with it. We'll stop.' And we did. And Dan said, 'Great.' And so at that point, his people and our people got together, and suddenly the issue was resolved."

Boeing shut down its kill vehicle backup operation in early 2000 under an agreement that called for Boeing and Raytheon to join forces in devising a new "best of breed" kill vehicle for a later phase of the missile defense program. But while top executives from both firms expressed a renewed sense of teamwork, antagonisms lingered at lower corporate levels. "Relations are better, but we'll never be able to turn archenemies into friends," Kadish observed. "You've got to understand that between Raytheon, Boeing, and a few other major defense firms, they're the only big players left in the business, and they're competing with each other over everything—not only in this program but across the board. So it's a real challenge, it's a constant struggle, to keep everyone on track. It's like the Army, Navy, and Air Force. There's built-in institutional conflict."

Having won the contract for the kill vehicle, Raytheon found itself in no small bind. What had been a rather small-scale, experimental endeavor for the company—run largely by a bunch of highly educated engineers out of a basement laboratory complex—was suddenly mushrooming into an urgent, high-visibility Pentagon program. Raytheon was neither staffed nor equipped to manage the adjustment well at first—and it promptly fell months behind schedule.

Early evidence of these problems surfaced when company engineers made a mess of a prototype of the kill vehicle in their Tucson laboratory in January 1999. The incident occurred during a "blowdown" test meant to power up the kill vehicle. A thermal battery had been wired backward as a result of a mismarked connector. It overheated and ruptured, spewing paraffins and hydrocarbons throughout the lab. The underground rooms filled with smoke, forcing an evacuation. Smoke particles also were sucked into the compartment containing the infrared sensor, clogging the filters and contaminating the sensor. Another blow-down test a month later also ran into trouble, this time with leaky plumbing fixtures in the kill vehicle's coolant system.

Such mishaps pushed the date for the first flight intercept test back from the summer to the autumn of 1999. They also raised the first of what were to be persistent questions about Raytheon's quality control and its processes for assembling the kill vehicle.

Because the program had been little more than a research effort for much of the 1990s, the company had no production facility with stan-dardized assembly procedures on which to rely for building the kill vehicle. The program lacked spare parts and was woefully understaffed for the accelerated effort on which it had embarked. There were plans to set up a production facility by renovating a section of a large manu-facturing plant on Raytheon's sprawling Tucson site. Spares would be ordered, and new workers hired. But in the meantime—that is, during the first year, probably a second one, and possibly even a third—the company would have to piece together painstakingly each kill vehicle for each planned test, making do with the parts it had and training freshly recruited personnel on the fly.

"We had a mix of Ph.D.s and scientists and very young people right out of school," said Ron Meyer, the deputy program manager. "We were staffing up quickly, so we had a big problem mentoring the younger ones. They had all the enthusiasm, all the basic knowledge, but they didn't have the experience of working in this type of laboratory or preproduction or flight-test environment, so we had an awful lot of training to do."

Operations in the lab were in desperate need of experience, authori-ty, and discipline. For instance, in the early months of testing many of the cables linking a test harness to the kill vehicle were found to be either misconnected or mislabeled. So the company, using a color-coded scheme, painstakingly marked all cables, putting red tags on wires that had not been completely checked and tested, white tags on those that were certified for use, and yellow tags on ones that had been used previously without incident but still had not been tested and checked. Training sessions were held to teach procedures for checking every connection.

"We were putting in a firm process, training and sensitizing folks that they were working on a $25 million space vehicle," Meyer said.

But even with all its growing pains, the missile defense program was ready for its first flight test in October 1999. Although the test was about three months behind schedule, it had still been only a year and a half since the selection of the LSI, and less than a year since the choice of Raytheon as producer of the kill vehicle. This test would focus on the kill vehicle—other elements of the system, including a new booster, the X-band radar, and the battle management computer network, would be incorporated later. And the test course—which involved firing a target missile from Vandenberg Air Force Base in California and an interceptor from Kwajalein Atoll in the Marshall Islands—was familiar to anyone in the missile defense business. It had been used by the Pentagon as far back as the 1970s during development of the short-lived Safeguard system.

Even so, achieving an intercept would be no small feat. The combined speeds of the kill vehicle and mock enemy target were estimated to approach 15,000 miles per hour—fast enough to go from Washington, D.C. to New York in a minute. More often than not, past intercept attempts involving other prototype interceptors had failed.

Adding to the jitters this time was a glitch that had occurred during the final practice run three days before launch. Test team members had been in the control room seats they would occupy on launch day, reciting the commands, pushing the buttons, and reporting the readouts on their instrument panels, as they would during the real event. Vandenberg had just reported the simulated launch of the target missile.

All of a sudden, the system shut down. No computer. No more simulation.

For several hours, technicians worked frantically to figure out what had gone wrong. They had never been confronted with such a problem before. That it arose so late in the preparations was especially disturbing. After several hours, someone noticed that Vandenberg had used fractions of a second to report the launch time. In all previous dry runs, the launch time had been entered using only round numbers. But Vandenberg had not participated in those runs, and they were accustomed to using fractions. That had confused the computer and triggered the shutdown.

So Vandenberg was instructed to stick to whole seconds. Then, to be sure the glitch had been fixed, a mission control team cycled the countdown clock through thousands of potential launch times—using every second of every day for a two-week period. Everything worked. Finally, the green light came from the Pentagon to launch.

At 7:02 P.M. PDT on October 2, an unarmed Minuteman rocket took off from Vandenberg and headed west. Right on schedule nearly twenty-two minutes later—and about 4,300 miles away—the interceptor shot up. It found the target and scored a direct hit 140 miles above the central Pacific Ocean. Despite intermittent cloud cover over Kwajalein, a high-resolution ground camera that had been focused on the projected hit point captured the brilliant, unmistakable flash of the collision.

"Very few people had given us much chance of being successful on the first flight test," said Major General Bill Nance, the Pentagon's program manager, who was in Kwajalein for the event. In fact, Nance himself had not fully believed it in the first minutes after seeing the flash. While everyone else in the control room was clapping and celebrating, Nance wanted to be sure that what the kill vehicle had struck was actually the target and not some other object. He was just that kind of person—the kind who, in his words, "likes to see the actual proof." So he checked the data flow from as many as nine different sensors monitoring the test to verify that the target had indeed been obliterated. Only then did he phone Kadish at the Pentagon and confirm a hit.

The intercept, pitting one speeding missile against another, provided critical validation of the hit-to-kill concept. It had come on top of a string of recent hits by two Army prototypes—the PAC-3, which was an advanced version of the Patriot used in the Persian Gulf War, and the THAAD system, an even more powerful weapon that finally broke a run of six failed intercepts against intermediate-range targets. The Vandenberg-Kwajalein demonstration had involved still faster speeds, greater distances, and higher altitudes.

The success of the first intercept attempt gave the Pentagon and the contractors a tremendous confidence boost. But controversy soon arose over whether the test had been rigged to ensure a hit—in particular, whether the presence of a large, bright decoy had actually helped to guide the kill vehicle to the target more than to deceive it.

The nature and number of decoys factored importantly in assessing the realism of the test. Russia and China were known to have incorporated such objects as balloons, jammers, chaff, and warhead replicas into their missiles to confuse or deceive potential interceptors. How quickly rogue nations would be able to develop and attach such devices to their missiles was a subject of some dispute, but there was no quarreling over the eventual need for a U.S. antimissile system to have the capacity to discriminate between the decoys and actual warheads—especially a system aimed at the midcourse phase when decoys are deployed.

In briefings before the flight test, Pentagon officials had highlighted the ways in which test conditions had been tightly circumscribed. After all, this was just the first intercept attempt, and the main objective, officials said, was to show whether the kill vehicle could find the target under benign conditions. This would not be a test of the whole system, they stressed. The focus was on the kill vehicle alone, with other system elements represented by surrogates or prototypes.

Defense officials were mindful that tests of hit-to-kill systems in 1984 and 1991, which succeeded in striking the targets, were faulted by congressional investigators years later for making the outcomes appear more effective than they actually were. This time, they noted, the target set had been simplified to include fewer decoys than had been used in two fly-by tests in mid-1997 and early 1998—down from nine to one.

And both the target and the kill vehicle, they said, were to be maneuvered onto a collision course in the test's early minutes with the help of global positioning satellites—although once released from its booster, they added, the kill vehicle would receive neither external assistance in distinguishing the dummy enemy warhead from the decoy balloon nor any help closing in on the warhead.

After the test, however, Pentagon officials neglected to explain just how the kill vehicle had found its way to the target, thus leaving the impression that the acquisition and discrimination processes had gone flawlessly. In fact, there had been a glitch. This fact emerged the following January when the *New York Times,* acting on a tip from the Union of Concerned Scientists, published a story disclosing that the interceptor had drifted off course and initially picked up on the decoy balloon rather than the warhead. Only after steering toward the balloon, the story said, did the kill vehicle spot the real target and adjust course toward it. Without the large, bright balloon, the story suggested, the test might not have succeeded. The story implied that the test was flawed and a dubious demonstration of the feasibility of antimissile technology. It quoted critics who believed the test showed that the system could easily be fooled.

Pentagon officials acknowledged that the process used by the kill vehicle to identify and lock on the target had been a bit circuitous, but they offered an entirely different interpretation of the significance of the test events. They said the kill vehicle had drifted off course as a result of an improper calibration of its inertial measurement unit, which is used for guidance. So when the kill vehicle opened its eyes in space, it did not find the target in its field of view at first. Consequently, it widened its view through a procedure known as a "step stare," which scans methodically across a larger area.

The first object to appear was the Mylar balloon decoy, which was about eight times brighter than the smaller, cooler mock warhead. Then the kill vehicle saw the canister—called a "bus"—that had carried the decoy and mock warhead into space before their release. The vehicle continued to search for the warhead, which was just about to come into view when this scanning phase timed out and the vehicle switched modes to begin discriminating between the objects that it had identified. But a second or two later, before the vehicle had fired any thrusters to steer toward the decoy, it picked up on the warhead and proceeded to home in on that as the proper target.

Officials maintained that the resulting hit proved that the system could work even under adverse circumstances, when not all of its components were functioning properly. If there had not been a balloon, they said, the interceptor would have continued to look in the assigned space and most likely still would have found the target.

Even so, the fact that the test's anomalies had not been made public cast suspicion on what the Pentagon had portrayed initially as an unqualified success. For some weeks too even senior BMDO officials seemed at a loss to explain just how the kill vehicle had identified the target, or more generally, how the vehicle's algorithms were programmed to deal with just the kind of confusion encountered in the test. Finally, BMDO and Raytheon produced a series of charts and diagrams that dissected the process used by the kill vehicle down to fractions of a second.

Program officials also found themselves on the defensive about the decision to use only a single decoy in the intercept test, particularly since earlier fly-by trials had involved nine decoys. Was the Pentagon dumbing down the tests to ensure a hit?

In fact, the question of whether to go with one decoy—and a rather large, bright one at that—had stirred considerable argument among those involved in planning the test. The decision was made by Nance, the National Missile Defense program manager, after a meeting with aides in Huntsville in the summer of 1999.

By then, the greatest number of decoys under consideration had dropped from nine to three. Gone were the replicas of the warhead and various other cone-shaped objects used in two earlier fly-by tests. The remaining candidates included one large balloon and two small ones. The reduction had reflected a decision by the Clinton administration to keep the system limited to hitting just rogue-nation missiles, not the more sophisticated Russian or Chinese ones. But even three struck some experts as too many for an initial intercept attempt.

An outside review team, headed by retired Air Force General Larry Welch, had advised Nance to eliminate all decoys in the first few tests. Welch had been called in to conduct a second assessment, following his "rush to failure" report the year before. Hitting an unobstructed speeding target in space was hard enough, Welch and his panelists argued, and they urged Nance to demonstrate such capability before complicating matters with decoys.

At the Huntsville meeting, test engineers for both Boeing and the Pentagon's program office recommended eliminating all decoys. But the Raytheon representative, confident of the capacities of his company's kill vehicle, pushed for raising the number back to nine. The government's chief program engineer and Boeing's program manager favored going with one, just to collect data on how the kill vehicle would behave. Nance thanked all for their views and spent the weekend mulling over what to do. On the following Monday, he announced he would take the small balloons off but leave the large balloon in order to, as he put it at the time, "exercise the algorithms."

"I left the balloon in there because the intelligence guys said it was representative of what we might see and because I wanted to check the ability of the kill vehicle to discriminate," Nance said. "But I also didn't want to overstress the system the first time we flew it. I wanted to see if we could do the basic functions that the kill vehicle was supposed to do. It never entered my mind that we were doing something that might be seen as cheating."

The decoy controversy would prove to be among the more frustrating aspects of designing not just the first test but others to follow. Program officials always felt on the defensive in explaining why the target set had been limited to one decoy. "We were damned if we flew with a decoy and damned if we didn't," said Keith Englander, the program's technical director. "It was a no-win situation."

The October test had shown that under carefully controlled and limited conditions the kill vehicle could locate and steer itself toward a small, fast-moving target in space. This was no small achievement, given the difficulty experienced by past attempts in getting even to this point. But the October test was still far from demonstrating a workable system under the stressful circumstances of real combat. At the time, Pentagon officials also were mulling fresh warnings about the program's break-neck pace and the abridged testing plan that served as a cautionary counterweight to the euphoria over achieving an intercept the first time around.

The warnings had come in the second report by Welch. Although the study had been completed a month before the October launch, Pentagon officials kept its circulation tightly restricted for more than two months, and then only quietly sent copies to Capitol Hill. This time there were no headline-grabbing quotes like "rush to failure." Welch's findings were not even presented in complete paragraphs but rather in brief sentences or sentence fragments under a collection of summary charts. Nonetheless, the forty-page report constituted a stinging critique.

It said that the program remained plagued by inadequate testing, spare parts shortages, and management lapses. Continued delays in test-ing and development, the report said, had jeopardized Clinton's plans to make a deployment decision by the summer of 2000. If further delays arose, it advised, the decision should be postponed. And in any case, there would not be enough information by mid-2000 to make a judg-ment about "deployment readiness." Instead, it suggested, the presi-dent's review should be converted into simply a "feasibility assessment."

The report also faulted government and contract officials for exhibit-ing "a legacy of over-optimism" about their ability to develop a reliable interceptor. Citing troublesome management gaps, the report said that government program managers lacked clear authority, and it faulted Boeing for being too focused on the overall integration of the system at the expense of overseeing the development of the kill vehicle and other components. It urged Boeing to audit the ground testing of the kill vehi-cle more closely. "Instead of unusual clarity, there is unusual fragmenta-tion and confusion about authority and responsibility," the report said.

Missile defense proponents were quick to draw what encouragement they could from the panel's blunt assessment. Senator Thad Cochran, the Mississippi Republican who had sponsored the new law requiring the deployment of an antimissile system as soon as "technologically possible," said that, while the panel had rated the program's inherent risks as high, they did not appear "unacceptably" high. He argued that the risks were outweighed by the urgency of getting a system in place to defend against potential attacks from rogue states. "We don't have the luxury of time," Cochran said. "Because of the threat, we have no choice but to accept a high-risk program."

But critics of the antimissile program saw in the panel's findings fresh cause to urge a go-slow approach. "In our view, delaying the program until there's more certainty of success is a reasonable course," said Steve Young, deputy director of the Coalition to Reduce Nuclear Dangers, a nonprofit group of seventeen arms control organizations. "This is something you don't want to get wrong, because the consequences, if it fails, can be disastrous."

Welch himself, since his days as Air Force chief of staff during the Reagan administration, had been branded as an opponent of missile defense. He certainly was a skeptic, but it would be wrong to categorize him as dismissive of the idea. He believed under Reagan, and continued to believe under Clinton—as his reports reflected—that the fastest way to deploy an antimissile system would be to follow a disciplined, step-by-step approach through essentially three levels of proof. First, the system would have to demonstrate it could achieve a hit. Second, it would have to show it could hit repeatedly. And third, it had to present a path for growth to deal with more sophisticated missiles.

What he and his panel concluded in the autumn of 1999 was that even after the program's restructuring a year earlier, significant schedule slips had occurred between January 1999 and September 1999, prompting program managers to look again for ways to cut corners and place things back on schedule. Welch's intent was to deliver a very strong message that no shortcuts to missile defense existed, that the quickest way to field a system was to dispense with politically driven calendar deadlines and instead set milestones based on performance.

In fact, Kadish and his BMDO team agreed with Welch's recommendations for additional tests, more hardware, and tighter oversight. For Kadish, the panel's findings were simply confirmation of the kind of tumult to be expected when efforts were being made to invent a complex weapon system in record speed. It was hardly surprising that there would be a hardware shortage, given that the program had only just been kicked into high gear that year. As an example of how pressed the program was for backup parts, the first flight-test vehicle had also served as the first laboratory prototype and had more than two hundred hours of test time on it when it flew.

"The Defense Department generally takes ten to fourteen years to invent something," Kadish observed at the end of 1999. "Right now, we're capitalizing on a lot of investment made in past years, but we're being expected to go from start-up to deployment in six or seven years. To get to 2005, we have to do things simultaneously to some degree, where you'd rather do things serially."

Kadish and Nance both had substantial experience developing military weapon systems. Kadish had managed a successful turnaround in the Air Force's once-troubled effort to produce a new cargo jet, the C-17. Nance had helped pioneer several of the Army's newest long-range artillery systems. Neither man was especially talkative or demonstrative. But each had a reputation for being among the toughest-minded, most thoroughly informed acquisition officers in their respective military branches. They knew how weapons got built, how to deal with contractors, and how to organize a development program.

Calling Nance and himself "hard-nosed realists," Kadish emphasized that the two of them were under no illusions about the difficulties involved in national missile defense. "It certainly rivals anything we've done as a nation—Minuteman, Polaris, Atlas," he said, citing major offensive missile programs. "If you go back in history to the Manhattan Project, it is certainly on that level."

Chapter 10

MISSED

Having scored a hit in October, the Pentagon headed into another intercept test in January with high hopes. Under a requirement that the Pentagon had set for itself, the kill vehicle would have to achieve a second intercept before officials could declare that the system had "demonstrated functionality." That rather arbitrary standard had been established the previous May—and not without vigorous dispute within the Defense Department. It represented a compromise between program officials, who had argued that a single hit should be sufficient, and two other Pentagon agencies—the Operational Test and Evaluation group and the Program Analysis and Evaluation (PA&E) office—which had pressed for a more stringent standard of at least three successes.

The matter was complicated by a reduction in the overall number of tests scheduled before the president was supposed to make a deployment decision in the summer of 2000. Originally, the program had scheduled four tests by then. But kill vehicle production delays in 1999 had postponed the initial flight test several months, pushing it into October and leaving time for only two additional tests ahead of

Clinton's decision. In view of the schedule slippage, Welch and his panel had suggested putting off the presidential review. But administration officials were reluctant to do so, fearing a political backlash and allegations that they were ducking a judgment.

In a split-the-difference call, the Pentagon's acquisition office set the hurdle for going forward with the program at two successful tests. "I just told them that this is a tough job and it wasn't realistic for them to expect we'd hit a perfect three out of three," Bill Nance said. "So we had more discussion and reached agreement on two."

Then the question arose: What if the program scored only one intercept? Would that be enough to justify any even partial steps forward?

To keep alive the option of fielding the system by 2005, Pentagon officials added a caveat that in the event of only one hit, the program office could at least award contracts for initial construction and long-haul communications and approve purchase of "necessary long lead hardware." But site construction could not actually start until a second intercept.

One other qualifying phrase got tucked into the Pentagon's self-imposed rule for proceeding. Of the two required intercepts, officials decided, one had to involve an "integrated system test." Just what constituted an "integrated system," however, was not specified, and the question became a point of heated controversy within the Pentagon in the run-up to the January test. That test was designed to be more complex than the October trial, which had focused essentially on just the kill vehicle. Now, Defense Support Program satellites would detect the target launch and provide a quick alert message. An early-warning radar in California would confirm the track. An operator sitting at the Joint National Test Facility in Colorado would replicate the military's main missile monitoring operation in Cheyenne Mountain. And for the first time, the battle management system would be incorporated to generate the targeting instructions for the interceptor. Finally, the kill vehicle would again have a chance to show its ability to acquire the target and smash into it.

Even so, the test would fall considerably short of a run-through of the complete system. For one thing, a number of key elements—particularly the booster rocket and the X-band radar—had yet to be produced, so surrogates or prototypes were being used. Initial targeting information that in the actual system would come from the X-band radar was simulated by data from a range radar in Hawaii and a transmitter onboard the target warhead itself. And the target and interceptor would fly the exact same geometries as before, with the target soaring out of Vandenberg, the interceptor taking off from Kwajalein, and the collision of the two expected at about the same place over the Pacific Ocean.

Because all the basic processes of the planned system were being worked into the event, Jack Gansler's staff argued that it should qualify as an integrated test. But PA&E officials strenuously objected on grounds that the interceptor was not really receiving its initial guidance from the X-band radar but rather from a simulated version of it. The argument raged until the day before the test, when Gansler finally ruled that the test could be considered an integrated one. "Reasonable people could come out on either side of the issue," said Bob Soule, the PA&E director. "But in the hallway afterwards, we were saying among ourselves, 'You know, we're never going to get to completely realistic testing.'"

Barely three months had elapsed since the last test when Nance and John Peller appeared before senior Pentagon acquisition officials to brief them on final preparations for the next one. With the launch several days away, the two program managers sounded confident about having things under control. "We felt, based on the success of the previous flight test, that this next one was going to be a virtual repeat of that," said Bill Carpenter, who managed the kill vehicle for Raytheon.

But Hans Mark, the Defense Department's head of defense research and engineering, was worried. Mark had a reputation for being hard to satisfy. A smart and accomplished aerospace industry veteran, Mark was especially proud of his own record—more than thirty spacecraft launches over forty years, including fourteen NASA space shuttle flights, and no failures. Listening to Nance and Peller give their report, Mark thought too many things were being overlooked and too much was being taken for granted in the wake of the first test's success.

An inveterate memo writer, Mark sent a confidential memo to Gansler after the briefing, complaining that Nance had appeared "too slick." Nance resented the allegation, taking personal offense at what he considered a none-too-subtle play on his first name, Willie. The general felt he had been as thorough as possible in preparing for the test. If the briefing had come across to Mark as a breeze-through, Nance thought that may have reflected the fact that it was indeed a condensed one-hour version of what had been planned as a three-hour presentation.

In any case, in contrast to the first test, which had been delayed for months by kill vehicle production problems, the second one went off as scheduled on January 18. A slight glitch occurred in the final countdown when a hunter was discovered on the Vandenberg test range, forcing a twenty-minute pause to ensure that the area was completely cleared. The target then took off and released its mock warhead and decoy balloon without a hitch. The interceptor lifted off flawlessly as well. Colored arcs representing the flight paths of the target and interceptor flashed across video displays on the ground, showing the two converging. A giant camera based near Kwajalein—the one that had caught the bright flash of collision during the first test—again focused its lens on the expected point of impact.

But the moment came and went without a flash. Both the target and interceptor were still flying, still broadcasting streams of telemetry. They had shot past each other about seventy meters apart, then continued falling back toward Earth until they disintegrated in the atmosphere.

Ron Meyer, Raytheon's deputy manager of the kill vehicle program, had been monitoring the interceptor's planned eight-minute flight from a computer terminal in a back row of the mission control room. In the final two minutes before the anticipated collision, he had noticed something wrong with the two infrared sensors that serve as the kill vehicle's eyes. They were seeing only a tiny fraction of the distance they were supposed to see.

Investigators were handicapped by one of the complications of space testing: the absence of any surviving hardware. But the kill vehicle had been extensively wired from top to bottom to send back data on virtually every moving part and action. The resulting avalanche of telemetry far exceeded the volume of data provided by, for example, the "black box" on commercial airliners.

Instrument readings from the kill vehicle clearly revealed that the infrared sensors had never cooled to the frigid temperature that allows them to operate. The sensors are supposed to be cooled to about seventy degrees Kelvin, or 334 degrees below zero Fahrenheit; being this cold enables them to track faint heat emissions from warm targets in the icy void of space.

Additionally, the vehicle had shown signs of an anomalous wobble. A krypton gas leak would explain both the lack of cooling and the wobble, so investigators initially blamed the failed flight on a gas plumbing leak. But further probing dismissed the wobble as insignificant and led to the revised conclusion that an obstruction in the flow of krypton had foiled the test.

The exact nature of the obstruction was never determined. It could have been moisture in the line that froze and clogged passage of the krypton gas. Alternatively, particles of brazing material from inadequately cleaned valves—or some other tiny debris—could have caused the clogging. Moisture was the favored theory because some water did get trapped in a valve that had been stuck open during gas-loading operations in the launch area. A Raytheon crew had tried to suction out the water, but investigators later found that the vacuuming procedures probably would not have removed all droplets.

To avoid another plumbing problem, Raytheon authorities scrutinized the cooling system in the kill vehicle already assembled for the next flight test—and did not like what they found. Replacement valves were ordered, fittings were redone, and some parts were redesigned. New assembly and cleaning procedures were put in place. Pentagon officials had little choice but to accept Raytheon's recommendation that a delay of several months be allowed for this work to be done and a few other glitches cleared up.

Although the kill vehicle has many redundancies built into it—two infrared sensors, for instance, as well as multiple thrusters and fallback electronic connections—the cooling system relies on a single plumbing network. It thus made the whole vehicle prone to what engineers call a "single point failure"—which is to say, if a plumbing line went down, the whole vehicle would shut down. "At the time the plumbing system was designed in the early 1990s," said Raytheon's Carpenter, "it wasn't considered to be a high probability of failure." Such a failure was still deemed unlikely, and adding a backup cooling system was ruled out, at least in the short term. But the program could not afford another plumbing problem.

It also could not afford to have the one technician at Raytheon who was most knowledgeable about the kill vehicle's plumbing off working on another project—but that is exactly what had happened, as Pentagon officials belatedly discovered to their amazement and consternation. John Karaba, the Raytheon engineer who had been the lead person in assembling the cooling system for the first test vehicle, had taken another job within the company and played no part in putting together either the vehicle that failed or the one about to fly. Raytheon believed in allowing its employees to shift to other company positions that interested them, and Karaba, who had worked on the kill vehicle project for four years, had opted for another assignment with greater responsibility. But the process for building the vehicle was still at an early stage: assembly was being done in a laboratory with many procedures far from firmly established, not on a standardized production line. (That was under construction.) Karaba had instructed his successors in how to arrange the plumbing, but assembly and installation still depended heavily on personal technique, particularly since some of the fittings went into very tight spaces. Karaba's successors lacked his touch.

Only after the test failure did Raytheon bring Karaba back to check out the plumbing assembly in the next kill vehicle getting ready to fly. He found that some joints and fittings were not assembled as he would have done them, and he expressed his concern about poor workmanship. One plumbing line, for instance, was pushed up against another in a way that was adding tension on the line and could restrict gas flow.

His judgment that the lines needed to be redone was central to the company's recommendation to the Pentagon to delay the next test and allow an overhaul.

It was this sort of down-in-the-details factor—the rather routine reassignment of a key company technician—that could make a momentous difference to the program as a whole and yet remain hidden from the attention of senior Pentagon officials until it was too late. Missing out on significant details was, of course, one of the risks inherent in running any big operation, but that did not make it any less frustrating for those on top. "If you've been around this business any time at all, this kind of thing is the maddening part," Ron Kadish remarked.

Kadish and Nance took some solace in the fact that so much had gone right in the second flight test. In fact, Pentagon officials had become somewhat notorious for claiming, even in the absence of an intercept, that a test had largely succeeded because a laundry list of other systems had worked. In this case, the battle management system had functioned as it was supposed to. The kill vehicle had received accurate targeting information, and it had separated cleanly from the booster and oriented itself correctly in space. It even had passed quite close to the target. But getting close was not sufficient, as those running the program knew. "We're not playing horseshoes," Nance said. "Close is not good enough in this game."

Nance was also well aware that the vehicle came as close as it did owing to happenstance as much as anything else. It passed close to the target because of the operation of a visible sensor on the kill vehicle— which was there alongside the two infrared sensors—and because of the way the target and its accompanying objects happened to have been arranged. The visible sensor, unlike the infrared ones, had not been given a role in discriminating objects in space or homing in on targets; its main function lay in determining the vehicle's position in space. Before the test, in fact, Raytheon and Pentagon officials had considered eliminating the visible sensor altogether in a redesign of the sensor package. But it was programmed during flight to keep as many objects as possible in its field of view, and in the test it did spot the target warhead, the balloon decoy, and the canister that had carried the warhead and decoy into space. As it happened, the warhead was in the middle of the three objects. So as the kill vehicle approached, the visible sensor stayed centered on the warhead as the other two objects fell out of its field of view. "If something else had been in the center, very honestly, it just as likely would have gone to that object," Nance said.

To strike the target, the kill vehicle still had to depend on the precision guidance of its infrared sensors in the final few seconds of flight. The high data rate of these sensors enables a vehicle to make minute course corrections and hit a target in a precise spot. But in the January test, when the kill vehicle checked for final homing data from the infrared sensors and found none, it shut down the terminal phase and went into a coast mode, gliding past the target. In the wake of the test, program officials started looking at ways of not only retaining the visible sensor but incorporating it into the final homing process.

Within a week of the failed flight test, Philip Coyle, the Pentagon's director of operational testing and evaluation (DOT&E), added censure to setback by issuing a report faulting the test program for an overly aggressive schedule and simplistic target scenarios. Studious and methodical, Coyle had served since 1994 as the Pentagon's chief assessor of the adequacy of test programs. Heading a team of forty experts, Coyle was responsible for sifting through the claims of contractors and the armed services to pass judgment on whether the weapons the Pentagon had under development were ready for mass production. His office tracked two hundred major acquisition programs and wrote about one hundred reports on them each year. He reported directly to the secretary of Defense, as well as to Congress, which created the testing post in 1983 as an independent voice commenting on the effectiveness of proposed arms.

Coyle prided himself on his technical intuition—the kind, he once said, "that gets built up over decades" of personal success and failure. His colleagues and admirers called him the technical conscience of the

Pentagon, although he knew he was not particularly popular. He told a story about once meeting a defense contractor at a conference who, suddenly recognizing Coyle's name, blurted: "You're the guy everyone hates!" Nevertheless, Bill Cohen, after taking over as Defense secretary in 1997, had made a conscious decision to keep Coyle. "He's a straight-shooter, and it's better to have someone like that in his job," said a senior Cohen aide involved in the decision.

Coyle was no stranger to missile defense. As test director in the early 1970s at the Nevada Test Site and at another test site in the Aleutian Islands, he had worked on nuclear warheads for the Safeguard system. Later, as an associate director at the Lawrence Livermore Laboratory in the 1980s, he had toiled on lasers for the Strategic Defense Initiative. He viewed missile defense as the most difficult effort the Department of Defense had ever undertaken. He thought the United States had the capacity to create an interceptor that could zoom reliably into space and smash a mock warhead to pieces. He saw it as largely a matter of engineering at this point. But what concerned him was that the testing program was unrealistic in terms of ensuring that the system could work in the combat conditions of nuclear war and enemy deception efforts.

By early 2000, Coyle was emerging as an influential counterpoint to those in the Pentagon pushing for an early deployment decision. Repeatedly, in public reports and private memos, Coyle contended that the early tests were by their nature limited. As a result, he argued, there would be insufficient information on which top administration officials could base a judgment by the summer of 2000 about the system's operational readiness. His concerns from inside the Pentagon echoed those expressed on the outside by the Welch panel.

In his February report, he noted that in the previous two years alone the National Missile Defense program had slipped twenty months behind schedule. The January flight failure, he added, would probably result in further delay. Since the deployment decision date had not been deferred, he said, program officials had come under "undue pressure" to meet "an artificial decision point." He said that schedule, not performance, was driving the program. He also wrote disparagingly about the structure of the tests. He said the unvarying Vandenberg-Kwajalein geometry, the relatively slow intercept velocities, and other aspects did not "suitably stress" the system, nor did the set of targets, which he described as unrepresentative of the decoys that U.S. interceptors might encounter.

By historical standards, the number of actual flight tests planned during development of the national antimissile system was paltry. The schedule was significantly shorter than those of previous missile and satellite programs, raising questions about whether enough tests would be conducted to ensure that the system actually worked. It amounted to less than half the 40 to 60 flight tests conducted during the first intercontinental-range ballistic missile and submarine-launched ballistic missile programs, and only a fraction of the number of flight tests of air defense systems. In the initial series of the Navy's Standard missile, 88 tests had been conducted, the original air-defense version of the Patriot had been run through 114, and the Advanced Medium-Range Air-to-Air Missile had undergone 111. The testing program for Safeguard had been consistent with these air-defense programs, with 165 tests conducted. By comparison, 19 intercept tests had been planned for the Clinton administration's proposed antimissile system.

One reason for the ambitious schedule was a sense of impending threat from rogue nations. But another important motivation for adopting this new approach was cost. A national missile defense flight test pitting a single interceptor against a single missile was priced at between $80 million and $100 million, including all the pre-test and post-test activities and analysis. Such high costs had driven the Pentagon to rely more on sophisticated simulations and ground tests to complement a sharply reduced number of flight tests. Technological advances in computers and ground tests also had facilitated this trend. Increasingly, flight tests were used essentially to establish certain critical data points to help adjust the computer models and simulations on which Pentagon authorities then based a majority of their performance assessments of new weapons systems.

But Coyle had expressed concern that ground-testing of the National Missile Defense system also had slowed. Besides, he argued, the planned system was more complicated than most weapons and therefore warranted more flight tests. Nor was it clear that the National Missile Defense program had the proper simulation and ground-test facilities. Although the Pentagon had built several significant simulation and ground-test facilities over the previous twenty years, none of the major antimissile systems under development—medium-range as well as long-range—had properly designed ground-testing programs, a point made by the 1998 Welch report.

Program officials defended their testing approach as sound. Kadish was fond of saying the system needed to learn "to walk before it could run." Privately, he and other senior program officials knew their tests were constrained, their flight envelope limited, their target sets oversimplified. But this was the plan that the department's senior staff had signed off on. Even Coyle had participated in approving the testing master plan several years before (although he would argue that the plan had been overtaken by events). What began to irritate the team running the program was a sense that the ground rules set for them were changing under their feet. Everyone had known, for instance, that only a few flight tests would occur before the president faced a deployment decision. They had known—or could easily have found out by reading the plan—what types of targets and decoys would be in those tests and how much reliance there would be on surrogates or prototypes for some major components.

Now all that was starting to be reassessed and second-guessed. Looking back months later, Keith Englander, the chief program engineer, would wonder whether the success of the first test actually worked against the program—by increasing the prospect that the administration might have to go forward with a deployment. "I think they were banking on a miss, which would have allowed them to say, 'Well, these guys certainly aren't ready,'" Englander said.

Preparing for the second flight test in January, program officials had felt the heightened scrutiny. They were asked to do a series of briefings for the senior acquisition staff headed by Gansler, the senior policy staff headed by Slocombe, the senior test analysis group headed by Coyle, and the senior analysis group headed by Soule. Englander laid out in considerable detail—"step by step, second by second"—what was expected to happen in the upcoming test, what the data sources were, which components were surrogate, and which were real.

"The department woke up and said, 'Gee, we have to start taking this program seriously and understand what they're doing,'" Englander said. "They really hadn't gotten in there before to ask the questions, to learn it, in order to give us the feedback that we needed. Even though they had attended all the test planning meetings, all the targets meetings, they never said, 'Gee, we're really going to have to make a decision.' And once they started waking up to that fact—you can see it in all the briefings we had to start giving—all of a sudden there was an interest in the program. And there were a lot of people who said, 'Gee, we didn't know that.'"

In addition to demonstrating that the system could work as advertised, the Pentagon had to get a better handle on what it would cost. Estimates had varied across a wide range, from $20 billion to more than $60 billion.

Just how decisive a factor cost would be in a final decision on deployment remained questionable. Given a grave enough threat, the system would be deployed even at great expense, particularly in times of flush federal surpluses. But if the threat was regarded as marginal, or the strategic rationale as questionable, the program could easily fall victim to congressional cuts or the competition for other resources within the Pentagon. That was what happened in the 1970s to the Safeguard system, which, once built, survived only several months after being deemed too expensive and too likely to be ineffective. More recently, in 1996, a joint effort by Senator Bob Dole and House Speaker Newt Gingrich to legislate deployment of a missile defense system never made it to the House floor after the Congressional Budget Office pegged the cost of the numerous antimissile system elements in the bill at up to $60 billion.

Almost invariably, costs rise in major acquisition programs, and the extent of the increases varies by type of system. Ships tend to have the lowest rates averaging roughly 15 percent, while tactical munitions and vehicles have the highest, doubling in price on average. The average cost growth for other types of systems falls somewhere in between, with most in the 20 to 30 percent range. That applies to strategic missile and space programs, which are most similar to missile defense programs and can be taken as a cost benchmark for them since only one national anti-missile system was ever actually built.

But the jumps for the Clinton administration program had been much more significant than that. The price tag for the three-plus-three system, introduced in 1996 and projected for fielding in 2003, was estimated initially by the Pentagon to be just short of $8 billion. By early 1999, when the administration rolled out its revised three-plus-five schedule, and after more than $4 billion already had been spent in the previous three years, defense officials projected another $8 billion would be needed through 2005. A year later, the estimated price tag through 2005 had risen to $11.4 billion, and the total acquisition and operating cost through 2015 for the C-1 system was set at $25.6 billion.

Some of this increase represented additional content. The August 1999 decision to go with an initial phase of one hundred interceptors in Alaska meant a several-fold jump from twenty, which had been budgeted at the start of the year. But much of the budget increase also could be attributed to overly optimistic assumptions about technology and costs—an optimism that was generated by the crash program mentality, combined with a desire to do the job as cheaply as possible. Together, those factors led to an acquisition program that included little of the usual funding for such essential activities as system engineering and risk reduction measures. Moreover, some of the assumptions that had been made about the amount of completed work that Boeing would inherit proved unrealistic. In addition, at the urging of Welch and his panel, the program was beginning to budget for more flight testing, ground testing, and other features common to acquisition programs of such complexity. These add-ons, while prudent, meant hundreds of millions of dollars in extra expense.

Finally, some of the increase could be blamed simply on poor estimation, reflecting the poor understanding early on by senior planners of some aspects of the program. For example, the projected cost of the system's communication links had skyrocketed from $800 million to $2.2 billion.

In the autumn of 1999, Kadish approached Deputy Defense Secretary John Hamre about the prospect of adding billions more to the National Missile Defense account. The reception was chilly. "Look, you may be a sacred cow," Hamre told him, "but that doesn't mean you get to eat everybody's oats. You're going to have to figure out how to absorb that yourself."

So BMDO officials went off to work the problem. They discovered that one factor driving up costs was the need for extremely high reliability along all communication links. To ensure secure connections, program engineers had planned to pull new undersea cables to Britain and Greenland as well as to Shemya in the Aleutian Islands—all sites of key radars. And just one new cable to those sites would not be sufficient. The design called for two to each location.

But by running only one fiber-optic cable to Alaska and none to Thule, relying instead on satellite communication links, program officials found they could save about $1 billion. They could trim another $300 million by leasing existing commercial fiber links to Britain instead of laying a dedicated cable. Further, by cutting back plans to "harden" cables and retrofit some ground facilities to withstand the electromagnetic shock wave effects of high-yield Russian and Chinese nuclear warhead detonations, defense officials came up with another $400 million in savings.

But there was a catch to switching from the ultra-secure optic cables to satellite links. The change would reduce the estimated reliability of the system by a fraction—a tiny fraction, but a fraction nonetheless. This was serious, because the reliability requirements for national missile defense, as for other weapons systems, were all but immutable. Each element of the system had been assigned a reliability number it had to meet—in some cases, down to four or five decimal points to the right of zero. Factored together, these numbers had to compute to an overall figure of 95 percent. That is, they had to show with 95 percent confidence that the system would be at least 95 percent effective. In that context, the communication links had been assigned a reliability requirement of 0.99999. Reducing the number of fiber-optic cables and relying more on satellites would drop the reliability to 0.9999 for the British and Alaskan radar sites and 0.999 for the Thule facility. This reduction, in turn, would lower the overall system reliability to 94.6 percent.

Kadish and Nance were incredulous that dropping the kill probability by four-tenths of a percent could result in such a substantial savings. It was not up to them, however, to authorize the change. Any reduction in the system's mandated reliability would have to be approved by the group that set the requirement in the first place—the Joint Requirements Oversight Council, a panel made up of the vice chiefs of the military branches and chaired by the vice chairman of the Joint Chiefs of Staff. As a sign of how gravely the military commanders take the parameters they set for new weapons systems, the council refused to lower the requirement for the National Missile Defense system. It did grant a temporary waiver for initial deployment of the system but insisted that program officials find a way eventually of reaching 95 percent.

The episode offered a lesson in the price the nation would be paying not simply to get the antimissile system to work but to get it to work to near-perfection. The former goal was hard enough to achieve; the latter struck some even in the Defense Department as unrealistic.

"Ideally, you'd like to have the civilians and the military working requirements together in a way that's sensitive to both political and military needs," a senior defense official said. "But that's not the way the system works; the military right now basically produces the requirement. When you look at our requirement for national missile defense, it's silly; I mean, it calls for a very high probability for no leakers. That makes extraordinary demands on the elements of the system."

What was clear was that senior officials were only slowly getting around to understanding some of the costs that had been designed into the system as a result of the requirements set. "Most new acquisition programs begin with a lot of work on what the requirements will be— what kind of airplane do we need, for instance, or what kind of tank— before any metal starts bending," another Pentagon official said. "But with national missile defense, things were happening in a compressed time period. I suppose a lot of people in BMDO had been working on what the various elements had to do to achieve a certain assurance of an intercept, but that had not received a lot of external scrutiny. Even more fundamental was the fact that the whole probability requirement was extremely ambitious, particularly if compared to any other system. So many things could go wrong with this system. It was extremely unlikely that a probability in the high nineties could be achieved."

The cable episode reflected differences over not only how reliable the system needed to be but also who the enemy really was. The administration had presented the system as focused on defending against a limited attack by a rogue state. But in introducing the original three-plus-three plan in 1996, Defense Secretary William Perry had also noted it would possess "some capability against a small accidental or unauthorized launch of strategic ballistic missiles from more nuclear-capable states," meaning Russia and China.

Just how capable the system would be against Russia or China had become a sensitive point, since any capability at all could appear to undercut the U.S. claims that the defensive weapon was not aimed at these countries. But the fact was that the system would have some inherent capability against a small accidental or unauthorized launch (AUL) of a Russian or Chinese missile. Moreover, since the early 1990s the military requirement for a U.S. national missile defense system had included a provision to engage a small number of warheads from any direction.

In early 1999, the Pentagon's civilian leadership tried to make the

AUL provision as incidental as possible. A classified guidance memo issued in January by Walt Slocombe, the undersecretary for policy, and Jack Gansler, the undersecretary for acquisition, stipulated that the system's primary objective was to combat a rogue-state launch. They noted that the system, by its nature, also would have some residual ability to counter a very small accidental or unauthorized launch from Russia or China, but they made clear that any move to enhance this aspect would have to be cleared first with the department's civilian authorities. BMDO officials were asked to assess periodically the added cost of enabling the system to deal with accidental or unauthorized launches.

Still, senior military officers balked at adopting similar language in the formal document spelling out the system requirements, which continued to make defense against AULs sound as significant as guarding against rogue-nation threats. "The military's view was, 'We're not sure of this rogue-state threat, but we're spending billions of dollars on this system and we want it to be able to handle a threat that might be just as plausible as a rogue-state threat, and that would be the threat of a small launch from a nuclear power,'" a senior defense official said.

The risk of a freak nuclear firing had been underscored in January 1995 when a scientific rocket launched from Norway was initially interpreted by Russian military authorities as a missile heading for Russian territory. While the Norwegian government had given the Russian Foreign Ministry advance notice of the launch plan, word had not reached the Russian Defense Ministry. The Russians ultimately refrained from sending off some nuclear missiles of their own, but not before sounding a nuclear alert that caused President Boris Yeltsin and General Mikhail Kolesnikov, chief of the Russian general staff, to open their nuclear control briefcases and consult each other via hot line.

The AUL issue became an area of civilian-military contention in the Pentagon in the spring of 2000 as the vice service chiefs moved to update operational requirements for the National Missile Defense system. Aides to Slocombe made several arguments for playing down the question of accidental or unauthorized launch. First was the diplomatic argument. "We had been telling the Russians and Chinese all along that the system wasn't directed against them, but if we have an operational requirement saying the system had to be configured—uniquely in some respects—so that it could shoot down Russian or Chinese warheads, then we were going to get ourselves into a little bit of an inconsistency," the official said.

Second, the intelligence community had provided no plausible size for an AUL. Was it a single ICBM, or a submarine commanded by a rogue officer and loaded with two hundred warheads? There was no way the system could counter the larger scenario without expanding well beyond existing plans. "So our point was, because we couldn't get a precise threat from the intelligence community, it didn't make sense to require this system to counter an AUL," the official said.

And third, the design of a system pinned to dealing with AULs would have to be much more expensive than that of a system whose only requirement was to counter a simpler rogue-nation threat. Just how much more expensive was uncertain. It was known, for instance, that the primary mission of one of the five upgraded early-warning radars planned for the system—the one in Clear, Alaska—was to look for Russian launches. But no study had been done to identify any other components in the National Missile Defense plan that were there just to deal with an accidental or unauthorized launch.

Officials continued to argue over the issue for several more months while BMDO produced a study showing what the differences in cost were between designing only to the rogue-nation threat and designing to the rogue-nation-plus-AUL threat. But the vice chiefs were reluctant to give up the AUL requirement, and so it remained in the system's Operational Requirements Document, even after a revision in mid-2000.

In the meantime, the issue of cost growth continued to plague the program. The Congressional Budget Office issued a report in April 2000 pegging construction and operational costs of the C-1 system at $29.5 billion through 2015. For the complete C-3 system, with two sites, 250 interceptors, nine X-band radars, and six upgraded early-warning radars, the cost would be $48.8 billion, CBO said. This estimate did not include the price of twenty-four SBIRS-Low satellites, due sometime after 2010 but not dedicated solely to national missile defense. The estimated price tag for the satellite network was $10.6 billion, raising the total system cost to nearly $60 billion.

In February 2000, the Navy's top officer challenged Pentagon plans to rely solely on land-based interceptors to shield the United States against missile attack, urging that ship-launched interceptors also be used to knock enemy warheads out of the sky. In a confidential memo to Defense Secretary Cohen, Admiral Jay Johnson, the chief of naval operations, argued that ships would make the proposed antimissile system more effective.

This marked the first time a Navy leader formally had pushed for a role in national missile defense, although Navy authorities had argued privately for nearly a decade that ships would provide a cost-effective substitute—or at least adjunct—to basing interceptors on land. In fact, the idea of a sea-based defense had been a favorite among some of the most ardent missile defense backers. The case for such a system had been presented in a 1995 Heritage Foundation study headed by Henry Cooper, who had directed the Pentagon's missile defense efforts in the first Bush administration.

But using interceptors aboard ships to guard U.S. skies would violate the ABM Treaty and posed all sorts of technical challenges, so the Clinton administration had done little to pursue the option. Johnson, worried that Clinton would decide on a deployment plan in the summer that would effectively preclude a Navy option, warned in his memo that such a step "would not be in the best long-term interests of our country." The admiral made clear he was not pressing to include sea-based interceptors in the near term, but rather in some later expansion of the antimissile project. He suggested that the sea-based option be considered as "complementary to, not a replacement for," the Pentagon's land-based scheme.

His pointed warning, which caught senior Pentagon civilians by surprise, added another controversial element to the escalating political debate over whether to construct a national interceptor system and, if so, what kind and how fast. What had prompted the Navy's top officer, known for his reserved demeanor and nonconfrontational management style, to call into question the Pentagon's antimissile planning? Some interpreted the move as a bid by the Navy for a piece of a major Pentagon growth program. But Johnson said the memo grew out of concern that some in the Pentagon bureaucracy, particularly on the civilian policy side, were inclined to choke off further consideration of the sea-based option. With 2000 shaping up as a pivotal year, Johnson, who had just several months remaining before retirement, felt emboldened to confront Cohen and argue the Navy's case. He worried that if the Navy did not put down a marker before the summer, it ran the risk of being dealt out. And if that happened, getting back in the game would be all the more difficult.

Whatever money the Navy had gained just for its theater antimissile program, Johnson felt, had required a persistent struggle. The lion's share of Pentagon funding for theater missile defense had gone toward the land-based alternative, the Army's THAAD program. He recalled that in his first months as chief, the Navy program was dismissed as a "science project" by Lieutenant General Mal O'Neill, the Army officer then in charge of BMDO. And while BMDO's current leadership seemed more receptive to a Navy role in both theater and national missile defense, Johnson considered the battle far from won.

Most of the support for a sea-based option came not from within the Defense Department but from Capitol Hill—from Republican Senators Jon Kyl, Bob Smith, John Warner, and others. Navy officers—most notably Rear Admiral Rodney Rempt—had aggressively pushed the Navy's case in Congress, to the consternation at times of some senior Pentagon civilians. In fact, another of Johnson's motivations in speaking out was to keep Rempt from following through with plans to retire by setting up a new Navy organization focused on missile defense and naming Rempt to head it.

Much of the appeal of the sea-based option rested in the idea that the Navy could convert some of its existing fleet of Aegis-equipped cruisers and destroyers into missile defense launchpads. By taking advantage of these floating assets, proponents said, the United States could greatly expand its antimissile coverage at relatively low cost. Ship-launched interceptors were being recommended both for doing midcourse hits, like the proposed land-based system, and for striking enemy missiles in their boost phase shortly after lift-off. But these notions were largely untested and filled with technical challenges. Although the Navy was close to fielding a short-range antimissile system for defending ports and coastal facilities, a more powerful medium-range version was not due for delivery until 2010. And even then, substantial advances would be required to produce a weapon capable of going against long-range missiles.

Just how much cost and effort a sea-based system would entail also was hotly disputed. Usually the price tag for one of these capabilities was advertised by proponents to be around $2 billion, but that figure neglected many of the requirements for developing an actual system. According to the Pentagon, a sea-based system aimed at midcourse intercepts would cost $16 billion to $19 billion. It would require a larger, more capable kill vehicle than the one being developed for the Navy's theater antimissile system, as well as a larger booster, which in turn would require modification of the ships to carry them. The Navy challenged the Pentagon estimate, insisting that the price would be much less because a sea-based antimissile system could plug into the same satellites, land-based early-warning radars, and battle management systems being developed for the ground-launched interceptor.

With so much argument over the numbers, and several variants for a sea-based system being bandied about, it was difficult at times to know what or whom to believe. "We've had this constant problem of people talking about something without really knowing the details," said one exasperated senior aide to Cohen.

To sort out the facts, Congress had ordered several studies from the Pentagon since 1997. The preliminary results of another one were due in March 2000. In fact, a thirty-six-page draft had been completed jointly by the Navy and BMDO, and the Navy was eager to send it to Congress. That idea was quashed, however, by Gansler and other senior defense officials, including Soule, Coyle, and deputy undersecretary for policy James Bodner, at a March 29 meeting in Gansler's conference room. They voiced concern that the study, which included a number of assertions about the value of a sea-based antimissile system, lacked supporting analysis and read more like the prospectus for a major new Navy program than a balanced, dispassionate assessment of the cost effectiveness and operational feasibility of ship-based interceptors.

"In the end, we as a department were unclear on a crucial question," Soule said. "Were we talking about the possibility of doing some kind of sea-based system after the land-based, C-3 phase? Or, as the Navy saw things, was anything beyond the initial C-1 plan of one hundred land-based interceptors up for grabs?

"There were two points of consensus at the meeting," he recalled. "First, everyone agreed we would have to analyze various sea-based options because of the Navy's interest and congressional interest, and because it's a fair question. But second, several officials stressed the need to be very careful about how we handle the issue because of the delicacy of the negotiations with the Russians."

The Navy and BMDO were instructed to revise their draft substantially. The Navy would live to fight another day.

Chapter 11

THE DECOY DILEMMA

Historically, the U.S. scientific community has shied away from involvement in major public policy battles. For one thing, taking political sides can jeopardize research grants. For another, it can cloud the image of researchers engaged in the pursuit of unbiased scientific truth. But missile defense has been an exception, with scientists playing an important and vocal role in challenging both the theory and practicality of building a national antimissile system.

As far back as the first missile defense debates in the 1960s, scientific critics pointed to the technical complexity of missile defense and the inherent advantages enjoyed by the side firing the missiles. In the 1980s, President Reagan's notion of space-based lasers and associated systems was deemed by a large segment of the scientific community to be fanciful and technically unachievable. Studies by the American Physical Society, the Union of Concerned Scientists, and the congressional Office of Technology Assessment explored the range of technologies under consideration by the Pentagon at the time and found them lacking. Dramatizing such scientific opposition was the refusal of thousands of academic physicists, computer scientists, and engineers to solicit or accept missile defense–related research grants.

So it was with a sense of some tradition that three Massachusetts physicists set out at the start of 1999 to mount a scientific assault on the Clinton administration's proposed antimissile plan. Their aim was to show that apart from all the political and diplomatic arguments for dropping the scheme—fears, for instance, that it would generate a new arms race and cost exorbitant sums—its technical shortcomings provided another compelling reason to turn away from the plan. The main problem, in the view of these scientists, was that enemy decoys and other deceptive measures would fool the U.S. interceptors.

The three—Lisbeth Gronlund, David Wright, and George Lewis—had met in their graduate student days at Cornell University in the early 1980s and embarked on careers of security policy analysis. By the early 1990s, they had ended up in Cambridge, Massachusetts—Lewis as associate director of the Security Studies Program at the Massachusetts Institute of Technology, and Wright and Gronlund splitting their time between research for the MIT program and work for the Union of Concerned Scientists (UCS). As they watched the renewed controversy over national missile defense in Washington, the three had little doubt that the Pentagon's tests would ultimately succeed in validating the basic principle of hit-to-kill. No scientific breakthrough was needed, they knew, for a speeding missile to strike another speeding missile, at least not under simple conditions. But they also were convinced that the U.S. interceptors would prove incapable in the long run of distinguishing between actual enemy warheads and the cloud of decoys sure to accompany them. In their view, the planned system risked becoming obsolete even before it could be fielded as a result of such deceptive techniques, known in the trade as "countermeasures."

The contractors and defense officials involved in developing the system certainly were aware of the countermeasures challenge. For decades, it had loomed as the biggest hurdle to perfecting a midcourse intercept weapon. But the initial C-1 phase of the antimissile system had been given the equivalent of a bye in this area, based on predictions from the U.S. intelligence community that North Korea and Iran were unlikely to employ any but the simplest of countermeasures with their first crop of long-range missiles. Consequently, to the extent that decoys had been included in the nineteen flight tests planned over the next six years, they were simple and few, consisting mostly of Mylar balloons of various sizes.

The judgment that countermeasures would not pose a threat with the first rogue ICBMs rested on a view that mastering decoy technology was considerably more difficult than producing long-range missiles. The techniques involved were said to be more challenging and information about them not as available from public sources as rocketry data. But this was vigorously disputed by some missile experts, who argued that any country capable of making an intercontinental-range missile could produce decoys to go with it or possibly obtain help from Russia or China.

Even more critical than this argument was the question of whether the planned U.S. system would ever be capable of defeating such countermeasures. Here too technical judgments differed sharply. The defense officials and contractors responsible for designing the system insisted that it would gain such a capability, certainly by the second phase of deployment. But what the three UCS and MIT physicists were intent on explaining was that even in its expanded form—with more interceptors and X-band radars and a new generation of tracking satellites—the system would remain vulnerable to deception, given the technical difficulty of distinguishing decoys from the real thing.

Informed debate about the decoy dilemma was handicapped because details about the antimissile system's performance parameters remained classified. But the system's components were known, as were their general physical properties, so outside experts had at least something to go on. Then too only certain properties about an object moving through space could be measured, and only certain types of technology were available for doing so. Temperature changes and emissivity area, for instance, could be read by the infrared sensors on the kill vehicle or on orbiting satellites. Speed, attitude, rotation, wobble, and other dynamic features could be detected by an X-band radar on the ground.

From this rush of data, characteristic traits could be computed and matched against the stored memory of images of known foreign warheads. In this way, a kill vehicle could distinguish a real warhead from a fake one, or so the system's designers argued.

But this process depended on U.S. defense and intelligence authorities having detailed knowledge about the shape, size, makeup, and other features of the enemy warheads the system was likely to encounter. Moreover, much of this purported discriminating capability had never been tested in flight. Nor was there any plan to do so ahead of Clinton's scheduled decision on deployment in the summer of 2000. Because the proposed system was a much-reduced version of Reagan's Star Wars vision, Gronlund, Wright, and Lewis worried about a general tendency to assume it was more likely to work. Although recognizing that yet another technical study on missile defense might be easily lost on the general public, they had in mind a target audience that was more narrow, focused mostly on journalists, Washington policymakers, and other scientists.

Several other factors also suggested the time was right for a scientific review. By 1999 the administration had defined its plan for the first time. It had settled on a specific Alaska-based architecture, stipulated where the interceptors and radars would go, and decided how many of each would be deployed in which phases of development. Enough information had become available for outside experts to make reasoned assessments of the system's operational effectiveness.

To add expertise and lend credibility and weight to their study, the Cambridge three set up a panel of more prominent scientists and missile experts. To chair the group, they enlisted Andrew Sessler, a senior scientist at the Lawrence Berkeley National Laboratory and a past president of the American Physical Society who had participated in the APS study in the 1980s. They also recruited Sherman Frankel, a retired University of Pennsylvania physics professor and expert on electronic countermeasures; John Cornwall, a UCLA physics professor and Pentagon consultant; and Bob Dietz, a systems engineer at Lockheed's missile division who had worked on submarine-launched ballistic missiles and studied countermeasures and foreign missile development. Four others on the panel—Kurt Gottfried, a retired Cornell University physics professor and UCS cofounder; Steve Fetter, a physicist and arms control specialist at the University of Maryland; Theodore Postol, a professor of science and national security studies at MIT and author of many reports on antimissile systems; and Dick Garwin, fellow emeritus at IBM's research center and onetime Rumsfeld commission member—had long been identified with the arms control movement and opposition to missile defense.

At a kickoff meeting in March 1999, panel members figured the study could be prepared and published within six months. But it took nearly twice that long as the authors expanded the scope of the study and made frequent revisions. Out of a list of about a dozen conceivable countermeasures, the group selected three for extensive examination. The idea was to pick measures simple enough to convince people that countermeasure technology could be easier to master than building the missile itself. The chosen three were: placing a nuclear warhead in a lightweight balloon of aluminized Mylar and releasing it along with a large number of similar but empty balloons; covering a nuclear warhead with a shroud cooled to a low temperature by liquid nitrogen; and releasing one hundred or more bomblets filled with chemical or biological agents shortly after liftoff.

Devoting a chapter to each countermeasure, the study described in great detail how it could be produced and why it would be likely to overcome the planned antimissile defense system. In the first two cases, the study argued, the warheads would elude the system's infrared sensors and radars. In the third case, simply the number of submunitions would overwhelm the system. Although the scientists had no access to classified aspects of the system, they were confident that such information would make little difference to their assessments. They simply assumed the sensors would work as intended. As good as these sensors might be, the study concluded, basic physics dictated that they could not detect some things.

Halfway through its effort, the group found encouragement in a new U.S. intelligence assessment acknowledging that the countermeasure problem was more imminent than previously thought. The revised National Intelligence Estimate said that rogue states "probably would rely initially on readily available technology" to develop countermeasures and could do so "by the time they flight-test their missiles." The NIE noted that Russia and China had developed countermeasures and "probably are willing to sell the requisite technologies." Among the technologies that rogue states might acquire, the report said, were balloon decoys, chaff, low-power jammers, radar-absorbing material, and "booster fragmentation" (blowing up a booster after release of its warhead in order to create lots of space debris).

The intelligence community appeared to be coming around to the notion that countermeasures required attention sooner rather than later, although privately, defense officials described the intelligence in this area as so sketchy and hedged as to be essentially useless. The practical problem was that the system proposed by the administration remained based on earlier estimates that had played down the countermeasure threat and so had little proven capability against it. This point was driven home by the UCS/MIT-sponsored panel, which released its report at a Washington news conference in April 2000. It was the first independent technical evaluation of the U.S. plan. "Available evidence strongly suggests that the Pentagon has greatly underestimated the ability and motivation of emerging missile states to deploy effective countermeasures," the report said. If true, it added, the system would fail in the real world. The report went on to criticize the planned testing program as inadequate to assess the operational effectiveness of the system.

Entitled simply *Countermeasures,* the 175-page study provided a densely packed but readable compendium of information about the decoy dilemma. Filled with charts, diagrams, photographs, and complex equations, it appeared extensive and authoritative and gave missile defense opponents a technical footing from which to criticize the administration's plan. It noted that most of the ideas for countermeasures were as old as ballistic missiles themselves. It said there was little reason to think rogue states would not develop such devices, particularly since major nuclear powers had done so and many of the measures required a lower level of technology than that required to build a long-range missile.

"This is basic physics, and it's airtight," Gronlund declared at the news conference. "It's as close to a proof as you can get that the system won't work." Added Sessler, "A defense that doesn't work is no defense at all."

The Pentagon initially offered only a muted, cursory response. For months, no high-ranking official came forward to rebut the report. Official comment was confined largely to statements by a spokesman for the Ballistic Missile Defense Organization, who merely reasserted his agency's confidence that the system would work against any challenges or countermeasures that could be mustered by the year 2005. But privately, senior defense officials were starting to take a harder look at the issue, especially now that the intelligence community also had assigned it greater urgency. Other developments too would conspire to keep the countermeasure issue in the news and continue to call into question the capability of the Pentagon's planned system.

A month earlier, a related scientific attack had emerged in a legal case involving development of an alternative kill vehicle—the one that Boeing had pursued before the disqualifying incident with Raytheon. Although the equipment was no longer a factor in the Pentagon's program, allegations that TRW faked tests and computer evaluations in developing software for the vehicle had raised disturbing questions about the conduct of contractors and defense officials.

The allegations came from a former TRW employee, Nira Schwartz, who was on the company's antimissile team in 1995 and 1996 helping to design computer programs meant to enable the kill vehicle to distinguish between incoming warheads and decoys. In test after test, she alleged, the software proved too feeble and failed to identify the warheads. But her superiors insisted that the technology performed adequately, refused her appeals to tell industrial partners and federal

© AFP/CORBIS

On August 31, 1998, North Korea launched a three-stage Taepodong I missile, attempting to put a small satellite into orbit. While the satellite never made it, the launch demonstrated that North Korea's skill at long-range missile technology had advanced further than U.S. intelligence analysts had predicted. Stunned by the event and under pressure from Republican lawmakers to accelerate work on a national antimissile system, the Clinton administration began to look more seriously at deploying a weapon to protect all fifty states from missile attack.

AP/Wide World Photos

The military chiefs were wary of national missile defense. From left, members of the Joint Chiefs of Staff: Army General Dennis Reimer, Navy Admiral Jay Johnson, Army General Henry H. Shelton (chairman), Air Force General Michael Ryan and Marine General Charles Krulak.

© Reuters NewMedia Inc./CORBIS

Secretary of Defense William Cohen, the only Republican in President Clinton's Cabinet, had long advocated a limited missile defense since his days in Congress and continued to push for one after joining the administration.

Samuel "Sandy" Berger, the president's national security adviser, cobbled together a compromise on national missile defense that balanced political, military and diplomatic considerations.

© AFP/CORBIS

Courtesy of Congressman Thad Cochran

Courtesy of Congressman Curt Welson

© Wally McNamee/CORBIS

Republican Senator Thad Cochran (*top left*) of Mississippi authored a bill that called simply for deployment of a limited national missile defense system "as soon as technologically possible." Republican Representative Curt Weldon (*top right*) of Pennsylvania drafted a similar measure. Both bills passed overwhelmingly in March 1999. Although the Senate's leading Democratic voices on foreign and defense policy—Joseph Biden (*lower left*) of Delaware and Carl Levin of Michigan (*at right with Air Force General Joseph Ralston, vice chairman of the Joint Chiefs of Staff*)—voted for the legislation, they later opposed a deployment decision.

Courtesy of the White House

Meeting with his senior national security advisers in the Cabinet Room of the White House on August 18, 1999, President Bill Clinton authorized a plan to place the first missile interceptors in Alaska by 2005 and to begin talks with the Russians about amending the ABM Treaty to permit such a system. From left around the table: Robert Bell, senior NSC specialist on defense policy and arms control; James Steinberg, deputy national security adviser; Strobe Talbott, deputy secretary of State; Madeleine Albright, secretary of State; Sandy Berger; William Cohen; General Joseph Ralston; and Leon Fuerth, the vice president's national security adviser. Seated in the back row: Steven Andreasen, an NSC defense expert; and James Bodner, a senior aide to Cohen.

Department of Defense

The Pentagon put two of its most experienced acquisition officers, Lieutenant General Ron Kadish (*right*)and Major General Bill Nance, Jr., in charge of supervising development of the National Missile Defense system.

Department of Defense

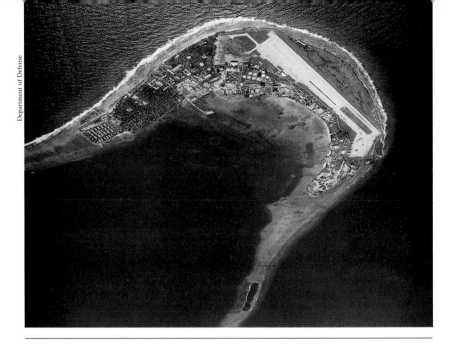

Kwajalein Island has been the site of interceptor tests for various prototype missile defense systems since the 1960s.

Courtesy of Raytheon Company

The "kill vehicle," made by Raytheon, has the makeshift appearance of a jet-propelled telescope.

Department of Defense

Boeing's John Peller (*left*) and Major General Nance posed for a handshake in front of the interceptor missile just prior to the first intercept attempt in October 1999.

AP/Wide World Photos

AP/Wide World Photos

AP/Wide World Photos

Deputy Secretary of State Strobe Talbott (*at left with Russian Foreign Minister Igor Ivanov*) led the U.S. effort to reach a deal with Moscow. But Russian President Vladimir Putin (*at right in photo with Clinton*) made clear at a summit meeting in June 2000 that he had no intention of agreeing to amend the ABM Treaty. Four months later, Secretary of State Albright traveled to Pyongyang to explore a possible agreement with North Korean leader Kim Jong Il, who expressed a willingness to halt all missile exports and discontinue development of longer-range missiles in exchange for financial assistance.

AP/Wide World Photos

President George W. Bush, in a speech in May 2001, called for moving "beyond the constraints of the thirty-year-old ABM Treaty" and replacing it "with a new framework that reflects a clear and clean break from the past, and especially from the adversarial legacy of the cold war."

© Reuters NewMedia Inc./CORBIS

Secretary of Defense Donald Rumsfeld (*left*) designed an aggressive testing program to pursue a broad range of potential antimissile weapons, while Secretary of State Colin Powell (*right*) offered assurances in the early months of the administration that the United States would not act precipitously to abrogate the ABM Treaty.

Courtesy of Boeing

In July 2001, a prototype interceptor lifted off from Meck Island in the central
Pacific and soared skyward, where it found and collided with a target warhead
launched from Vandenberg Air Force Base in California. It was the second hit in
four attempts. Among the devices recording the intercept was an airborne Infrared
Instrumentation System that showed the hit through four sensors of different
wavelengths.

Department of Defense

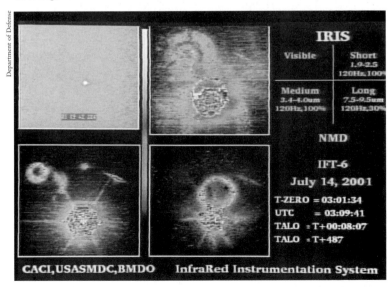

patrons of its shortcomings, and then fired her. Schwartz said that her protests precipitated her dismissal from the company. Her claims of TRW wrongdoing extended beyond the time she left the company, to a June 1997 "fly-by" test of the kill vehicle against a target warhead and nine decoys. The Pentagon had hailed the test as a success, but Schwartz said that the results had been doctored after the kill vehicle's computer had selected a decoy instead of the target.

Her story became public in March 2000, after filings in a federal lawsuit she started in 1996 seeking to recover damages for the government were unsealed and she spoke to the *New York Times*. She also had filed a wrongful termination action against TRW. The company, for its part, vigorously denied misrepresenting or improperly manipulating flight test data and defended the performance of its kill vehicle software. The Pentagon, also denying that any test results were rigged or doctored, pointed to two studies that it said had discredited Schwartz.

Reading about the case, Ted Postol, the MIT professor who had worked on the countermeasures study, was intrigued. Schwartz's attempt to defy a big defense contractor and accuse it of fraud recalled Postol's own much-publicized efforts a decade earlier challenging the Army's claims of success for its Patriot antimissile system during the Persian Gulf War. Patriot interceptors, Postol said, had in fact destroyed no Iraqi missiles at all. Though the Pentagon and Raytheon, the manufacturer of Patriot, at first denied his assertion, they later conceded that initial reports of the Patriot success had been grossly exaggerated and reduced Patriot's claimed success rate by half.

Now Postol saw a chance to learn more about the proposed National Missile Defense system by getting to know Schwartz and her case. So he phoned her and arranged for her to attend a one-day workshop in Boston with a group that included himself, Garwin, and several other scientists.

By outward appearances, Schwartz and Postol can come across at times as prickly or combative personalities, driven to right wrongs where they see them. But both have extensive technical training and impressive scientific credentials. She is a naturalized U.S. citizen, born in Israel, with a Ph.D. in engineering from Tel Aviv University and eighteen U.S. patents, including ones involving computerized image analysis and pattern recognition. He is a bona-fide rocket scientist—a Ph.D. in nuclear engineering from MIT, a former adviser on ballistic missiles to the Navy's top-ranking officer, and a onetime analyst of the MX missile for the congressional Office of Technology Assessment. He also has a knack for infuriating Washington officialdom, as the Patriot controversy made clear.

Postol came away from the meeting with Schwartz persuaded that she knew what she was talking about. He started poring over the documents she had collected for her lawsuits, focusing particularly on data from the 1997 fly-by test. He concluded that the way the warhead and decoys appeared to the kill vehicle's infrared sensors—as twinkling signals of light—were random and insufficiently differentiated for the interceptor to be able to tell them apart. In May he laid out his findings in a letter to John Podesta, the White House chief of staff, claiming that he had found a major flaw in the Pentagon's antimissile plan. The brightness readings from the mock warhead and decoys, he told Podesta, "fluctuated in a varied and totally unpredictable way," revealing no feature "that could be used to distinguish one object from the other." He accused the Pentagon and its contractors of engaging in an "elaborate hoax" to cover up the failure and called on the White House to appoint a high-level scientific panel to investigate.

Pentagon officials and outside experts hired by the Pentagon disputed Postol's conclusions, saying that he had misread the data. Raytheon authorities also stressed that the algorithms used in their kill vehicle worked differently from those in the TRW software, relying on another form of statistical reasoning. But it was the allegations of fraud that infuriated Podesta and top Pentagon officials. Postol's accusation that the antimissile plan was not only based on faulty science but guilty of outright dishonesty made his attack more personal and insulting. This was not out of character for him. For all the laser-like precision he brought to his work as a scientist, Postol often behaved publicly with what one journalist called "the subtlety of a Gatling gun."

Still, the Pentagon made matters worse for itself by moving to classify Postol's letter and four appended reports as secret—after they had been sent to Podesta and widely publicized. The action, prompted by BMDO officials—who noted that the documents contained some classified information that inadvertently had been cleared for release— caught the White House by surprise. Learning about it in the newspapers, Podesta phoned Rudy de Leon, who had replaced Hamre as deputy Defense secretary, and commented on the absurdity of classifying a letter already disseminated. But even top Pentagon officials could not seem to stop their own bureaucratic wheels from taking another turn or two and making Postol look bullied and abused. A few weeks later, three agents from the Pentagon's Defense Security Service arrived unannounced at Postol's office, saying they had come to point out the classified material that had been contained in Postol's letter. Postol refused to go over the documents with them, complaining publicly afterward that the visit may have been an attempt to intimidate or entrap him into talking about classified information. A security service spokesman insisted the visit was merely advisory and entirely proper.

After giving some thought to providing an extensive rebuttal to Postol's scientific arguments, the Pentagon settled for a brief statement in late May, suggesting that security considerations prevented them from saying much more. "The criticisms he's made are not new and have been reviewed by independent experts, including those at MIT's Lincoln Labs," the statement said, referring to the defense laboratory responsible for most of the government's research on countermeasures. "The information upon which he based his claims is incomplete and his conclusions are wrong." As an example, the statement said that Postol, in focusing only on the kill vehicle, had failed to take into account the discriminating capabilities of ground-based radar and space-based infrared sensors.

Postol himself was not finished. A month later, he was the source of another front-page *New York Times* article, this one highlighting the fact that the Pentagon had revised its target set for future flight tests, substituting simpler and fewer decoys that would be easier for the antimissile weapon to recognize. Postol portrayed the move as a

Pentagon attempt to hide the system's inability to distinguish a warhead from a decoy. Other technical experts quoted by the paper also spoke disparagingly of the plan, noting that none of the tests addressed what one called "the reasonable range of countermeasures" that an enemy would use to try to outwit an antimissile weapon. But defense officials maintained that the changes had been made to reflect more accurately the kinds of targets the system was likely to face in its initial phase.

In the article, Postol got personal again. He accused Pentagon officials of "systematically lying about the performance of a weapon system that is supposed to defend the people of the United States from nuclear attack." Even some of Postol's own colleagues regarded the accusations of fraud and deceit as, at a minimum, impolitic. "Ted is a sound, original, and energetic scientist who has done great work in this field. But diplomacy and Ted Postol are an oxymoron," Garwin told the *New York Times*. "Many, many times I have told him, 'You will be more effective if you are more controlled and edit your language.'"

Asked by fifty-three members of Congress to investigate Postol's charges of fraud and cover-up, the Federal Bureau of Investigation eventually cleared TRW Inc. of allegations that it had manipulated the test results. The FBI closed the case in February 2001, saying Postol's charges were "a scientific dispute and Postol's attempts to raise it to the level of criminal conduct had no basis in fact." Nonetheless, as a technical critique that focused attention on the discrimination shortcomings of the planned system, Postol's arguments resonated in the scientific community.

In contrast with Postol's brash tactics, Garwin preferred a quieter, more diplomatic approach. The two men were friends and collaborators, but while Postol was shunned by the administration for his offensiveness, Garwin was invited by officials at the Pentagon, State Department, and White House to brief them on his concerns and ideas.

A much-honored physicist and onetime boy wonder now in his seventies, Garwin had the patient, earnest air of a professor lecturing students when explaining missile defense. After all, he had been making some of the same arguments critical of national antimissile systems for more than thirty years. "It's like teaching freshmen," he remarked once, when asked how he felt addressing generation after generation of government officials. "They never learn."

Although he had served on the Rumsfeld commission, he did not believe that its warning about a growing missile threat necessarily favored construction of an antimissile system. In his view, there were other ways of dealing with the problem—and other potential threats to U.S. security that warranted more immediate attention. Still, if the United States wanted to build a national defense against missile attack, he would not lobby against the idea. What he would do was campaign energetically for what he considered a more sensible system than the one the administration was pursuing. Instead of going after enemy warheads during their midcourse phase in space, he argued, it would be better to hit them shortly after launch during their boost phase.

This was not a new concept. Many experts had long asserted that the optimal defense would be to strike missiles while they were still burning and on their way up. The Pentagon had begun researching boost phase technology in the 1950s in its BAMBI program, and much of the missile defense research under Reagan had focused on developing laser boost-phase weapons. The Air Force was still funding work on a boost-phase intercept system called the airborne laser, although that program was targeted against shorter-range battlefield missiles, not ICBMs. A separate space-based laser program run by BMDO was considered more than a decade away from proving itself.

After making the case for a boost-phase system before the Senate Foreign Relations Committee in May 1999, Garwin became one of the most outspoken and authoritative proponents for it on the national scene. His notion, however, did not involve lasers. He envisioned stationing interceptors on cargo ships a safe distance off the coast of North Korea or on land in Russia as part of a joint operation. Satellites would provide launch notice directly to the interceptors, which would fly out to catch the enemy missile before its motors shut off and it reached maximum speed.

The advantages to this approach, Garwin argued, were obvious and compelling. For one thing, a missile still burning as it climbs into space is many times more visible than the cooled warhead in space. For another, an ascending missile has not released the decoys and other countermeasures that become problematic in the midcourse phase. Further, a boost-phase system could be designed to operate over an area small enough to be effective against an emerging threat like North Korea without itself posing a threat to missiles deployed in the vast expanses of Russia or China.

At the Pentagon, program officials met with Garwin several times during the year to discuss how such a system might work, but they concluded that it would be more challenging and potentially more costly than the existing plan. They would have to devise a new interceptor, they said, one that was considerably bigger, faster, and more maneuverable than the Standard Missile onboard Navy Aegis ships, or even than the land-based interceptor being planned for Alaska. Although the United States had built high-velocity boosters before, as well as high-acceleration ones, it had never built a booster that achieved both the high velocity and rapid acceleration needed for boost-phase intercepts. And assuming such a booster could be built, the missile would be too large and heavy to launch from current Navy combat ships. Besides, program experts doubted whether a kill vehicle would be able to identify and home in on an enemy warhead through the bright, encompassing flame of a booster rocket. Distances, too, posed a problem. While U.S. ships conceivably could get within striking range of North Korea's coastal launch site, inland launch pads, particularly in large countries like Iran, were another matter. Moreover, the boost-phase approach would require a nearly automatic response to a missile launch, eliminating time for consultations with political authorities in Washington.

"There's a basic approach by opponents of the Pentagon effort, which is to say, 'Your plan is to do "A," but "A" won't work, what you really want to do is "B,"'" said a senior defense official, voicing some suspicion of Garwin's motives in pushing the boost-phase idea. "And if they ever succeed in killing 'A' and getting us to move to 'B,' then they'll suddenly say that 'B' isn't going to work, what you really want to do is 'C.'"

For his part, Garwin continued to insist that the Pentagon could overcome its technical concerns about the boost-phase concept more quickly and cheaply than it could build what the administration had in mind. "It's like when I worked on automobile fuel economy and exhaust emissions in 1970," he said. "There was a clean air amendment, and the Detroit car companies were going out of their way never in the same car to demonstrate catalytic converters to treat the exhaust gas and advanced carburetors to have an appropriate mixture. Because once they did that, the law said this would be taken as a demonstration of technical feasibility and they would be mandated to adopt that. Meanwhile, the Japanese had no such problem. Instead of paying lawyers, they paid engineers and went ahead and did it. So in the case of missile defense, it would be faster to do a boost-phase system than redoing all these radars worldwide and building the new interceptors and the radar in Alaska."

Garwin had urged Kadish to award "some small contracts" to test aspects of the boost-phase idea, including the question of whether a kill vehicle could home in on an enemy rocket enveloped by a bright burning plume. But the general concluded there was little likelihood he could get administration support and funding for yet another experimental missile defense project. Garwin considered such resistance shortsighted. "They feel that the countermeasures challenge will wait until they have a land-based, midcourse system deployed," he said. "I think they're absolutely wrong. They ought to be doing the boost-phase approach first. If you think missiles will have countermeasures, then why even build a midcourse system?"

Other scientists also were weighing in, challenging the technical viability of the system. In May the American Physical Society, the world's largest group of physicists, with more than forty-two thousand members, faulted the Pentagon's test program as "far short" of what the

president needed to make an informed decision about whether the proposed weapon could actually shoot down enemy warheads. The group
zeroed in on what it called meager testing of possible ways an enemy
might outwit the proposed weapons system. The United States, it said,
should make no decision to build a national missile defense unless it
was shown "to be effective against the types of offensive countermeasures that an attacker could reasonably be expected to deploy with its
long-range missiles."

In early July, fifty Nobel laureates signed an open letter to Clinton
urging him to reject the system, saying the plan would be wasteful and
dangerous. "The system would offer little protection and would do
grave harm to this nation's core security interests," they wrote.
Although Nobel winners had formed other informal groups occasional-
ly to address issues, it was unusual for so many to do so; the signatories
included about half of all living American science laureates. Hans A.
Bethe, a Nobel winner in physics and a main architect of the atom
bomb, helped write the letter and was the first to sign. In the late 1960s,
he and Garwin had coauthored one of the first public scientific critiques
of national missile defense. The new single-page letter noted that scientists independent of the Pentagon had long argued that foes could outwit or overwhelm any such attempt at defense.

Scientists were not the only ones taking issue. The revival of a
national missile defense plan had galvanized the arms control community, whose ranks had dwindled after the end of the cold war. For much
of the 1990s, in the absence of the kind of grassroots antinuclear campaigns that had opposed Reagan's policies, activists had focused their
efforts inside Washington, lobbying administration officials and members of Congress on a handful of matters, including the nuclear test
moratorium, the Comprehensive Test Ban Treaty, and extension of the
Nuclear Non-Proliferation Treaty. By 1995 more than a dozen arms
control groups had banded together loosely under the Coalition to
Reduce Nuclear Dangers, which became a main coordinating body for
efforts to block deployment of an antimissile system. The coalition
included such groups as Peace Action, the Union of Concerned

Scientists, Physicians for Social Responsibility, the Natural Resources Defense Council, the Arms Control Association, the Center for Defense Information, and the Council for a Livable World. Although these groups were generally opposed to missile defense, they did not consider it realistic to campaign for outright termination of the Pentagon program. Rather, coalition members united around the objective of forcing a postponement of the deployment decision and avoiding a U.S. withdrawal from the ABM Treaty.

On Capitol Hill too some Democrats who had backed the administration's plan began urging Clinton to postpone a decision. In the wake of the failed January flight test, Senator Joseph Biden, the ranking Democrat on the Foreign Relations Committee, was the first of his party's congressional leaders to call for delay. Significantly, some Republicans started doing so too—notably, Senators Chuck Hagel of Nebraska and Gordon Smith of Oregon, both members of the Foreign Relations Committee—although the Republican leadership continued urging a quick decision to deploy.

Even some former high-ranking members of Clinton's national security team who had once backed the land-based, midcourse intercept approach began having second thoughts. By late spring, several had decided to speak out against the proposed architecture. Two former deputy Defense secretaries under Clinton—John Deutch and John White—joined with Harold Brown, President Jimmy Carter's Defense secretary, in calling the administration's scheme misguided. Writing in the journal *Foreign Policy*, they suggested an alternative: an interim sea-based system that would use available Aegis ships, while the Pentagon took time developing more powerful interceptors. This approach, they said, would be more cost-effective, more tailored to the immediate threats from North Korea and Iran, and more responsive to the concerns of Russia, China, and the NATO allies.

William Perry, who as Defense secretary under Clinton from 1994 to early 1997 had been the father of the three-plus-three missile defense plan, also had doubts about the administration's course. Long skeptical of missile defense on technological grounds, the former Pentagon leader concluded that the time was not right to proceed because the initial flight tests had gone poorly, the potential threat had yet to materialize, and the risk of ruptured relations abroad was too high. Following a conference on missile defense hosted in Washington by the Carnegie Corporation, Perry joined Garwin, Gronlund, and twelve others in signing a letter to Clinton in June that urged deferral of the deployment decision. "We are convinced that significant unresolved issues remain concerning the costs, technology, and especially the security and foreign policy implications of a national missile defense system," the letter said. Among the other signers were retired General John Shalikashvili, former chairman of the Joint Chiefs of Staff under Clinton, and former Senator Sam Nunn, the Georgia Democrat who had chaired the Armed Services Committee.

Although such seeming acts of desertion by onetime top Clinton aides suggested trouble for the administration, at least some of those who broke ranks did so with a wink and a nod from their erstwhile colleagues still on the inside. Perry and Deutch, for instance, each got the impression in separate contacts with senior presidential advisers that by taking their concerns public, they would be helping the administration back away from a deployment decision, which was where most of those around Clinton—and the president himself—were increasingly inclined to go.

"I didn't think the administration's proposal had been genuine," Deutch said. "I don't believe they put it forward because they really wanted to see it developed. I think they put it forward because they wanted to take the issue out of the 2000 election and sort of quiet it down." He added that he had received encouragement from two senior administration officials to write the journal article. Similarly, Perry said that when he spoke to a high-ranking White House adviser of his concerns and his intention to speak out, the adviser seemed sympathetic. "They could look at what we were doing as giving them a way out," Perry said.

• • •

Officially, the administration's position through the first half of 2000 was that Clinton would stick to his timetable. The military would conduct another missile test—originally scheduled for April but delayed until July to allow for investigation of the January flop and retooling of the kill vehicle. The Pentagon then would present to the president its assessment of the system's readiness to deploy. And the president would make his decision in the summer.

But as criticisms of the plan picked up pace in the spring, the Pentagon seemed surprisingly unprepared to address the technical issues raised by the UCS/MIT report and other scientists. For several months, defense officials had little to say publicly about the decoy dilemma. Finally, in late June, ahead of the third intercept attempt, Kadish and Gansler appeared at a Pentagon news conference to defend the system. Even then, they were hesitant to venture beyond general assurances that the weapon would be able to distinguish between warheads and decoys, citing security concerns about revealing too much. In the first place, they said, a rogue state could not acquire countermeasures as easily as suggested in the UCS/MIT report. Moreover, they went on, the system had "inherent capabilities" that would be developed in time to deal with whatever countermeasures an adversary might devise. The capabilities involved "multiple discrimination techniques," but to get any more specific in public, they said, could tip off potential adversaries.

At about the same time, a twelve-page assessment of the USC/MIT report began circulating in the Pentagon—written not by the department or even a U.S. contractor but by Uzi Rubin, onetime head of Israel's Arrow antimissile program. The Arrow is a shorter-range defensive system with a very different technical configuration than the proposed U.S. system, but it also faced a countermeasures challenge. Rubin had an interest in assuring himself and others that countermeasures did not pose the insurmountable hurdle indicated by the scientists' report. His paper began by actually commending the UCS/MIT report as "perhaps the most exhaustive study ever published in the open literature on the issue of countermeasures," calling it well researched and based on very detailed calculations. But he took issue with its basic assumptions, which he described as biased against the system. He said that countermeasures were more difficult to produce and test than the report suggested, and he stressed the U.S. system's potential for growth and improvement. The UCS/MIT panel issued a lengthy rebuttal of Rubin's points.

Behind the scenes, White House officials mulled the question of how to make sense of the conflicting technical assessments. Here was a prime example of one of the most frustrating aspects of missile defense: scientific experts could not agree. One side said that any country capable of building an ocean-spanning missile could build sophisticated decoys to go with it. The other side said that while missiles require rocketry and mechanical skills, countermeasures require electronics and photonics—a markedly different set of disciplines. One side said that no known sensors could adequately discriminate some countermeasures. The other side said that the kill vehicle's infrared and visible sensors, combined eventually with infrared satellites and X-band radars, could provide enough data on a potential target from various angles to determine whether it was a warhead.

Who was right? Despite their antagonism toward Postol, White House officials did give serious thought to his call for an independent technical review. Podesta, who was particularly interested in the scientific aspects of the system, prodded Berger about commissioning an outside study. But earlier in the year, Cohen already had requested a third review from Welch, who intended to look at the countermeasures issue. Based on the reception given Welch's previous two efforts, the retired Air Force general appeared to have the respect and trust of those on both sides of the missile defense argument. "At some point, you can't have an independent panel to review the independent panel," a senior White House official said, explaining the decision not to appoint another one. "Our feeling was that Welch and his group had proven they were pretty legitimate, and all we'd do was undermine the process if we got another independent panel."

The truth of the matter was that senior Pentagon authorities had not focused very closely on the countermeasures problem. They had assumed the system would have whatever discriminating capability it needed when it needed it, but just what that would be had not received much high-level scrutiny. Since 1992, BMDO had run a countermeasures program at a remote Air Force research laboratory near Albuquerque, New Mexico. Using small teams of young military personnel with backgrounds in such fields as physics, math, electrical engineering and aerospace technology, the program attempted to build the kinds of decoys and other deceptive devices that emerging missile states were thought capable of constructing. The teams were restricted to using only commercial off-the-shelf technology and to gaining information only from publicly available sources. They worked out of a machine shop equipped simply with a few lathes and drill presses and such basic electronics gear as oscilloscopes, voltmeters and desk-top computers. But the effort, called the Countermeasures Hands-On Program—or CHOP shop for short—had remained a relatively low Pentagon priority, hardly scratching the surface of the challenge posed by countermeasures. The program was oriented largely toward theater missile defenses; only three of the fifteen projects pursued in its first eight years had applications to national missile defense. Additionally, team members tended to have no specific training in missile defense and served at the CHOP shop for just a year or so—the typical length of a single project—before moving on to other military assignments.

Soon after taking over as head of BMDO in mid-1999, Kadish had started a project to improve the discrimination capabilities of all the Pentagon's missile defense efforts, both national and theater. Called Project Hercules, the effort aimed to draw together the nation's leading academic and industry experts on sensor technology and statistical analysis. But this was a long-term undertaking. No specific plan existed to ensure that adequate capabilities for dealing with complex countermeasures would indeed be incorporated into later, expanded phases of the National Missile Defense system.

BMDO officials were bothered by allegations that they had dumbed down the flight tests to cover up weaknesses in the system. While the number and types of decoys had indeed been simplified, several justifications were offered. The changes were said to reflect earlier intelligence predictions that rogue states would not be capable of any but the most unsophisticated countermeasures for years—namely, balloons and radar-absorbent materials. Also, keeping things uncomplicated during initial flight testing made good technical sense. And it made political sense, since the simpler the system's demonstrated capabilities, the less threatening they would appear to Russia and China.

But the choice of decoys was not the only area in which the Pentagon was vulnerable to allegations of ignoring more difficult potential threats. The mock warheads being used in the tests also were questionable, differing from warheads that North Korea, Iran or other rogue states might field. The test replicas were smaller and patterned after advanced Soviet designs. This in some ways actually made them harder to detect. But in one very important way, they presented easier targets for U.S. sensors because they spun like tops, a feature common to high-precision warheads. The rotation, known as spin stabilization, increased accuracy by maintaining the direction of the warheads as they sped through space. By contrast, a rogue state might decide to forego such spinning and fire off cruder warheads that would tumble toward their targets, sacrificing accuracy for ease of design—and also complicating the job of U.S. interceptors by obscuring the often subtle differences between warheads and decoys. BMDO officials said that to obtain testing targets in a hurry, they had been compelled to rely on an inventory of Soviet-model replicas. Some rogue nation-specific replicas were on the drawing boards for future tests. But plans showed the first set of these also were to be spin-stabilized.

Pentagon officials found the whole countermeasures debate frustrating, filled with too much hypothesizing and too few hard facts. One high-ranking official described it as "mushy." On the one hand, he said, projections of rogue nation capabilities provided by the intelligence community were rather vague. On the other hand, every critic seemed to have his own favorite countermeasure. "We call it the 'threat du jour,'" another official said, exasperated. "You could keep chasing the threat, but at some point you need to put a stake in the ground—the engineers need something to build to, the testers need something to test to, everyone needs some stability in the process."

Still, the notion that the Pentagon would embark on an antimissile system without being assured that it would be effective against a countermeasure threat that everyone agreed was coming—even if they could not agree on when—struck critics as ludicrous. Such a plan would be irresponsible, not least for risking the waste of billions of dollars on a potential white elephant.

The severe politicization of the missile defense issue made the task of sorting through scientific truths all the more difficult. Each side could present experts armed with impressive credentials and arguments that supported its view of the workability—or not—of national missile defense. It was not an environment in which doubts or questions about program elements could be admitted or discussed publicly, because they would be seized on by critics determined to discredit the program as a whole.

"There's only a limited number of people knowledgeable enough to form an independent opinion," said one Pentagon civilian who was supervising development. "The others have to rely on the advice of experts—theirs or someone else's. Consequently, it's very easy to get misled and to choose what to believe or what not to."

Even within the Pentagon, analysts charged with sorting through the conflicting claims clashed with BMDO officials. As early as his 1998 report to Congress, Phil Coyle, the Pentagon's chief test evaluator, had warned that the National Missile Defense program was "building a target suite" that "may not be representative" of decoys and other countermeasures. He attributed much of this limitation to "the lack of information about the real threat." Throughout 2000, Coyle and his staff continued to press concerns that the test plan was not doing enough to take countermeasures into account. Bob Soule's Program Analysis and Evaluation staff also bore in on the countermeasures issue, pressing BMDO officials for more details about how the kill vehicle actually picked its targets, how the discrimination algorithms worked, and what could fool them. But they suspected program officials of holding back.

IV

Decision

Chapter 12

NUMBERS GAME

On a late winter day in March 2000, Ted Warner, a senior Pentagon civilian involved in setting nuclear arms policy, traveled from Washington to a three-story brick building on an air base in Omaha where targeting plans for America's nuclear arsenal are prepared. Warner had come to see Admiral Richard Mies, the four-star Navy officer who as head of U.S. Strategic Command had considerable say over the size and use of the nation's deadliest force.

The two men, while both members of a very small group in the U.S. government privy to the nation's nuclear hit list, were not very familiar with each other. They had spoken only a few times before. Warner, a Naval Academy graduate and onetime Air Force officer with a doctorate in political science from Princeton, had worked as a senior analyst at the RAND Corporation and joined the Clinton administration in 1993 to deal with strategic policy issues. Mies, also a Naval Academy graduate with degrees from Oxford and Harvard, had spent most of his military career in or around submarines.

Both men had a reputation not only for intelligence but for a strong will. Both also had a tendency, as one Pentagon official who worked with them put it, to be "on transmit a lot more than they're on receive." They were about to find themselves at loggerheads over a critical question: By how much should the nation's nuclear arsenal be reduced?

Talks with the Russians were approaching a decisive juncture as officials in Washington and Moscow looked ahead several months to a planned summit meeting in June between Bill Clinton and Vladimir Putin. The new Russian leader had shown little sign of a readiness to deal, but the U.S. administration was still exploring the possibility of constructing a big arms control package, hoping that a "grand bargain" could be struck that would gain Russian acquiescence to changes in the ABM Treaty in return for U.S. agreement on deeper cuts in nuclear weapons. Clinton had approved the notion of such a trade the previous August. But at the time, the president's top national security advisers had skirted a decision on just how much of a strategic arms reduction they would be willing to accept.

The nation's military chiefs had no great fondness for nuclear weapons. In the aftermath of the 1991 Persian Gulf War, the primacy of superior conventional technologies had come to dominate conceptions of U.S. global military strategy. The costs of maintaining strategic and theater nuclear forces competed with new requirements for improved conventional capabilities, especially at a time of shrinking defense budgets. With the notable exception of the submarine community, there appeared to be increasing military support for an explicit strategy to deemphasize nuclear forces in favor of more vigorous investment in advanced conventional weapons.

But the chiefs also were wary of reducing the arsenal too quickly. They did not fully trust the staying power of the changes in Russia. And the potential spread of nuclear technology to rogue states posed a worrisome new danger. Mindful that nuclear weapons, once cut, were hard to restore, military leaders looked on the reduction process with great caution.

The number of warheads that the United States and Russia had agreed to keep had already dropped substantially since the end of the cold war, from about 11,000 in 1991 to a range of 3,000 to 3,500 under the START II accord signed in 1993 (although still not ratified seven years later). At a spring summit in Helsinki in 1997 between Clinton and Yeltsin, the two leaders signaled a readiness to go down to 2,000 to 2,500 warheads in prospective START III negotiations. Then, in the autumn of 1997, the Russians—eager to save money and avoid the need to deploy a new single-warhead missile to preserve their numerical parity under START II—proposed an even deeper cut to 1,500 warheads.

Mies saw problems with going lower. An inherently cautious, firmly principled officer, Mies believed that America's ability to maintain its nuclear deterrent—in accordance with policy guidance issued by Clinton as recently as 1997—would be dangerously weakened by dipping below the Helsinki numbers. He had presented his concerns to Clinton's top national security advisers at a meeting in the White House Situation Room on August 16, 1999. The issue had been set aside then for further review.

By early 2000, however, it was back—and with greater urgency.

The president's national security adviser, Sandy Berger, had pressed Defense Secretary Bill Cohen about the prospect of Pentagon willingness to accept a new U.S. ceiling at 2,000 warheads. Cohen sounded generally receptive, but the job of building a case for going lower—and confronting Mies—fell to Warner, the assistant secretary of Defense for strategy and threat reduction.

Warner took on the task energetically. He was excited by the notion of a grand bargain and wanted to be ready if push came to shove with the Russians on a settlement. For some time, he had sensed that lower numbers were possible; it had only been a question of when to make the effort to tackle the problem in detail.

During his meeting with Mies in Omaha, Warner sought a better understanding of the admiral's concerns. Mies ran through some of them. He worried that further cuts would leave the Pentagon unable to deter war by holding at risk certain targets in Russia and China. Going lower, he argued, might also jeopardize the nation's ability to maintain a "triad" of land-based Minuteman III missiles, submarine-launched Tridents, and Air Force B-52 and B-2 bombers. The Air Force in particular could be left short of needed conventional capability, since its B-52 fleet was responsible for conventional as well as nuclear missions. And Mies fretted that the reductions in long-range nuclear weapons being contemplated did not address the continued existence of as many as 13,000 shorter-range nuclear devices that Russia had withdrawn from the field in the early 1990s but could still use.

Warner regarded all these concerns as legitimate, though not necessarily insurmountable. He left the session with Mies determined to explore whether it would be possible to shrink the U.S. arsenal further while still meeting the deterrence policy set by the president and subsequent targeting guidance issued by the secretary of Defense.

But Mies, for his part, found the meeting strained. He told others afterward that he had resented Warner's professorial manner, which came across as lecturing. Beyond the personality clash, Mies was suspicious of the Clinton administration's interest in pursuing deeper reductions, fearing that the president's aides were in too much of a rush to seal a deal with Russia. He believed it was the job of civilian authorities to set the overall guidelines for how they wanted to deter nuclear war, then leave it to Strategic Command to figure out what forces were necessary to fulfill the guidance. What he did not want to see was a U.S. government simply picking a bottom-line nuclear arsenal number and then adjusting its forces to achieve it.

Clinton and his top national security aides were to wait through the spring, hoping the split in the Pentagon, represented by Warner and Mies, would sort itself out. Such sharp divisions between the Defense Department's civilian and military ranks over nuclear policy were unusual. The START II and Helsinki reductions had been achieved with the joint concurrence of the civilians in the department's policy branch and the military officers at Strategic Command, although Mies was not yet in charge. But the differences this time ran especially deep and broke along a fundamental fault line—namely, the question of how much was enough to hold Russian targets at risk and maintain a strong U.S. deterrent.

The demise of the Soviet Union in 1991 had opened new opportunities for arms control measures that were quickly seized by the first Bush administration. With the threat of global war receding, the rationale for maintaining extensive nuclear arsenals had been severely undercut. Fresh concerns about the security and safety of Moscow's stockpile had given added impetus to the task of whittling down the estimated 24,000 tactical and strategic weapons that were dispersed across Soviet territory and tightening controls on what remained.

The last of the cold war arms treaties, START I, had been signed in 1991 after nine years of negotiations. It required the United States and Russia to reduce their respective nuclear forces to no more than 6,000 accountable nuclear warheads, a cut of almost half of deployed weapons for each side. The agreement set an ambitious schedule for dismantling launchers under intrusive verification and accounting procedures. It also required three former Soviet republics—Belarus, Kazakhstan, and Ukraine—to eliminate nuclear weapons on their territory and accede unconditionally to the Nuclear Non-Proliferation Treaty as nuclear weapons–free states, a process completed in 1995.

In addition, the Bush administration ordered a series of unilateral actions to prompt reciprocal measures by the Russian government. The United States withdrew all ground-launched, short-range nuclear weapons and removed all tactical nuclear weapons from naval ships and submarines. It took land-based Minuteman II missiles off alert by reconfiguring safety switches to prevent ignition of the motors and reduced the alert status of bombers as well. It also stopped development of mobile systems for the MX and Midgetman missiles and canceled the SRAM-II short-range nuclear attack missile.

This was a heady time, and developments moved fast. President Bush asked Moscow to meet his unilateral actions with similar moves of its own. First Mikhail Gorbachev in late 1991 and then Boris Yeltsin in early 1992 responded with announcements of cuts in tactical nuclear forces. The U.S. aim was to consolidate as many arms control gains as possible. Capping his accomplishments in this area, Bush, just days before leaving office in early 1993, signed the START II agreement with Yeltsin to eliminate two-thirds of the strategic nuclear warheads on both sides by 2003 and ban altogether multiple-warhead, land-based missiles, such as Russia's SS-18 and America's MX. Such weapons had long been viewed as the most destabilizing strategic arms because an attacker might be tempted to use them to carry out a preemptive nuclear strike.

Clinton thus inherited major nuclear force reduction agreements still to be implemented and a series of innovative cooperative ventures with Russia and the former Soviet republics. Many observers assumed that Clinton would continue and perhaps accelerate these initiatives. The announcement of a Nuclear Posture Review by Defense Secretary Les Aspin on October 29, 1993, suggested that a thorough reassessment of nuclear policy and doctrine would soon be under way.

The Bush administration already had taken some major steps toward revising the targeting plan for nuclear war, known as the Single Integrated Operational Plan, or SIOP. A review ordered in 1989 by Defense Secretary Richard Cheney and General Colin Powell, chairman of the Joint Chiefs of Staff, led to a new SIOP several years later that reduced the war plan by about 3,000 targets, down to a total of about 5,900. The drop reflected a reduction in former Soviet forces and also more efficient approaches to targeting. Now the Clinton administration would be embarking on what arms control advocates hoped would result in a series of novel initiatives—perhaps significantly deeper force cuts, an end to virtual hair-trigger alert, or an agreement not to use nuclear weapons first. But when the findings of the administration's review were released in September 1994, they represented little significant change. Forces would go no lower than START II levels, and there would be no serious alteration in U.S. operational policies.

At the time, the administration's top priority in the field of nuclear weapons was preventing "loose nukes"—that is, ensuring the removal of nuclear weapons from former Soviet republics and keeping Russian warheads from being sold or smuggled out of the country. There was little high-level interest in a sweeping revision of U.S. nuclear posture or in the debate such change would have entailed. What was the continuing sense of a doctrine of nuclear deterrence in the post–cold war world? What role should nuclear weapons still play in confronting Russia and China? What mission should they have in dissuading rogue states from acquiring or using nuclear, chemical, or biological weapons? These were all relevant questions certain to evoke sharply different views among policymakers and experts. But the administration did not want to go there, preferring to hold to traditional strategic doctrine and the established START process, for the time being at least. Another agreement in 1994 ostensibly had committed Moscow and Washington to stop aiming missiles at each other's territory. But the retargeting agreement was largely symbolic. Neither side was interested in undertaking steps—such as changing the target sets programmed in the missiles' computer memories—that would have obviated the ability to retarget the missiles in a matter of seconds.

In 1997 Clinton did take two important steps involving nuclear weapons—one doctrinal, the other numerical. The doctrinal move was to issue new presidential nuclear policy guidance, something no president had done since Reagan in 1981. The Reagan directive had called for U.S. nuclear forces to "prevail" in a prolonged nuclear war, an aim that defense officials had long since dismissed as so vague and excessive as to have little pragmatic significance. Nonetheless, it was still on the books, even in the wake of the sweeping revisions of the nuclear targeting list accomplished during the Bush administration's reign.

Clinton's new guidance, known as Presidential Decision Directive 60, was a relatively brief if highly classified document shorn of much of the political rhetoric of its predecessor. It shifted emphasis away from winning a nuclear conflict, recognizing that no nation would emerge as the victor in a nuclear war. Instead of planning for a long nuclear battle, it said, the Pentagon should aim the nation's nuclear forces to deter war by threatening a devastating response to nuclear attack. One consequence of this policy change was to provide the basis for a cut in the number of nuclear weapons held in reserve for possible use after an initial nuclear exchange.

But in other ways, the new directive did not depart much from the past. It did not fundamentally alter the categories of Russian capabilities to be held at risk or the overall philosophy for targeting that had informed U.S. strategy for decades. It called for U.S. war planners to retain a triad of nuclear forces and long-standing options for nuclear strikes against the military and civilian leadership and nuclear forces in Russia. It also continued to permit the United States to launch its weapons after receiving warning of attack—but before incoming warheads detonated—and also to be the first to strike. Additionally, it reincorporated Chinese sites in the SIOP after years on a reserve list. And it contained language allowing U.S. nuclear strikes in response to enemy attacks using chemical or biological weapons, an idea hotly debated by independent arms control experts.

The other important action that Clinton took in 1997 was to outline a possible START III deal with Yeltsin that would drop warhead levels to a range of 2,000 to 2,500. The Russians had balked at ratifying START II, concerned in part about the financial burden of committing to maintain forces at the accord's stipulated level of 3,000 to 3,500 warheads. So Clinton sought to induce Moscow to affirm that deal by offering to reach quick agreement afterward on even lower numbers. Meeting in Helsinki in March 1997, the two leaders reached an agreement on a proposed framework for START III, which, if realized, would represent a 30 to 45 percent drop in the number of deployed strategic warheads permitted under START II and a more than 65 percent cut in the number of warheads permitted under START I.

Two years after the Helsinki meeting, with START II still not ratified by the Russian parliament, the Russians were back pressing again for even deeper reductions in nuclear forces, down to 1,500 warheads. The 2,500 ceiling that had been tentatively set in Helsinki was the threshold below which U.S. military leaders had asserted at the time they could not go and still be expected to meet the demands of nuclear deterrence.

Warner and other senior defense officials who were sympathetic to the idea of shedding more nuclear weapons figured at the outset that there was little chance of achieving a consensus within the Pentagon for dropping as low as 1,500. In their view, the more realistic objective was 2,000, although even that number would be a huge stretch for some of the Pentagon's nuclear planners. They assumed that the Russians, while asking for 1,500, would accept a range of 1,500 to 2,000. But how were they to move from 2,500 to 2,000 without violating presidential instructions to destroy what the Russians valued most in the event of war? That was the question that confronted Warner.

Part of the solution could be found in changing the ways in which nuclear launch platforms were counted—or discounted. Not all nuclear-equipped bombers and submarines are always available for action should war break out. At any given time, some of the force will be undergoing extended overhaul for months in dry docks or bomber depot maintenance facilities. Taking these out-of-action "phantom" warheads out of the count would reduce the number by about 250 without actually cutting forces and target coverage, but to drop another 250 beyond that, down to 2,000, would require more than a change in counting rules. It would involve the removal of some sites in Russia from the U.S. nuclear hit list. It was on this point that the discussion within the Defense Department got interesting—and heated.

Drawing up the U.S. targeting map is an intensely secretive and elaborate process. There are strict procedures through which political authorities indicate their intentions to military planners and military targeters seek approval for needed changes. The president retains formal authority for setting strategic policy for employing nuclear weapons. But this guidance, contained in a document less than ten pages in length, is very broad.

More detailed guidance flows from time to time from the secretary of Defense in the form of the Nuclear Weapons Employment Policy. Still greater detail comes on a biennial basis from the Joint Chiefs of Staff, which issues the Joint Strategic Capabilities Plan. On the basis of this guidance, the U.S. Strategic Command—known in Pentagon parlance as STRATCOM—annually develops a detailed operational plan for the employment of strategic nuclear forces. The determination of target coverage and alert levels needed to uphold strategy has long resided with the nuclear command.

Still, the question of how much nuclear destruction is enough is as much art as science—and contains no small measure of bureaucratic politics. It is based on the notion that Russian aggression can be deterred by holding at risk those assets that Moscow "values most." But the value attached to those assets has never been strictly quantifiable. It involves judgments not only about military capabilities but about the political, economic, and even cultural motivations of foreign leaders. U.S. presidents have traditionally set several broad categories of targets, including not just nuclear forces but conventional forces, Russian leadership facilities, and war-supporting industries (itself a category encompassing a wide variety of enterprises, including manufacturing firms, petroleum facilities, and power plants). But these broad categories have allowed for considerable latitude in interpretation by Pentagon officials.

In recent years, the SIOP has consisted of a still-lengthy list of targets in Russia and a shorter list of targets in China. Three former republics of the Soviet Union—Belarus, Kazakhstan, and Ukraine, where strategic nuclear forces were deployed until the mid-1990s—were dropped from the plan in 1997. Even so, the target list has grown instead of contracted; it was up about 11 percent by 2000. Russian targets totaled about 2,260, according to a generally reliable tally published in 2000. Of these, 1,100 were nuclear arms sites in Russia, 500 were conventional—that is, non-nuclear-related—military buildings and bases; 160 were "leadership" sites, including government offices and military command centers; and 500 were war-related factories and economic assets.

To critics, the SIOP appears out of sync with the realities of a Russia whose economy and armed forces are weak and have little prospect of substantial recovery for many years. The numbers of weapons thought essential to provide requisite target coverage seem well in excess of any conceivable nuclear contingency, reaching absurd levels. Even top Pentagon officials tend not to believe that a cold-blooded, deliberate nuclear strike by either Russia or the United States is even remotely plausible. But in the view of the small group of Pentagon experts who think about nuclear war, the arsenal remains a critical hedge against an uncertain future—particularly the prospect of a resurgent evil Russia. The probabilities of this happening may be small, these experts concede, but the consequences would be catastrophic if insufficient attention were paid to the nuclear deterrent.

"We don't think about having to deter Yeltsin or Putin; nuclear weapons are practically irrelevant in that context," said a defense official involved in nuclear targeting. "We're thinking about how to protect against the Russian transition going south on us—and then try to figure what a new Stalin in Moscow might value most."

Moreover, the number of nuclear weapons actually available at a moment's notice is considerably smaller than the full arsenal would suggest. Planners differentiate between "day-to-day" forces, which would be available for immediate retaliation, and "generated" forces, which would take some hours or a few days to ready for attack. Today, while U.S. ICBMs could be launched within minutes and some submarine-born missiles could fire soon after receiving orders, no bombers remain on alert. So figuring out how many weapons the United States would need in a nuclear war involves considerations of the number that might be used to retaliate quickly and the number that could survive an attack and be available for later use.

To get a more precise handle on STRATCOM's planning, Warner dispatched his deputy, Frank Miller, to Omaha in April to study the targeting plan. Miller was a career civil servant with strong ties to the military's nuclear community. In the 1980s, as the director of strategic forces policy at the Pentagon, he had become the first civilian to delve into core details of the SIOP. His skeptical views of Russia's conversion to democracy and arguments for preserving a strong nuclear deterrent had given him a reputation around the Pentagon as the last remaining nuclear warrior.

During the day he spent in Omaha, Miller discussed the targeting plan category by category, subcategory by subcategory, with the STRATCOM targeting staff. Back in Washington, Miller—along with Warner and several others in Warner's office—reviewed the plan. Together, they came to doubt the military value of some targets. Ultimately, they concluded that there were two hundred or so targets that could be dropped without weakening U.S. ability to deter attack from Russia or China in accordance with existing presidential policy guidance. Many of the targets that seemed most questionable fell in the category of "general purpose conventional" military sites rather than leadership or economic infrastructure facilities. But each of the policy staff members had his own preferred list of targets that could be dropped.

"It didn't get down to a main difference over a single set; it didn't tend to be that clear, it was usually two or three different sets," a defense official said.

The traditional nuclear triad of forces—bombers, submarine-launched ballistic missiles, and land-based ICBMs—has required a sharing of roles in the country's strategic arsenal. This split mission has had at least as much to do with competition among the military branches as with any articulated military imperative. But as intent as Warner was on finding a way to shrink the nuclear force, he was mindful of preserving the triad. The issue was which of the three legs would end up trimmed more than the others.

Under START II, the United States was counting on a force of five hundred single-warhead ICBMs, fourteen Trident submarines with 120 warheads per boat, seventy-six B-52 bombers with up to twenty warheads apiece, and twenty-one B-2 bombers with sixteen warheads each. Nuclear planners were most reluctant to reduce the number of submarines, which are considered the delivery system most able to survive—or "ride out"—a nuclear first strike against the United States and hence the one most likely to be available to retaliate. More vulnerable to cuts was the land-based ICBM force, currently distributed among three bases; closing one base would eliminate at least 150 warheads. But the bulk of reductions would have to come from the Air Force's fleet of aging B-52 bombers. This prospect naturally alarmed the Air Force and presented a larger problem for the Pentagon, because the B-52 performed important missions as a conventional bomber as well. It had become the aircraft of choice, for instance, for firing cruise missiles. It also was essential for unloading tons of "dumb" bombs in saturation runs, such as those flown against Iraq during the 1991 Persian Gulf War or over Kosovo during NATO's 1999 campaign against Yugoslav forces.

Warner met directly with General Michael Ryan, the Air Force chief of staff, several times through the spring to explore ways of reducing the B-52 nuclear count while preserving its conventional capacity. Initially, Warner floated the idea of designating about thirty or forty bombers for nuclear missions and segregating them from the rest of the force by stationing them all at a separate base, say Minot, North Dakota. The rest of the fleet could then be earmarked solely for conventional missions and stationed at another B-52 facility in Barksdale, Louisiana. But the Air Force rejected that idea, since it wanted to close the base at Minot, not add aircraft there.

Instead, Warner and Ryan focused on finding a means of reducing the number of warheads that would be attributed to each aircraft—down to fewer than ten air-launched cruise missiles for the B-52 and eight bombs for the B-2. Under START II, the United States was permitted to attribute whatever number of warheads it wanted to each launching platform. The catch was that, whatever number was chosen, the United States would have to convince the Russians that it was based on the number that each bomber was operationally equipped to carry. If the attribution number for the B-52 were reduced from twenty to fewer than ten, and the B-2 number went from sixteen to eight, the aircraft might have to undergo some structural changes to justify the smaller numbers. Warner and Ryan pored over bomber designs, examining how external pylons on the wings and rotary launchers in internal bomb bays could be adjusted to make the numbers fit. The modifications could be expensive and elaborate. And there was another problem: some of the nuclear-armed, air-launched cruise missiles closely resembled the conventionally armed cruise missiles.

After several weeks of intensive study, Ryan suggested something that Warner thought might do the trick: the Air Force would limit itself to the nuclear-armed, stealthy cruise missiles that were easily distinguishable from the conventionally armed cruise missiles. This would allow all seventy-six B-52s to retain internally mounted rotary launchers that could carry the conventionally armed missiles but could not be used for the larger, nuclear-armed, stealthy missiles that have to be mounted on external pylons. Whether the Russians ever would accept that move, Warner and Ryan could only speculate. But the two senior Pentagon officials were satisfied that they had at least found a reasonable proposal to put on the table.

Mies, meanwhile, was still having trouble seeing his way below a U.S. force of 2,500 warheads. In fact, he was uncomfortable even with the 2,500 figure that had been agreed on at Helsinki in 1997. The admiral thought that his predecessor, Air Force General Eugene Habiger, had signed off on the deal without adequate analysis, particularly regarding the impact that the drop from 3,500 to 2,500 would have on conventional bomber capability—a claim that Habiger vigorously disputed. Mies also regarded the strategic environment as having changed significantly since the Helsinki summit. Russia appeared to be placing increased reliance on nuclear weapons employment in its declarative policies, and recent statements by senior Chinese officials were pointing toward an increased role for nuclear weapons in their security policy as well.

Additonally, Mies worried about Russia having a substantially larger capacity to produce nuclear weapons than the United States, which had stopped its production. This reflected contrasting approaches in how the two countries maintained their nuclear forces. The United States preferred to extend the life of its weapons by replacing components as they aged, while Russia kept its manufacturing plants going in order to remake whole plutonium pits whose impurities limited their life spans.

Also of concern to Mies was Russia's large stockpile of tactical nuclear weapons, which went beyond any plausible military requirement for them. In Mies's view, the Russians would be able to compensate for strategic reductions by bringing theater weapons into a strategic role, using shorter-range weapons to cover threats in Europe, for instance. Mies wanted greater accountability and transparency of these warheads, which were estimated to number anywhere from 6,000 to 13,000. Although U.S. officials had raised the need to get a handle on such tactical weapons, it was unclear whether they would be a part of the START III accord.

Mies could appear quite sensitive about others outside STRATCOM getting involved in nuclear force planning. He was said by some who worked closely with him to be bothered by Warner's direct dealings with Ryan and the other chiefs. He also was described as suspecting the White House of trying an end run by seeking to enlist support for deeper cuts from General Ralston on the Joint Staff, who was sympathetic to the idea. Indeed, the NSC's Steve Andreasen had set up a small group that included representatives from the Joint Staff and from the Pentagon's civilian policy branch to look at the question of new counting rules that would exempt submarines and bombers in extended overhaul from the START totals. The group even drafted language for a possible START III treaty based on a range of 1,500 to 2,000 warheads.

Another of Mies's persistent arguments was that any reduction below 2,500 warheads would essentially make it impossible for the Pentagon to carry out existing policy guidance for deterring nuclear war. This point struck senior Pentagon civilians as essentially a smoke screen to buy time. They acknowledged that directives would indeed need to be changed, but in the targeting guidance provided by the secretary of Defense, which could easily be done, not at the more general level of presidential guidance. Mies insisted that even the presidential guidance would need revision—a process that could take months of review. The last full-scale guidance review a few years earlier had lasted eight months. There simply was not that much time ahead of Clinton's planned summit meeting with Putin in June.

This argument resonated with the Joint Chiefs, who were clearly uneasy about being pressed to consider lower cuts without benefit of a thorough review of U.S. nuclear policy. They worried that the administration might be in too much of a hurry to give up nuclear weapons and make concessions to the Russians in the interest of reaching a deal on an antimissile system that the military leaders were not enthusiastic about in the first place. They presumed too that no matter what the United States did, Russia's missile force would drop far below the size that START II allowed, simply because the country could no longer afford to replace existing weapons as fast as they deteriorated. Seeming to bolster this point, Alexei Arbatov, deputy chairman of the Russian Parliament Defense Committee, had said at a Washington conference in May 2000 that if Russia were forced to keep Start I levels (6,000 warheads), it would cost $33 billion over ten years; at Start II levels (3,000 to 3,500 warheads), $26 billion over ten years; and at Start III levels, $14 billion over ten years.

Mies presented his case to the military chiefs at a meeting on May 9 in the secure Pentagon conference room known as the "tank," normally reserved for the most sensitive discussions of the service chiefs and other senior military commanders. A week later, the military leaders discussed the issue with Warner. Cohen joined the discussion in the tank the next day, May 18. Warner was given the task of outlining both his position and the admiral's, with firm instructions to be evenhanded. But Mies was not pleased with Warner's presentation and asked to see him in his Pentagon office the next morning, May 19.

Warner had thought he had been accurate in his presentation and faithful to Mies's views, while adhering to instructions to keep his talk as succinct as possible. But Mies complained that the briefing had failed to convey the breadth of STRATCOM's analysis. He showed Warner some charts about alternative force mixes developed by STRATCOM's staff, and Warner acknowledged that he had not seen them before. Still, Warner thought his presentation had come close to crystallizing the main differences between them. The meeting was tense, the exchanges sometimes bitter. At the end, Mies walked out gruffly. Warner emerged looking grim.

If only Habiger were still commanding STRATCOM, and not Mies. That thought nagged at several of Clinton's top aides—among them, Berger and Talbott. After all, Habiger had worked so cooperatively in finding a way to get to 2,500 at Helsinki in 1997. He surely would have been less resistant than Mies about going even lower. The president's aides were struck by how much influence over U.S. strategic nuclear policy was vested in the commander of STRATCOM. There was no other command, it seemed to them, where the incumbent was more decisive in setting policy. In this case, Mies's resistance was critical because the administration lacked the time to rewrite the defense guidelines and establish a new consensus behind them.

Berger had told Cohen clearly that if there were to be a deal with the Russians, it would have to involve agreement by the United States to drop to the 2,000 level. Now it appeared that doing so would require the Pentagon leader to overrule his STRATCOM chief. But the administration had not come to that point. There was no deal yet in the offing with the Russians, which in some ways was fortunate for Clinton. He had no desire to pick a fight with the military chiefs during an election year.

The resistance of the chiefs broke into the open in mid-May with articles in the *New York Times* and the *Washington Times* reporting divisions between administration civilians and the chiefs over whether further reductions would damage national security. Senator John Warner, chairman of the Armed Services Committee, then summoned the chiefs to a public hearing on May 23, a week before Clinton's planned summit with Putin. Senior administration officials viewed the move as a preemptive Republican strike against any efforts by Clinton to use lower nuclear arms cuts with Russia to win agreement on a limited missile defense system. "We were very nervous about what the chiefs would say and whether they'd leave any room for us to go down a bracket, if we determined that's what it would take for a deal," said a senior State Department official.

On Capitol Hill, Republicans and Democrats were sharply divided over the proposed deeper reductions. Many Senate Democrats, including Carl Levin of Michigan and Bob Kerrey of Nebraska, believed the U.S. arsenal could shrink substantially without compromising U.S. security, but Republicans chastised the administration for bowing to political expediency and insisted that further cuts should await a more formal Pentagon review and ratification of START II. By law, the United States was required to maintain at least Start I levels until Start II went into effect. Although the Russian parliament had approved the accord in April, it had made its ratification contingent on U.S. Senate approval of the ABM Treaty protocols on succession and demarcation concluded in 1997 but opposed by Republicans. Earlier in May, the House Armed Services Committee rejected, forty to seventeen, a proposal to permit the Pentagon to reduce unilaterally its strategic forces to Start II levels before the treaty actually entered into force. The Pentagon had supported the measure because of the high cost of maintaining the weapons. (The price, for instance, of keeping the fifty Peacekeeper missiles due for retirement under Start II was $1 billion through 2007.) So for the time being, the arsenals remained far from START II numbers. As of 2000, the United States had 7,519 nuclear warheads on missiles, submarines, or bombers, while Russia had 6,464.

In a statement at the start of the Senate hearing, Senator Warner noted the president's constitutional responsibility to consult the Senate on international agreements limiting U.S. forces and weapons. He chided the administration for having "attempted often to circumvent the Senate in a range of arms control agreements." A series of questions he then put forward about the prospect of further nuclear arms cuts were phrased in ways that echoed Mies's concerns: How would new reductions affect the triad? What counting-rule changes would be required to maintain conventional capability? What about the implications of all those tactical nuclear weapons still held by Russia? With low force levels, could the United States still deter regional adversaries as well as Russia, and would U.S. cuts encourage countries other than Russia to seek nuclear peer status with the United States? Should the Senate be concerned that the United States might be pressing ahead toward a START III agreement without having conducted a recent comprehensive review of its strategic nuclear policy doctrine and requirements?

Mies attended the hearing with the chiefs and outlined his reservations. His bottom-line appeal was for "more comprehensive consideration" before the U.S. government agreed to move toward lower levels. The admiral suggested that lower numbers could mean less security, not more. "There is a cushion in larger numbers," Mies said. "And there is a tyranny in smaller numbers that reduces stability in certain situations. And so the issue is not parity in numbers as much as it's stability that ought to be the criteria. And that has to—that requires a fair amount of dynamic analysis and war-gaming to convince yourself that your forces are postured and their capabilities are such that you can deter under a wide variety of situations."

Each of the chiefs, queried individually by the senators, expressed similar concerns. Although none flatly opposed going lower, each said he would not be "comfortable" with numbers below the Helsinki range without further analysis. "Clearly, whatever we do ought not to be driven by a predetermined number of systems to be agreed to," said General James Jones, the Marine Corps commandant. "We should also recognize that the journey is more important than the destination, since strategic force reductions are essentially irreversible."

The questioning by the senators was deferential and nonconfrontational. Everyone already knew where the chiefs stood, and no one seemed interested in drawing them into a public debate. Only at the end of the hearing did one of the senators, Max Cleland, a Democrat from Georgia, take a stab at the root issue of how much destructive power is enough—and he did so in the form of a statement, not a question. He noted that the answer to the question of how large a nuclear arsenal the United States should maintain had shifted over time. But he suggested that the U.S. arsenal was still too big.

"I understand," he said, "that we have the destructive power—if we unleashed an attack on the Russians and the Russians unleashed an attack on us—we have a combined destructive capacity to unleash, within thirty minutes, on each other 100,000 Hiroshima bombs in effect. So the destructive power here has dramatically increased. I understand one missile out of one Trident submarine—and the Trident submarines are based in Georgia at Kings Bay—carries destructive power equivalent to all the bombs dropped on Western Europe in World War II.

"I am surprised that the number of targets in the nuclear war plan has increased. So I think we're a little schizophrenic here," the senator added.

In June, several weeks after the hearing with the chiefs, the Senate convened a closed-door session with Democrats and Republicans to receive a classified briefing on the SIOP, the nuclear targeting plan. Among those most eager to attend was Bob Kerrey, whose own advocacy for deeper reductions had led him on a frustrating search for more information about the nuclear planning process. But he emerged no further informed from the briefing, which was delivered by Mies and the Pentagon's Walt Slocombe.

"The briefers were unable, or unwilling, to give us the kind of specific information about our nuclear forces and plans we need to make the decisions required as elected representatives of the people," Kerrey complained on the floor of the Senate. "In fact, when asked for detailed targeting information, we were given three different answers. First, we were told that they did not bring that kind of information. Then, we were told there were people in the room who were not cleared to receive that kind of information. Finally, we were told that kind of information is only provided to the Senate leadership and members of the Armed Services Committee. Because members of the leadership and the Senate Armed Services Committee indicated they had never received such information, I can only surmise there must be a fourth answer.

"We find ourselves in an uncomfortable and counter-productive Catch–22," the senator from Nebraska went on. "Until we as civilians provide better guidance to our military leaders, we are unlikely to affect the kind of changes needed to update our nuclear policies to reflect the realities of the post–cold war world. Yet, providing improved guidance is difficult when we are unable to learn the basic components of the SIOP. . . . In truth, it is important for citizens, armed only with common sense and open-source information, to reach sound conclusions about our nuclear posture and force levels."

Chapter 13

━━━━━━━━━━━━━━━━━━━━━━━━━━━━━━━━━━━

LOOPHOLES

Of all the places that Pentagon planners could have picked to host the new X-band radar, Shemya Island was among the bleakest, most remote, and least hospitable. It is nearly at the end of the Aleutian chain, located about 1,500 miles southwest of Anchorage—and about 500 miles east of Russia's Kamchatka Peninsula. Arctic winds blast across this desolate speck of land most of the year, and for the few months of the summer when they ease, the fog slides in so thick that planes are often unable to land for several weeks at a stretch. Ships have their own problems navigating the swells and maneuvering into the island's single dock, which even on good days commonly has waves washing over it. As a warning of the dangers, the beached remains of a fuel barge have sat deteriorating on the southern shore since 1958, and the skeleton of a crab boat that fell victim to strong winds and high seas is tilted on the rocks near the pier.

But history has a way of taking what seem to be the most insignificant outcroppings and making them strategically crucial, which was the case with Shemya. During World War II, it served as a frontline staging ground for U.S. forces attacking fortified Japanese positions on neighboring islands. For several decades afterward, the Air Force used the island as a base for air operations over Asia and the Soviet Union, and it renovated dormitory and recreational facilities there just as the cold war was ending, then decided in the early 1990s to shift its operations ashore. The Navy still maintains an intelligence-gathering outpost on the island. Also operating there is a missile-tracking radar called Cobra Dane, built in 1977, and an aviation-related transmitter and radar beacon.

Under the Clinton administration's National Missile Defense plan, this two-by-four-mile hunk of grass-covered dirt and rock, bordered on the north by the Bering Sea and on the south by the Pacific Ocean, was in line again to play a vital strategic role as the site of the system's first X-band radar.

Why Shemya? Because it is about as close to North Korea as the Pentagon can get and still be on U.S. territory. The closer the radar, the earlier and better it can detect and track a missile launch. The interceptors themselves also were to go in Alaska, but on the mainland, in the central part of the state near Fairbanks, where snow and ice presented their own construction challenges, although nothing quite like what engineers were going to face on Shemya.

To illustrate just how short the construction season on the island would be, the Air Force Combat Climatology Center prepared a bar graph showing that even in July, the calmest month of the year, the number of days on average that the wind blew at less than ten miles per hour was under one—0.7 to be exact. Most construction would have to be done between May and September. After that, the winds were just too fierce.

But weather and remoteness were not the only challenges. There was something else to worry about: earthquakes. One of the world's major fault lines runs under Shemya. The radar would need to rest on seismic springs, and other reinforcement measures would have to be taken to keep it standing in the event of shock waves.

As tough as conditions appeared to be on Shemya, things could have been worse. Pentagon officials had actually considered putting the X-band radar on Alaska's north slope. That site was ruled out as *absurdly* difficult. Shemya, on the other hand, was regarded as *merely* difficult.

In a normal climate and under normal soil conditions, constructing an X-band radar might take a little more than a year. In Shemya's case, defense officials told the Clinton administration to allow at least four years—that is, four summers—for construction. A fifth year was slated for pre-operational testing. So if the idea was to have the whole National Missile Defense system up and running by 2005, work on the radar would have to begin no later than 2001. This constraint, in effect, made the radar the driving element in the schedule. Construction of the interceptor silos and other facilities could wait until 2002; the radar could not.

There would be major diplomatic implications if the start of radar construction was going to mark the beginning of deployment of the missile defense system, because at some point in the construction the United States would find itself in violation of the ABM Treaty. But just what was that point? Administration officials had been proceeding on the rather loose understanding that sometime around the actual start of construction—either when ground was broken or the first concrete was poured—the treaty would be breached. Under the current schedule, both events were due to occur in the spring of 2001.

This did not leave much time to negotiate amendments to the treaty. Because of a treaty provision requiring six months' advance notice of withdrawal, the United States effectively had until the autumn of 2000 to strike a deal or else notify the Russians it was pulling out. Unless, that is, one of two things happened: either the Pentagon found a way to compress the construction time, which BMDO's Kadish determined was unlikely, or the actual threshold of violation was determined not to be the start of radar construction but some later stage in the building process.

Senator Carl Levin, a leading Democratic skeptic of missile defense—and one of the keenest legal minds in Congress—was intrigued by the second possibility. After all, he reasoned, digging a hole and pouring concrete do not in themselves necessarily constitute radar construction. What if the determining moment was when an element unique to radar—like the antennae mount—started going on the concrete base? The platform for the radar was scheduled to go up in 2001, but construction on the track and yolk structure for holding the dome was not due to start until the spring of 2002. If the United States could wait until then to give notice, it would buy a year, and building could thus get under way in Alaska without immediately triggering conflict with the ABM Treaty.

Levin was ambivalent about this timing issue. He worried that extending the decision time could actually work against his goal, which was to carry out a deployment constrained by a modified ABM Treaty. On the other hand, there was a certain pragmatic appeal in the idea of gaining the United States months of additional leeway, allowing it to keep all its options in play. That way, the administration might be able to proceed with construction while also avoiding a decision on whether to withdraw from the ABM accord.

But would it be legal?

In a conversation with Strobe Talbott in late February 2000, Levin asked whether any formal legal analysis had been done about the breach issue. Talbott produced a brief written statement essentially saying that the treaty would be violated at the onset of construction but providing little legal rationale to back that up. "That isn't a legal paper," Levin responded to Talbott, who himself had been taken aback by how little his own State Department staff had come up with. Returning to his office, Talbott requested a conclusive legal opinion based on a review of the negotiating record and deliberations by the Standing Consultative Commission, the U.S.-Russian group charged with handling treaty issues and implementing the treaty.

The review was conducted by an interagency team of State Department, Pentagon, and NSC lawyers. They had little precedent to go on. The treaty itself did not define the actions that would constitute the actual start of a deployment of a national antimissile system. So the lawyers looked at two sets of related agreements worked out in previous years by negotiators in the Standing Consultative Commission. One involved a 1976 protocol governing the dismantling of radars and other missile defense components that were being replaced. It defined when a facility could be deemed dismantled and off-the-books—sort of the flip side, administration lawyers figured, of when something might be considered under construction. Another agreement in 1978 on the construction of new test ranges also seemed to have some relevance to the Shemya question. It stipulated notification procedures based on various stages of radar construction.

Both the 1976 and 1978 accords made clear that the act of excavation itself did not constitute construction. Determining what did depended in part on the kind of radar at issue. Three categories were spelled out in both accords: a radar with its antenna array integrated into the side of a building, like the old Safeguard radar in Grand Forks, North Dakota; a radar with an antenna pedestal support, like a once-controversial Soviet radar called Flat Twin in Kazakhstan; or a third catchall category, which would seem to apply to the X-band radar planned for Shemya. "All this was written for what would be allowed," one participant in the legal meetings said. "What we were hunting for was the threshold for doing something—namely, putting up a national antimissile system—that was not allowed."

The lawyers had expected to complete the review in a few weeks, but it ran for several months as they peered at schematic drawings of the Shemya radar, compared them to other radar structures, and debated when, in effect, a radar became a radar. Ultimately, they agreed on three possible thresholds at which the Shemya radar could be deemed "under construction" for treaty purposes. First was the laying of an initial concrete slab. Second was the pouring of another slab on top of the foundation slab and a set of seismic stabilizers. Third was after completion of a tall concrete base and installation of a circular rail on which the radar's supporting frame would rest—a point that, according to the Pentagon's planned schedule, would not come until nearly a year after ground was broken, meaning well into 2002. A fourth option involving placement of the antenna inside the supporting yolk was considered but dismissed as unreasonable.

The lawyers did not indicate a preference among the three thresholds, figuring their responsibility was limited to outlining options. But in their view, options one and two would be easier to defend than option three. If they had to argue for option three, they could try to make a distinction between construction of the radar's antenna and construction of supporting elements. They could say that because the Shemya radar required an exceptionally large concrete base for seismic protection, construction of the base should not be considered the start of construction of the actual radar. But privately, they were the first to acknowledge this would be a weak case. After all, the idea that the United States might start building in Shemya for what everyone knew was a radar, yet claim that the legal definition of construction would not be crossed for a year or so, seemed on its very face to defy common sense.

Nonetheless, once the interagency legal review group listed option three as a possibility, the administration split over it. Senior Pentagon officials seized on the option as a way around an all-or-nothing choice for Clinton. They viewed it as a perfectly defensible way of allowing the United States to move forward with initial construction, thus preserving a potential 2005 deployment date, while avoiding a sudden confrontation with the Russians over the ABM Treaty. The legal reasoning behind the option, they argued, might be a stretch, but it was consistent with prior U.S. positions.

By contrast, officials at the State Department and the White House regarded option three as too devious. The whole approach struck them as reminiscent of Judge Abraham Sofaer's highly contentious legal arguments in the mid-1980s attempting to interpret the ABM Treaty so broadly as to allow for development and testing of space-based weapons. Around the State Department, officials referred to option three by the initials LRA—for Lawyers Run Amuck. "There was a snicker factor for any of us who had been through the Abe Sofaer thing," one senior State Department official said. "We joked about the

Abe Sofaer Memorial Prize for whoever came up with the most ingenious new interpretation of the ABM Treaty."

The Pentagon-State split was played out in exchanges between Walt Slocombe, the Pentagon's policy chief, and Strobe Talbott. Slocombe's view of the third option was that it was worth exercising if it would prevent a diplomatic blowup and allow the United States to proceed with the military program. Besides, he argued, the Russians themselves might appreciate some flexibility in the system. They did not have to go along with the U.S. interpretation, he said, but if the United States proceeded with it, they could simply add it to an existing list of more than a dozen pending treaty-compliance disputes. Talbott, on the other hand, readily showed his disdain for the whole idea by labeling it "Philadelphia lawyering." To which Slocombe once replied helpfully that "pettifogging" might be a more appropriate term.

The two men had known each other for years. In his previous career as a *Time* correspondent and author of several books on arms control, Talbott had written about events involving Slocombe, who had held a number of influential national security jobs over three decades. They respected each other's intelligence and commitment. But the closer Clinton got to a decision, the more they clashed over what decision the president should take.

Slocombe remained unshaken in the belief that the United States needed to move forward with construction of a limited antimissile system to protect against the rogue-state missile threat. In contrast, Talbott came increasingly to doubt the course on which the administration had embarked. By early 2000, he had begun to question just how imminent the rogue-missile threat was, how feasible the defense technology was, and how understanding and accepting the allies and Russians would be. He took some degree of personal responsibility for having awakened late to the whole missile defense issue and having allowed the administration to get into a position where it seemed on the political defensive in late 1998 and 1999. In retrospect, he felt that he and his senior colleagues had been caught up in a kind of collective folly over missile defense, pressured by congressional Republicans and a surprise North

Korean launch into venturing down an all-too-risky path. He saw his job now as minimizing the potential damage in European relations and avoiding a diplomatic catastrophe with Russia.

As the administration's point man in talks with the Russians, Talbott tried to keep his own misgivings about the U.S. policy to himself and carry out his diplomatic instructions as straightforwardly as he could. His administration colleagues, including those at the Pentagon, gave him high marks for that. It was not that Talbott felt he was going against his convictions or betraying any personal principles. He believed the administration's arguments were defensible and was convinced that ultimately the ABM Treaty would need to change to deal with new missile threats. But he did think that he and his colleagues had gotten themselves into a difficult position, and he was not sure how they would ease themselves out of it.

In March, Talbott was authorized to raise the breach issue delicately with Georgi Mamedov, and he did so during a trip to Italy for another of his periodic meetings with the Russian minister. The two diplomats, accompanied by their wives, dined one afternoon in Spoleto, where they listened to a contest of sopranos. It was during the van ride back that Talbott mentioned to Mamedov the developing U.S. view on the range of options for determining just what construction activity would constitute a treaty violation. Suggesting that construction of the Shemya radar could begin without the United States immediately breaching the treaty, Talbott was careful not to appear to be putting an actual proposal on the table, not least because he personally had some real concerns about the idea.

Mamedov also responded very carefully. On the one hand, he did not want to give the Americans any reason to think such a proposal might be accepted. On the other hand, he did not want to stop U.S. officials from trying to develop a way out of the diplomatic impasse. He sounded noncommittal, referring to the option as the "escape clause." Talbott called it "managed slippage," a term he credited to Sandy Berger. Mamedov also took the opportunity to tweak Talbott about the troubles the U.S. administration was having with critics of the technical feasibility of the missile defense plan. Referring to Nira Schwartz, the former TRW employee whose claims of poor performance and falsifying documents were making headlines in the United States that month, the Russian minister said he hoped she would be hired to draft the next version of the Welch report.

On the question of whether Putin would eventually be willing to deal, the Clinton administration received mixed messages through the winter and spring of 2000. Albright met with the Russian leader in early February, soon after he took over from Yeltsin, and came away slightly encouraged. During their meeting, Putin expressed concern about Iran and North Korea, saying he worried about the willingness of the Pyongyang leadership to "behave in responsible ways" and about what would happen if the North Koreans possessed nuclear weapons. He voiced interest in finding an "acceptable formula" to deal with the threat and still preserve the "fundamental principles" of the ABM Treaty. Albright and several aides thought there was something significant in Putin's reference to preserving the principles, rather than the actual terms and conditions, of the treaty. But other U.S. diplomats, including Talbott, remained skeptical. "Unlike most Russians who say no when they mean yes, Putin says yes when he means no," Talbott said.

Later that month, Sergei Ivanov, head of Russia's national security council, also seemed to crack open the door to a deal. During a visit to Washington, he told reporters that Moscow was prepared to discuss allowing the United States to move its current authorized site for a radar and interceptors from North Dakota to another location. But at the same time, he expressed continuing opposition to modifying the ABM Treaty to permit deployment of a national missile defense system. And in private conversations with Ivanov, senior administration officials said they detected little softening of the Russian position.

Then in April, in his first legislative initiative as Russia's elected president, Putin won approval of START II from the lower house of parliament. The move came seven years after the signing of the accord, and the Kremlin portrayed it as a sign of Putin's interest in a constructive relationship with the West. But it also came with strings attached. Conditions had been written into the ratification resolution insisting that the U.S. Senate also ratify the ABM Treaty changes on succession and demarcation that were so controversial among Senate Republicans. So while the Russian parliament had finally acted on START II, it had done so in a way that still left the accord effectively stalled.

With a summit between Clinton and Putin scheduled for June, time was running out to lay the groundwork for a deal. Administration officials had continued to hammer against the main Russian argument that the U.S. system would threaten Russia's nuclear deterrent. But ominously, the Russians had started raising a new kind of nightmare scenario in their talks with U.S. officials, involving not only U.S. strategic nuclear weapons but also intense air strikes by high-precision conventional weapons against Russian forces. NATO's successful seventy-eight-day bombing campaign against Yugoslav forces in Kosovo the year before appeared to have rattled the Russians. This conventional bombing scenario was presented to Talbott and other U.S. officials at a meeting in March with Major General Nikolay Zlenko, a regular member of Mamedov's delegation. He portrayed national missile defense as part of some grand U.S. strategy to defeat Russia and break it up. "It was by far the most sinister, lurid case that I'd heard," said a U.S. official who was present.

Zlenko, a former military attaché in Cuba with a background in military intelligence, had a reputation as a tough, no-nonsense officer, but he was also known to be articulate and very professional. In late April, both he and Igor Ivanov, the Russian foreign minister, flew to Washington for further discussions. This time, the Russian general came with charts to drive home Russia's argument about the threat the U.S. system would pose to Russian nuclear forces. Presenting his briefing to U.S. officials in Talbott's seventh-floor State Department conference room, Zlenko insisted that American interceptors launched from Alaska or North Dakota could knock down Russian ballistic missiles. One chart said one hundred interceptors could engage thirty to fifty

Russian missiles; another chart said an expanded system with two hundred interceptors could knock down eighty to one hundred missiles; a third chart said that if the system were enhanced by placement of land- and sea-based theater defenses around the United States, it would be able to engage three hundred to four hundred Russian missiles.

The Americans regarded the presentation as fanciful and over-drawn, but it effectively registered the extent of Russian fears, which by that point administration officials were taking very seriously. In truth, what seemed to worry the Russians most was not phase one or even phase two of the planned U.S. system, whose limited scope could still be overwhelmed by Russian warheads. Rather, the Russians cited the system's "breakout" potential—that is, the prospect of it expanding relatively easily into a powerful defensive shield that would give the United States a clear strategic advantage over them.

After Zlenko finished, the Americans brought him and Ivanov to the Pentagon and into the secure "tank" conference room. There defense officials delivered their own briefing, highlighting the inherent limitations of the planned U.S. system. Previously, American negotiators had focused on the relatively small number of planned U.S. interceptors. Now defense officials, led by Walt Slocombe and Ted Warner, introduced a new point: they asserted that technical limits to the new X-band radar posed an even greater constraint on the system. This was something that Slocombe and Warner themselves had learned only recently during discussions with Keith Englander, the missile defense program's technical director, when they were probing for more information about how the system actually was designed to work.

The Americans told the Russians that the new radar, while providing greater resolution than early-warning radars that scanned for a Russian nuclear attack, would be able at any given moment to track only a limited number of objects. The reason, as Warner explained in one chart, was the "dwell time" it would take the radar to distinguish the actual missiles from the decoys. Dwell time referred to the need for the beam to remain focused on a "threat cluster" for a certain period of time in order to gather the information that the radar's algorithms then sorted out.

The exact number of missiles the systems could counter was classified, but the Russians were told that a radar system like the first one on Shemya Island would be able to track and destroy only twenty to thirty missiles "with high confidence." Simply expanding the number of missile interceptors would not significantly expand the effectiveness of the missile defense system since the new radar system would quickly reach this "saturation point." In theory, the officials said, the United States could build enough radar systems to overwhelm the Russian force eventually, but several thousand of them would be required, and even then, building so many radar systems would be politically and practically impossible because of the costs and the diplomatic sensitivities of placing radar installations in foreign countries.

During the briefing, the Americans also showed the Russians charts diagramming the trajectories of potential missile shots from North Korea, all of which arced far north. The point was to underscore the significance of locating the radar in Alaska. The trajectories were much the same as the paths that Russian or Chinese missiles would follow. But that was just the nature of the missile business, the Americans said. They stressed that if the United States were interested in designing a system that could plausibly threaten Russia's projected arsenal of 1,000 to 2,000 warheads, it would not be this system but an entirely different architecture—one, for instance, with many more space-based elements.

Albright, who was attending the session, offered her own analogy. "You know, there's a difference in housing habits here," she said. "You Russians are worried that we're going simply to keep building onto an existing house; we're going to have a two-story house called National Missile Defense and we're going to make it into an eight-story house. That's not the way Americans approach construction. When Americans are planning to build an eight-story building, they design an eight-story one. By contrast, the Russians start with one floor and then they see if they get enough money for another floor and ultimately they may end up with an eight-story building."

Just to tweak the Russians a bit, Albright had worn a pin, made by the wife of the U.S. ambassador to NATO, that was in the shape of a missile. Ivanov noticed the jewelry and asked why the secretary was wearing it. "It's an interceptor," she quipped.

Learning of the Pentagon's explanation of the limits of the X-band radar, some U.S. missile experts disputed it. George Lewis, the MIT physicist who had worked on the countermeasures study, said that the radar was designed to be upgraded by increasing the number of receivers and transmitters on its face. That could allow the radar to track warheads and decoys, he told the *New York Times*.

Indeed, there was little sign that Ivanov and Zlenko had been persuaded by their trip to the Pentagon's "tank." In a speech at the National Press Club, Ivanov publicly pushed the notion of some kind of U.S.-Russian cooperation on "nonstrategic," shorter-range missile defenses as an alternative to the administration's plan. U.S. officials dismissed this approach as inadequate for dealing with the kind of long-range missiles that North Korea and other rogue nations were projected to develop. Similarly, they regarded other alternatives put forward by the Russians—a new "global cooperation" system to control proliferation of missile technology and intensified "political and diplomatic work" with rogue states—as insufficient to deal with the problem.

While Ivanov was in town, Talbott invited Mamedov to his house for a dinner that was also attended by Berger. There the president's national security adviser pushed Mamedov hard on the argument that it would be in Moscow's interest to do a deal with the existing administration rather than wait for the next. This had been much the same line that Talbott had pressed for weeks, noting that Clinton's successor, once in office, might feel under even more time pressure to build a system and less inclined to maintain the ABM Treaty.

"There's a 75 percent chance you'll have a worse situation after the election, and a 25 percent chance it will be just the same," Berger asserted. "That is, with a Bush administration, you know you'll end up with a more robust system and an inclination to do away with the ABM Treaty. And with Gore, he also might do more. So if you want to lock in an architecture in the context of arms control, we should do it now."

Mamedov seemed to have gotten the message. "I think we succeeded in convincing him—more perhaps than we convinced ourselves—that we would do this in the face of a Russian *nyet,*" said a U.S. official who was present. "And that was a huge part of what we were trying to do, because we didn't want them to be complacent, to think that all they could do was just hunker down and keep saying no. Mamedov, in turn, was spending quite a bit of time saying, 'If you have to do it, do it, but don't overdramatize it. Find ways of softening the blow. Keep it from becoming a disaster.'"

A few weeks after Ivanov's visit, Berger went to Moscow for another round of high-level talks with the Russians and final preparations for the planned June summit. By then the administration had developed two plans, referred to simply as plan A and plan B. The A option envisioned a deal with Russia along the lines that had been outlined: changes in the ABM Treaty to permit construction of an Alaska-based missile defense system, coupled with a START III agreement reducing nuclear arsenals at least to Helsinki levels and possible lower. The fallback, or plan B, was no deal, just a statement of principles that, if the Americans were lucky, would affirm the principle of amending the ABM Treaty.

Berger met with Putin and found the encounter discouraging. The Russian leader still appeared to have a distorted notion of the nature of the U.S. plan, its architecture, and expected capabilities. As Berger described how the proposed system would work, stressing that it would pose no danger to Russia, he had the impression that Putin was hearing some of this explanation for the first time. It was as if the briefings to Foreign Minister Igor Ivanov and national security adviser Sergei Ivanov had never occurred, Berger thought. Whether Putin was feigning ignorance or simply was poorly informed Berger could not tell, but he could see that Putin had little intention of reaching an agreement with Clinton.

Berger and Putin had known each other for several years, since Putin's days as Yeltsin's national security adviser. In fact, Berger had joked that Putin was a role model for all national security advisers who might aspire to higher office. But Putin was proving as cool and cautious in dealing with the United States as Yeltsin had been effusive and bold. Arms control was clearly not at the top of the new Russian leader's agenda.

And why should it be? There were clear signs that any deal he might cut with Clinton would have a hard time in Congress. In a letter to Clinton a few weeks earlier, twenty-five senior Republican senators, including Majority Leader Trent Lott and Jesse Helms, the Foreign Relations Committee chairman, said they would oppose any efforts to negotiate amendments to the treaty that might limit the national missile defense system they favored. Helms himself went further in a speech on the Senate floor a few days later, explicitly ruling out backing for any kind of new arms accord that Clinton might negotiate with Russia during his final months in office. "This administration's time for grand treaty initiatives is at an end," Helms said, adding that he wanted no part in a "final photo op" to help burnish Clinton's legacy in the international arena.

With Putin, Berger nonetheless tried to underscore the risks of waiting to negotiate with the next U.S. administration. "No matter how this election comes out, you're not going to be better off," Berger told the Russian leader. "You've got a Republican candidate who says he doesn't care what you think. You've got a vice president who is at least as hawkish if not more hawkish than this president. So if I were in your situation, I'd deal with us."

Putin looked at Berger with a trace of a smile on his face. "You're a dangerous man," the Russian remarked. Berger interpreted that to mean, "You clever devil"—and took it as a disingenuous compliment. But Putin was not budging.

In fact, not only did plan A look like a washout, but plan B also appeared in jeopardy after Berger received from Sergei Ivanov a revised draft of the proposed principles document. The document had undergone repeated revisions since initially conceived by Talbott and Mamedov the previous autumn, and U.S. officials had thought the two sides were getting close to a mutually agreeable wording. But the new draft, in the words of one high-level U.S. official, marked "a big step backwards." Said the official: "There were a lot of people on the Russian side, particularly in the Ministry of Defense, who not only hated national missile defense, but hated the idea of negotiating with us about it—and hated the principles document, as we saw in the version that emerged when Berger went to Moscow. We thought we were going to lose the principles document."

To salvage the document and plan B, Talbott was dispatched to Moscow twice in the week immediately prior to the scheduled June 3 start of meetings between Clinton and Putin. "The statement really went down to the wire," said a White House official who participated in the drafting. In the end, it contained several phrases that allowed the Americans to claim some progress, however minor. The statement took note of the emerging missile threat and alluded to the possibility of amending the ABM Treaty in view of world strategic changes. But there was no agreement on the nature of the threat or a clear program for how the two countries intended to deal with it. In one of the most heavily negotiated paragraphs, the statement said simply that the two countries had agreed to develop "concrete measures" to maintain "strategic nuclear stability," but it provided no particulars. The Americans had hoped to include a date at the end of that paragraph for reaching agreement on amending the ABM Treaty and concluding START III. But none appeared.

Several days before the summit, in an interview with NBC's Tom Brokaw, Putin caused a stir by appearing to suggest a joint U.S.-Russian missile defense system as an alternative to the American plan. "We could put up these umbrellas above potential areas of threat," the Russian leader said. "We could jointly with this umbrella protect all of Europe. We have these possibilities both technically and politically. We would like to propose them, and we would like to discuss that issue with President Clinton."

U.S. officials were not sure what Putin was talking about. Some news accounts indicated that the alternative he had in mind resembled the boost-phase idea that Garwin and others had promoted. Russian officials had been talking privately to American negotiators about the idea in recent weeks. Andrei Kokoshin, a senior lawmaker who had been national security adviser under Yeltsin, told the *New York Times* that Putin's plan was aimed at exploring ways of adapting theater missile defense systems to conduct "boost-phase interception." But Putin's own comments in the NBC interview were sketchy and at times hard to interpret.

Clinton, stopping in Portugal en route to Moscow, created his own minor stir. In the Lisbon suburb of Queluz on May 31, the president tried to reassure European allies about the proposed U.S. missile defense program, saying the United States would share the technology with friendly, "civilized" countries if it decided to build such a system. "I don't think that we could ever advance the notion that we have this technology designed to protect us against a new threat, a threat which is also a threat to other civilized nations who might or might not have nuclear powers . . . and not make it available to them," Clinton said at a news conference outside an eighteenth-century palace where a U.S.–European Union economic summit was under way. "I think it would be unethical not to do so."

The president's comments came eight days after Republican presidential candidate George W. Bush, outlining his own missile defense vision, had called for a more extensive plan that would protect not only the United States but many of its allies. Clinton spoke only of sharing technology, not of extending a U.S. protective umbrella over other countries. The president said that such sharing had "always been my position," but White House aides could find no record that he had said so publicly before. Indeed, Clinton's remarks in Portugal had put him steps ahead of his own policy. Additionally, it was not clear from Clinton's comments whether he would include Russia among the nations sharing in the technology. "We've done a lot of information-sharing with the Russians," he said. "We have offered to do more, and we would continue to." White House spokesman Joe Lockhart, pressed on whether the technology would be shared with Russia, declined to answer definitively. "Much of this is premature, based on a decision [about a missile defense system] that has not been made," he told reporters.

From Portugal, Clinton traveled to Aachen, Germany—to receive the Charlemagne Prize, given by a German-based foundation to recognize services on behalf of European unity and peace—and from there to Berlin. On the flight from Aachen to Berlin, Clinton received a briefing from the NSC's Steve Andreasen that focused on the status of U.S. proposals to Russia for cooperation and also the limits of such cooperation. Andreasen made the point that the United States had to be careful about how much missile defense technology it shared because some system components contained sensing and homing techniques that U.S. experts considered secret. Some of the technology, Andreasen noted, was not even available to NATO allies yet. But Clinton expressed interest in doing as much as could be done with the Russians and pressed Andreasen on the reasons for the U.S. constraints. "I was never sure we convinced him we couldn't do more than what we were doing," Andreasen said months later. "He was always pushing the edge on this."

Clinton and Putin had met twice before in 1999 when Putin was prime minister, but this was to be their first meeting since the Russian had been elected president in March. They had a range of issues on their agenda—including Russia's unresolved war in Chechnya and international peacekeeping efforts in the Balkans—and the Americans expected arms control to be a central focus of discussion. But during a two-and-a-half-hour private dinner with Putin on June 3 in the presidential quarters of the Kremlin, Clinton got the impression that the Russian leader was not particularly eager to talk about national missile defense. The two men had finished the meal and covered a number of other topics, when Clinton brought the conversation around to missile defense.

He noted his efforts during eight years in office to work for a "strong partnership" with Russia. He acknowledged that missile defense was a "tough issue" for Russia, but he emphasized that the phase-one deployment plan would pose no threat to the country. He argued it was possible to preserve the principle of mutual deterrence with Russia while pursuing a limited defense against the new rogue-nation missile threat. It was in Russia's interest to do a deal on missile defense, he went on, for two reasons: first, it might make a more ambitious START III possible; and second, it would establish a precedent that any future U.S. missile defense moves be premised on agreement with Russia.

Clinton offered to work with Putin in developing the Russian leader's boost-phase proposal, but that technology would take time, he said, and the United States needed an antimissile defense sooner. "I've got to make a good-faith effort with you to figure out how to change the treaty, not to ruin it but to allow it to hold up for another thirty years," Clinton asserted. "Whatever I do, I promise I'll never support putting Russia in an untenable position with regard to mutual deterrence."

But Clinton did not want Putin to mistake his accommodating tone as a sign of wavering commitment, or to misjudge his promotion of missile defense as simply a political expedient. "Don't make the mistake of thinking that this is just about current politics or me protecting Al Gore," he told Putin. "This is a real strategic problem for the United States."

Putin responded that he actually agreed with Clinton's argument about the new threat of proliferation, "in contrast to some of my colleagues, including several in the Ministry of Defense." But he took issue with the need for urgent action, cautioning that a precipitous U.S. move could drive Russia and other countries into taking measures of their own. He warned Clinton to "do no harm" and to "think carefully before making a decision."

Putin also complained that the U.S. system was aimed not just at rogue states but also at Russia. Claiming that U.S. interceptors were being designed to go against submarine- and bomber-launched weapons as well as land-based missiles, he pointed out that Russia was the only other nuclear power with a triad of forces like the United States. This understanding of the system's capabilities was wrong, as Clinton tried to explain to Putin. But the episode reinforced for Clinton and his advisers an impression that the Russian leader was receiving a skewed picture of the technical capabilities of the proposed U.S. system. Clinton urged Putin to accept a more detailed briefing on the system's design from Defense Secretary Cohen, who was due in Moscow later in the month.

As for the issue of when, if the United States moved ahead, it might end up actually breaching the ABM Treaty, Berger and his Russian counterpart, Sergei Ivanov, discussed the topic on the sidelines of the summit. But the Russians ended up firmly rejecting any notion that construction could start on Shemya without violating the treaty. And the Americans dropped plans to make a public reference during the summit to any potential legal options.

The summit ended with mutual praise but, not surprisingly, without any narrowing of differences over missile defense. Overall, senior administration officials described the atmosphere between the lame-duck American president and the new Russian leader as correct and businesslike, though not necessarily warm. At a press conference June 4, held in the grand, gilt-trimmed St. George's Hall in the recently renovated, czarist-era Grand Palace of the Kremlin, Putin made clear he felt no urgency to change his position on missile defense. While acknowledging a "commonality" with the Americans in seeing new threats emerging, Putin said, "We're against having a cure which is worse than the disease." He also indicated that he was willing to take his chances with Clinton's successor. "We're familiar with the programs of the two candidates. . . . We're willing to go forward on either one of these approaches," he said.

The summit did produce two narrow but significant agreements. One was an undertaking by the United States and Russia each to destroy thirty-four metric tons of weapons-grade plutonium during the next twenty years, enough to make tens of thousands of nuclear weapons. The other called for setting up a Moscow-based center where the United States and Russia would exchange information on missile launchings. The purpose was to help the Russians compensate for gaps in their system of early-warning radars and reduce the chance that they might launch missiles on a false alert.

Although the summit dashed whatever lingering hopes Clinton may have had about obtaining a new arms control framework before his term expired, it at least lowered the fever level that had surrounded the missile defense issue. Those on the left who had worried about Clinton leaving office with arms control in shambles could take assurance that the ABM Treaty would remain intact, at least for a while longer. Those missile defense enthusiasts on the right who were terrified that Clinton would lock the United States into a deal that allowed only a system that would be too minimal also could breathe easier. But the Russians were hardly ready to let up.

Putin had barely said farewell to Clinton before he and top aides fanned out across Europe to continue campaigning against the U.S. antimissile plan by promoting Russian alternatives and playing on European fears that the United States had embarked on a risky course. In Rome he proposed working with Europe and NATO "to create an anti-rocket defense system for Europe" that "would permit in an absolute manner a 100 percent guarantee of the security of every European country." In Germany he suggested that Europe join the United States and Russia in creating an early-warning center in Moscow that would monitor missile launches around the world. At NATO head-quarters in Brussels, Marshal Igor D. Sergeyev, the Russian defense minister, presented a vague seven-step plan envisioning development of a joint European and Russian defense that would use theater antimissile systems.

In another surprise to U.S. officials, Putin also announced plans to visit North Korea—the first such trip by a Russian leader. The Korea visit would provide Putin with an opportunity to make his case that diplomacy, not missile defense, was the way to cope with an emerging missile threat from Pyongyang. He also could use the visit to challenge the U.S. image of North Korea as an unpredictable rogue state. "There is a public relations war over missile defense, and all's fair in PR wars," a Clinton administration official said of the Russian blitz. "Some of it has been sophisticated. Some has been crude. But we have really seen an activist Russian foreign policy."

U.S. officials labored to dismiss the Russian talk about an alternative system as largely empty rhetoric, although they also scrambled to learn more about just what the Russians had in mind. Much of this task in the first days after the summit fell to Cohen and his aides, who were on a trip to Brussels and Moscow. From what they could discern initially, the Russians seemed to have floated two notions. One was the development in cooperation with NATO of some kind of theater antimissile system to protect Europe. The other was a boost-phase system that could guard U.S. as well as Russian territory by shooting down missiles as they were launched. But getting details from the Russians about either approach proved frustratingly difficult, deepening U.S. skepticism that much of a plan really existed.

After meeting with Marshal Sergeyev in Brussels, Cohen dismissed the idea of a theater-based missile defense system as inadequate to deal with the threat of strikes by long-range rockets. A system limited to shorter-range threats, Cohen told reporters, would not shield the United States and most European nations from ballistic missiles launched in North Korea or the Middle East. Cohen also expressed doubts about the boost-phase approach, questioning whether such a system could be developed by 2005, the deadline for the first phase of the antimissile program planned by the Clinton administration. Among the technical difficulties posed by trying to hit a missile soon after launch, Cohen said, was getting the interceptor to distinguish quickly between the missile's flame and the missile itself which tended to be obscured by its own burning plume.

Still, Cohen traveled from Brussels to Moscow expressing interest in learning more about how far along the Russians were in mastering boost-phase technology. In Moscow, Sergeyev told him that the boost-phase system under development was essentially for theater missile defense. The two men then went to see Putin, who said the same thing and suggested that Russian experts compare notes with U.S. experts. Cohen was led to believe that more information would be forthcoming later in the month at a working-level meeting between Russian and U.S. officials. Representative Curt Weldon, the Pennsylvania Republican and ardent missile defense supporter, was with Cohen and met separately with senior Russian Defense Ministry officials. He said that what the Russians apparently had in mind was an upgraded capability of the

S–400, a medium-range antimissile system. He said Russian defense officials told him they had completed all the mathematical equations for a new S–500.

In any case, when a U.S. delegation led by the Pentagon's Ted Warner returned to Moscow later in the month expecting a briefing on the Russian boost-phase plan, they were told there would be no such briefing. And through the end of Clinton's term, none ever came.

Chapter 14

MISSED AGAIN

Giant tracking radars jut above clusters of palm trees. Missiles soar over tranquil lagoons. An entire small American town sits on a narrow, crescent-shaped band of Micronesian coral. Ever since the late 1940s, when the U.S. military began using the Marshall Islands to test nuclear bombs and missile defenses, such scenes have reflected the sharp contrast inherent in pursuing high-tech ambitions in a remote natural setting.

The Marshall Islands consist of 1,225 isles grouped in twenty-nine atolls and five stand-alone land masses scattered over a broad swath of the central Pacific Ocean. The United States wrested control of the islands from Japan during a hard-fought Pacific campaign in World War II, then administered the territory under a United Nations mandate. In 1986, the islands entered into a "free association" with Washington that requires the United States to remain responsible for defense and external security and to provide financial assistance. Since then, American officials have negotiated fifteen-year renewable agreements to secure use of the missile testing range on Kwajalein Atoll and a base on the atoll's largest island, also named Kwajalein. The atoll's ninety islets form the world's largest lagoon—a loop of coral reef enclosing about 950 square miles.

On Kwajalein, an old hulking missile control structure stands as a reminder of earlier missile defense programs, with names like Nike-Zeus, Sentinel-Safeguard, HOE, and ERIS. The island has in fact become a distant American outpost, replete with paved roads, TV sports, and a general store called Macy's. Only American employees, their dependents, and those with special permission are allowed to visit or stay there. And many enjoy the sense of security and community that life on the island affords. Everyone pedals around on bicycles, and nine-hole golf, tennis, basketball, and scuba diving provide recreation. Consumer goods are subsidized, keeping the cost of living down. Over the years, armies of defense contractors have come and gone, pushing the island's population to more than five thousand at times. Today about twenty-five hundred live there, all but about two dozen of them civilians working for the Army or for defense contractors. For a missile defense test, the population can swell by several hundred more.

It was there that launch crews began returning in the summer of 2000 for the most expensive thirty minutes in the military testing business. It would be the last flight test of the National Missile Defense system before President Clinton decided whether to authorize the start of construction. The Pentagon had hoped that the program's future would not rest on a single test. In fact, one of the truisms in the defense acquisition business is never to let a program get into such a position. But to the dismay of program officials, that is exactly what had happened.

Originally, the schedule had called for two more flights prior to the president's decision. But the January failure had prompted a delay as review boards pored over what went wrong and Raytheon reworked parts of the next kill vehicle due to fly. Political considerations had kept the Clinton administration locked into a self-imposed deadline for making a deployment decision by the summer; the president did not want to be accused by Republicans of ducking a decision, which would leave presidential candidate Al Gore vulnerable to charges of being part of an administration weak on defense.

So with only one hit and one miss going into the summer, BMDO's Kadish took to referring despairingly to the upcoming test, scheduled for July 8, as a "binary event": if it succeeded, Clinton would be more likely to authorize preparations to build the radar on Shemya, and if it failed, he probably would not, thereby effectively postponing deployment at least until 2006.

This test, like the others before it, would draw on the efforts of nearly six hundred people. It involved the biggest names in the defense industry, and it would cost about $90 million. Phil Coyle, the Pentagon's chief weapons tester, had flown out from Washington to be in the control room. So had Mitch Kugler, the aide to Senator Cochran who had conceived of the legislation mandating deployment of a national antimissile system "as soon as technologically possible." Back at the Pentagon, Kadish and other high-ranking defense officials would be watching the launch on a video feed—in the presence of the *CBS News* anchorman Dan Rather and a camera crew taping a segment for the program *60 Minutes.*

From the outside, the run-up to the launch appeared routine, with no glitches. But from inside, the preflight planning looked considerably more frenetic and fretful. Even after all the rehearsals and readiness reviews, after the energetic engagement of all those hundreds of technicians, mission controllers, range safety authorities, and other contractors, there still were surprises.

After the January test failure was blamed on an obstruction in the krypton gas line, Raytheon had gone to considerable effort to avoid another plumbing problem. It had replaced pipes and valves, modified fittings, and revised assembly procedures. But on June 3, a day after the kill vehicle was filled with krypton and nitrogen gases in preparation for launch, measurements revealed another leak. This time it was nitrogen.

Raytheon officials were incredulous; so were their Pentagon clients. Compounding matters, Raytheon's crew could not pinpoint the source of the leak. Without knowing the location or shape of the leak hole, officials could not determine the chances that moisture might be seeping into the system—moisture that might freeze and obstruct the flow of gas in flight.

Fortunately, the leak rate was very gradual, so program officials decided after several days simply to monitor it while considering other options.

Concern about the leak was still shadowing launch preparations a month later when Bill Nance, John Peller, and other senior test managers gathered for a review in Building 1009, a plain, one-story office structure beside the Kwajalein runway that serves as local headquarters for the National Missile Defense group. With six days to go, they were assessing the several dozen problems that had surfaced in preparation for this test. Each problem had been written up in a Test Incident Report (TIR). Before the launch could proceed, each TIR needed to be certified as resolved or inconsequential. Only a few appeared to be of any lingering significance to test officials, and most of them involved software glitches that were being addressed. Even the nitrogen leak appeared less menacing than it had in June. The gas tanks had been refilled, and based on various structural analyses, Raytheon officials had assured Nance and Peller that the probability of the leak worsening in flight was minuscule.

"The chance of any of these things happening is one in a million," said Dan Testerman, Boeing's deputy director for test evaluation, as the review droned on to cover the most esoteric of issues. But Nance wanted no irregularity left unexamined. A new problem had emerged that very morning, when a Lockheed Martin crew working on the booster discovered a loose power cable on the nozzle control unit. The cable would have to be replaced, but the spare was in Hawaii. And the Air National Guard C-141 plane that ferried cargo to Kwajalein several times a week had broken down. That night Nance asked the pilots of a surveillance plane that was in Kwajalein just for the test to spend the next day fetching the spare.

Another day, another review: on July 3, the launch team traveled by large catamaran to Meck Island, over on the eastern rim of the lagoon. Meck is just large enough to host a launch site on a man-made hill at one end, a small dock and short runway at the other, and, in the middle, an aging, five-story, windowless concrete structure that houses the flight test's control room and support offices. The building was erected for tests of the Safeguard system in the 1970s, when computers still occupied whole rooms and bore gold-plated circuit boards.

Nance began the review by noting the particular importance of the upcoming test, an implicit reference to the decision Clinton would be making. As the review proceeded, he invited comment from anyone who wished to offer a thought. This open approach was typical of Nance, to the mild annoyance of some associates, because it sometimes resulted in uninformed comments and meandering meetings. But the general did not want to overlook anything that could help the mission.

Between midmorning and late afternoon, the review covered everything from the condition of the kill vehicle to the weather forecast for launch day. One new problem intruded: a critical communications facility for sending target information to the interceptor while in flight had suffered a power outage during maintenance the night before. The facility, known as the In-flight Interceptor Communications System (IFICS), was making its debut with this test.

A troubleshooting team that morning had concluded that the outage was caused by humid air passing through open panels in the small IFICS facility and blowing across hot computer equipment. Nance ordered that greater care be taken during maintenance; from now on, he instructed, no one would touch anything without a procedure.

Nance had been wrestling with how to get the best handle on all the issues that had come up and their status ahead of launch. Now he directed staff members to devise charts that would lay out all the critical test events, so potential glitches could be spotted in the sequence in which they might emerge. Will the target launch? Will the radars pick it up? Will the interceptor fire? Will the kill vehicle identify the mock warhead and intercept it?

"What we're trying to get to," he explained to his team, "is whether we have any weak links with single-point failures"—failures caused by any element that lacked a backup or was of overwhelming significance in itself.

On the walk back to the pier for the return catamaran ride to Kwajalein, Peller mused that just scoring a hit was hard enough, but these early tests were even more demanding. There were data to be collected and test range safety to be maintained. And in real life, the United States would be able to fire a salvo of interceptors against an incoming warhead; in these tests, only one interceptor was being shot. "Testing," Peller lamented, "is actually a lot harder than operating a system."

Senior test team members spent the next day, July 4, compiling the charts that Nance had ordered. The general would use them later in the week in a final video-conference briefing to high-level Pentagon officials. That night Boeing hosted a beachside party with free-flowing margaritas and a view of fireworks shot from a barge.

With three days to go, it was time for the final full-scale simulation. Tradition called for corporate team photos on launch hill in front of the interceptor. The photo shoot went smoothly, but sorting out another tradition—the positioning of corporate decals on the booster—was not so easy. There just wasn't room enough for all the dozen or so decals to go on the missile's "front" side, the one that faces the cameras on launch day. Nance regarded the decal-placement decision as one of the most politically sensitive he had to make. He appointed a group to make a recommendation, then issued his verdict: put Boeing, TRW, Raytheon, and Lockheed Martin on the front, and post the others on the back.

Despite the glitches that had popped up, Nance and Peller were giving this test better odds than they had the earlier ones. Peller put the chances of success at greater than fifty-fifty. Nance pegged them at about 80 percent. But the simulation that day turned out to be more eventful than expected.

About fifteen minutes before target launch, a fire alarm went off in the building housing the control room on Meck Island. A 240-amp breaker had burned out, apparently from old age, causing an air compressor to shut down. This in turn allowed humid air to waft into the ductwork and trip the alarm. "A 25-cent circuit breaker is threatening to foil a $100 million flight test," said Jim Ussery, a Pentagon test analyst.

With less than five minutes to go, range safety officials declared a "red" condition, halting the countdown. A UHF transmitter that would send a destruct signal in the event of a misfire had gone down, owing to a faulty amplifier. Finally, the simulation was run. As team members occupied the same seats they would on launch day, computers generated mock launches of the target and the interceptor. Mission directors recited in-flight progress reports as if the events were real. A video screen at the front of the control room showed the trajectories of the simulated vehicles converging and, ultimately, colliding.

In a video-conference call with the Pentagon on July 6, Nance and Peller briefed Kadish and Hans Mark, the Defense Department's head of defense research and engineering. Nance and Peller knew that Mark was the official who needed the most convincing. They remembered his biting memo before the last flight test, which had called their pre-launch briefing "too slick." As the July launch approached, Mark had worried particularly about the nitrogen leak.

During the briefing, Raytheon provided assurances that the leak was under control and not likely to pose a threat to the flight. Peller showed the charts listing the critical functions, from launch to intercept, that had to go right for the test to succeed. About thirty potential problems were cited, along with what had been done to address them. Most were given a "low probability" of occurring in-flight.

Mark asked what "low probability" meant.

About one chance in one hundred, Peller replied.

Using a standard probability equation, Mark quickly calculated an overall probability of success of about 70 percent. "If you were selling lottery tickets, I'd buy one," he cracked.

But buying a lottery ticket and recommending an important launch were two different things for Mark. He still had reservations about proceeding with the test, although Nance and Peller came away from the briefing with the impression that Mark had no objection to launching on July 8. Mark knew that the probability calculation he had done was very sensitive to the guesses that had been made about the probability of each event occurring, and people had widely different estimates in some cases. He could not put his finger on any single item that would warrant scrubbing the July 8 launch date.

Hit or miss, each test of the National Missile Defense system had come to cost about $90 million, according to figures from the Ballistic Missile Defense Organization. The kill vehicle itself consumed $24.1 million. The booster—a refurbished Minuteman rocket—burned $11.4 million. What BMDO refered to as "checkout, execution, and post-test analysis" of the mated booster and kill vehicle totaled $17 million. The target missile, which included a mock warhead and decoy packed in a dispersing container, or "bus," amounted to $19.1 million. There also were rental charges for the use of Kwajalein and Vandenberg Air Force Base in California ($3.2 million) and payments for "radar and battle management support" ($9.6 million). Finally, $4.7 million went for "system-level planning analysis and reporting," which covered preflight mission scenarios and postflight studies.

The high price meant that what went into a test counted for even more than it used to, so the Pentagon's Coyle was especially interested in observing an intercept attempt up close. Coyle was the only senior Pentagon civilian to make the two-day journey from Washington to Kwajalein for the July test. Two weeks earlier, he had sent a memo to Jack Gansler, the Pentagon's top acquisition official, saying that the test, though the "most significant" so far, contained "significant limitations to operational realism."

The memo reiterated some of the concerns that Coyle—along with critics outside the Pentagon—had been voicing for months. Coyle was particularly critical of the use of a large Mylar balloon as the decoy. He described it as "not especially stressing" to the kill vehicle and "not a true decoy" since it could in fact help rather than confuse the interceptor by alerting it to the presence of the real target nearby. He had in mind the October test, when the kill vehicle strayed off course and fixed on the balloon initially, without seeing the dummy warhead. Coyle said continued use of the balloon "only invites further criticism from the academic community." Because the kill vehicle had already demonstrated that it could tell a warhead from a balloon in the first test, he observed, it was time for "progressively more challenging countermeasures."

He noted that all the major components of the system were still represented in the test by surrogates or prototypes, and that the final versions would in some cases differ significantly from these stand-ins. The ultimate booster, for instance, would travel several times faster—and shake more violently—than the refurbished Minuteman missile being used to power the kill vehicle into space in these early flight tests. Moreover, Coyle pointed out, the test was using the same flight geometry each time—the familiar Vandenberg-Kwajalein scenario. He urged launches from more operationally representative locations—out of Alaska, for instance—and intercepts at higher altitudes and involving multiple interceptors.

Another of the artificialities troubling Coyle—and resulting from use of Kwajalein—involved how the interceptor was first told where to fly to find the target. In the actual system, such initial guidance—known as a "weapon task plan"—was to depend on X-band radar tracking data. An X-band prototype existed on Kwajalein, but there it could not see a California-fired target in time to help the interceptor before launch; it could only play in providing updates after launch. So the test had to rely on a range radar in Hawaii, which in turn required placing a C-band transponder on the target.

Coyle knew, of course, that early developmental tests were often limited and somewhat artificial. This test program had never been structured to produce operationally realistic test results this early. But that was Coyle's basic point: even if they succeeded, these tests could not realistically support a deployment decision now.

Coyle had written the memo out of concern that some Pentagon and White House officials did not fully understand the significance of the tests. He considered it unfortunate that the Pentagon was proceeding with plans to hold a "deployment readiness review" in the summer. It was too early, he thought, to make any assessment of deployment readiness, let alone for the president to make any deployment decision. In his view, the early trials had revealed little about the ultimate viability of the planned system.

At the same time, Coyle thought the tests already had demonstrated considerable progress. Many of the system's core elements, which had not even been available a decade earlier—such as the kill vehicle's infrared sensors or the battle management computers that process data from the sensors and produce a target map for the interceptor—had been shown to be working. What remained uncertain for Coyle was whether these elements could work reliably in an integrated system.

Coyle had come to Kwajalein to get better, more precise answers to some questions. "You just get a different story from the guys here than you do in Washington about the way the system is supposed to work," he explained, standing in the control room. "I don't mean anyone has been trying to mislead us. It's just that they don't have the same detailed information at their fingertips."

Nance welcomed Coyle's presence. The general was troubled by the persistent doubts that Coyle and the outside critics had continued to raise about the value of the tests. Part of the problem, he felt, was that people were expecting too much too soon. He was convinced that these early trials had been about as difficult as they should be at this stage. After all, their basic purpose was just to demonstrate the principle of using a missile to obliterate another missile, not prove the complete operational effectiveness of hit-to-kill technology. But somehow this initial series of tests, originally intended as proof-of-concept demonstrations, had become the touchstone of a system deployment decision—and the blurring of this distinction, in Nance's view, threatened to distort the whole development process.

"The first problem is, we're being graded against what the expectation would be for an end-of-the-development cycle, full-operational test," Nance said during a break in the action one day on Meck Island. "This system will go through that, but not until 2004 and 2005. We're not there yet. We're still in the front part of the test program. Our objective is to learn as much as possible about the elements of the system, then move to the next phase and add a little more rigor."

He could not disagree with Coyle's argument that the initial tests were not operationally representative. They were not supposed to be. But he took deep offense at allegations by others that the tests had been rigged or simplified to ensure success. "My disappointment is that we don't put the test in its right context," Nance said. "The message that you get in the media is that this is a rigged test. It's not. We may know where the target is going to launch from and what is in the target array, but it's pretty damn hard to rig a test to ensure we're going to intercept when the test range is nearly 5,000 miles long, and the speed is greater than 15,000 miles per hour, and we're trying to hit something as small as this target."

One of the questions that had confronted Nance and his boss Kadish in preparing for this test was whether to change the number of decoys, given the criticism that had followed the earlier tests. Some Boeing authorities had urged the generals to remove even the Mylar balloon in order to squelch the argument over whether it was serving to deceive or orient the kill vehicle. But Nance and Kadish decided to leave the target set unchanged and preserve a standard base of comparison with the previous tests.

On the morning of the critical flight test, people started moving into position very early. The launch was not scheduled until 2:00 P.M. local time, but the first ferry to Meck Island left at 4:30 A.M. The next—and last—departed two hours later.

Prelaunch rituals abounded: Nance walked up the hill to the launch site and looked around silently. The mission control director took his customary launch-day bike ride along a lagoon-side path to the ferry. A Lockheed Martin marketing specialist put on extra lipstick and kissed the kill vehicle. Jerry Cornell, Boeing's site manager, brought two lucky charms: a palm-size stone engraved with an Indian thunderbird image, and a knife that had belonged to J. B. Coleman, a sergeant in the Second Texas Cavalry during the Civil War. "He went through several battles— Antietam, Gettysburg—and died of old age in 1910," said Cornell, who had owned the knife for twenty-two years. "He kind of represents the soldier, the user."

Then there were the team shirts. The kill vehicle crowd wore white with blue trim; the Lockheed Martin booster contingent showed up in green with white stripes; the X-band radar group favored black; and the battle management team had bright blue shirts with an island motif of billowy clouds and palm trees.

As for the Boeing group, it went loudly against convention—and superstition—by donning bold red shirts. "Historically, red has been a no-no on the range," said Jim Hill, the Meck site manager. "Red means stop, abort. It used to be that if anyone wore a red shirt on mission day, he'd not be allowed in the building and would have to go home to change it. Maybe Boeing was trying to do a reverse on us." The Boeing test official responsible for shirt acquisition said red had been the only color sufficiently stocked at the company store in Huntsville, Alabama.

By midmorning, about four hours before the mock warhead was due to be launched from California, everyone was settling in for the wait when Vandenberg reported a voltage drop in a battery on the target missile. The battery powered a transponder used to track the container that carried the warhead and decoy. Vandenberg officials quickly determined that the battery still had enough voltage to do the mission, but they decided, without consulting Nance—and to his later annoyance—to recharge it anyway. The action delayed the flight two hours.

The countdown resumed just past noon. Shortly before 2:00 P.M., a security camera focused on the launchpad showed a fiberglass skiff in the background, racing across the water toward the island. Inside the control room, incredulous officials stopped their preparations to watch on a giant video screen. They could not believe their eyes. Who could that be? Everyone connected with the test had been ordered indoors and in their assigned places hours earlier as the countdown entered its final stages. U.S. authorities had taken extra security measures, beefing up a force on Kwajalein and running air sweeps over the surrounding lagoon.

The skiff hit the beach; a man and a woman got out and started walking up a road toward the launchpad. They carried a banner reading, STOP STAR WARS, GREENPEACE. It was an incredible stunt—and it posed a sudden threat to the launch.

Amid warnings that Greenpeace would try to disrupt the launch and reports of protest activity in California, Nance and his test team had sought intelligence information from local authorities on potential trouble on Meck. But the intelligence had yielded nothing suspicious, and so U.S. authorities had not anticipated any assault on the island that day. Upon getting word of the approaching intruders, several blue-suited civilian guards on the island had fanned out to check the shoreline. But the two protesters already were well up the path and approaching the launchpad, which was unfenced.

Army Colonel Earl Sutton, Nance's test director, dashed out of the control room. Michael Bright, Lockheed's manager for the booster, ran after him. They commandeered a golf cart and caught up with the protesters about one hundred feet shy of the interceptor.

"You need to stop right there," Bright shouted. The protesters stopped. Sutton was uncertain of his powers of arrest. His basic aim was to avoid a struggle. The Greenpeace activists—James Roof of Missoula, Montana, and Meike Huelsman of Lübeck, Germany— refused to move at first, saying they wanted to exercise their right to protest. But eventually they were escorted peaceably down the hill, where they were held until after the launch. Turned over to Marshalese authorities, they spent nearly three days in jail, then were released and fined $100 each for trespassing.

In the Meck control room, it did not go unnoted that had it not been for the delay caused by Vandenberg's battery problem, the protesters would have thwarted the launch. They had appeared on Meck at precisely the original start time of the test. "This," Bright announced to the control room, "is probably the only time when a battery problem saved the mission."

At 2:18 P.M., the countdown resumed, with two hours remaining. Nance opened a fortune cookie that the battle management computer team had given him earlier in the day. The fortune read, "Time is a wise counsel."

At 4:18 P.M., the target missile lifted off from Vandenberg. The second and third stages ignited, then burned out on schedule. Four minutes into the flight, Vandenberg reported "trajectory nominal," meaning on course. The dummy warhead was confirmed deployed about two minutes after that.

Nance peered at the large video screen at the front of the control room, which traced the target's trajectory over the Pacific. A mission control checklist was on the table in front of him, showing the minute-by-minute callouts for a normal test run.

About eight minutes into the flight, right on schedule, the tracking radar in Hawaii reported picking up the target. But a confirmation that the balloon decoy had deployed did not come. About fourteen minutes in, the unit that monitors the target data being relayed to the interceptor advised, "You will not see large decoy in the target object map." In other words, the balloon would not be playing in this test. It had failed to inflate.

About eighteen minutes in, word came on the network that Altair, one of the giant range radars, had reported "a non-nominal complex, a few extra pieces." Evidently, some debris had broken loose from the container that carried the dummy warhead and decoy into space; so even without the balloon, the kill vehicle would be encountering more than just the dummy warhead.

About twenty minutes in, attention switched to conditions on the launchpad at Meck. Safety radars were reported "green," meaning ready to track the interceptor. Then came a general alert: "All stations, stand by for terminal count, for go for launch. We are armed."

The fifteen-second mark was called out, then the final ten . . . nine . . . eight . . . seven . . .

The Meck control room began to rumble slightly, and a muffled roar penetrated the concrete walls. A few hundred yards away, the interceptor's booster was firing, shooting off into partly cloudy skies. Bright's hopes soared with the rocket. Nance jabbed his fist into the air, and applause burst out around him.

"Sensor cooldown commanded," intoned the voice of mission control, indicating coolant gases had begun to flow around the infrared sensors, preparing them for their space hunt.

Bright stood in his customary spot in a back corner of the control room. From there he could observe the rush of data streaming into computer consoles. He also could overhear the chatter of technicians monitoring the interceptor's performance.

After about two minutes, the talk suddenly turned worrisome. Transmissions from the missile had become "noisy" with lots of static interference.

"Where's the cover eject?" someone called out anxiously. "We didn't get cover eject." The cover—a giant aluminum clamshell-like device—protects the infrared sensors on the kill vehicle until reaching space. Nor did a signal arrive confirming that the booster's second stage had stopped burning. This signal was necessary before the kill vehicle could separate from the booster and home in on the target.

"We're not going to separate," someone blurted.

Three and a half minutes into the flight, the mission control network crackled with word again from Altair, confirming the technician's gloomy forecast. "Altair reports no separation of KV from PLV"—the kill vehicle was still attached to the payload launch vehicle. Instead of maneuvering toward its target, it was likely to tumble back toward Earth.

The control room fell silent. An overwhelming sense of failure struck Bright, a huge deflation, like the air rushing out of a balloon. Second-guesses were streaming into his mind. What had gone wrong? Where did we make mistakes? What more could we have done? He just shook his head and walked away.

Nance folded his arms across his chest and stared at the screen, which still showed the target and the kill vehicle arcing toward each other. Perhaps Altair's report was a miscall, Nance thought at first. What if Altair had been fooled, its view obscured because the kill vehicle had separated and somehow gotten behind the booster? Or perhaps the electronic signal that the kill vehicle must receive from the booster before cutting itself loose had been delayed and would still come through? Or maybe the connector between the kill vehicle and booster had been jammed and the kill vehicle would muscle free on its own when its thrusters fired?

For the next five minutes, as a wall-mounted digital clock clicked down to the scheduled moment of intercept, many in the control room simply sat silently, their eyes on the tracking picture. But it was clear that the flight had flopped. No one heard the reports normally broadcast on the mission control network when the kill vehicle closes in on a target. A telemetry indicator on the video screen that signals "valid" when the kill vehicle separates never switched on.

Finally, Nance swiveled around in his chair to address the room. "You've got to take this in context," the general said. "This is the most complex mission that the Defense Department has had since the Manhattan Project or some early strategic system programs, and it is not going to come without flight test failures. Our job is to evaluate the results of this, learn from what happened today, and apply it to the next tests. You've got to remember that our mission hasn't changed. Our mission is to design and develop—and test—a capability to defend the nation against ballistic missile attack. And it doesn't change tomorrow just because of this test."

Staff members, as if welcoming any activity to stave off depression, quickly turned to reviewing the telemetry still pouring in from flight monitors.

In the weeks that followed, the most likely culprit was judged to be a defective part in the booster's avionics processor, a ten-year-old device with an excellent track record. The malfunction early in the booster's flight had shut down data transmission between it and the kill vehicle. The communications loss prevented the kill vehicle from receiving the command to separate from the booster, thereby dooming the mission. Some missiles have backup processors; this one did not.

At the Pentagon, Jack Gansler, the acquisition chief, wondered whether more attention should have been paid to checking the booster. Hans Mark, his deputy for research and engineering, thought back to the briefing he had received from Nance and Peller and the calculation he had done showing a 70 percent chance of success. The dozens of anomalies that had sprung up during preflight preparations had continued to worry Mark, even after the go-ahead was given for launch. "You can do all the calculations you want, but you have to depend on your gut," he said. "It can't all be calculation. It has to be to some extent a feeling about whether something might go wrong. I canceled space shuttle flights for no good reason other than I didn't feel right that day about a flight." He knew it was all too common in the testing business for judgments to be clouded by an eagerness to get on with any given test. Testers have a term for it: "They had launch fever," Mark said. "I've seen that. And you know what should happen when you have launch fever? You stop, you don't launch. Never mind the calculations."

But that retrospective assessment struck Nance and other senior program officials as gratuitous. In postflight interviews, they disputed the notion of having been in the grip of any fever. They felt they had been as thorough, deliberate, and extensive in their preflight checks as they knew how to be. Even Mark had been blindsided by the outcome. During the July 6 review, he along with all the other participants had glossed over one chart that in hindsight should have drawn more attention. "Will the kill vehicle separate from the payload launch vehicle?" it read across the top. Only two words appeared on the page below: "No issues."

Missile defense advocates found some consolation in the nature of the malfunction. It was the old technology that failed, not the new stuff. The glitch had occurred during the routine procedure of launching a payload, not in the more innovative elements required to knock down a warhead. Additionally, several important components of the missile defense system had functioned without incident, including the IFICS link and the X-band radar prototype, even if they had been denied the opportunity to perform fully.

But program officials knew that the test failure made it more likely that Clinton would decide against proceeding with deployment. After all, the Pentagon's own criteria for certifying that the system was ready to move forward required two intercepts, and so far the program was only one for three. The criteria did include a provision allowing for some initial construction contracts to be awarded after a single successful test, but the actual start of construction still would have to await a second intercept. The chances of another test being run in 2000 were virtually nonexistent, given the time necessary to investigate the July mishap and the need for exhausted program teams to take a break.

Senior administration officials were stunned by the test failure, so sure had they been that the Pentagon simply would not allow anything to go wrong this time. At the State Department, officials had joked that Pentagon policy chief Walt Slocombe himself was going to be mounted atop the kill vehicle, with little foot pedals and a joystick, making downright certain that it hit what it was supposed to hit. In the immediate aftermath, some at State and at the White House suggested that Clinton move quickly to decide against deploying the system, but Sandy Berger urged delay until the Pentagon could proceed, as planned, with a review of the program's overall readiness.

"When the test failed in such a fundamental way—even though it failed before any of the interesting parts of the system actually had a chance to perform—there was some initial thought that maybe we should put this thing out of its misery," Strobe Talbott said. "A number of people joked about euthanasia, that sort of thing. But the wiser course quickly prevailed, and that was to wait for the Pentagon's review. We didn't want to make it look as though we were so grateful to have a pretext for slipping the noose and running away from the program."

A personal consideration also played an important role in the delay. Clinton, Berger, and others wanted to give Bill Cohen the time they had promised to make his final pitch for moving ahead with the program. The lone Republican in the cabinet, he had joined the administration in its second term amid considerable skepticism over whether he could fit in or would want to. But he had earned the respect of the president and his cabinet colleagues for playing things straight, keeping differences out of the news, and appearing loyal. It was no secret that he still favored a missile defense system, although just what he would recommend in the aftermath of the latest test failure was unclear. He would be given the opportunity to prepare his case.

In the meantime, Clinton also heard from Vladimir Putin about the test—two days after it happened. The Russian leader phoned to discuss an upcoming meeting with Clinton in Okinawa, and one of his opening lines was, "I don't know whether to congratulate you or to commiserate with you." Clinton ignored the remark and moved on to talk about other things.

That same day, dozens of the civilian contract employees who had sweated and agonized over the test boarded the first commercial flight to have departed Kwajalein since the launch. Many of them looked weary and sounded glum. Naturally depressed by the test flop, they were worried about the program's future. They reached Honolulu at 3:00 A.M., only to find a shortage of taxis at the airport. One Raytheon employee cracked, "I can't help but think that if the test had succeeded, there'd be limos here waiting for us."

Chapter 15

COHEN'S LAST STAND

While other politicians were driven to embrace the idea of a national missile defense out of ideological bent, party loyalty, or constituent benefit, Bill Cohen first came to the issue largely out of intellectual interest. In the 1980s, as a U.S. senator from Maine with a seat on the Armed Services Committee, he gravitated to the subject and became a legislative player on matters of nuclear strategy and arms control. It was not a move that would gain him many votes back home, but Cohen found the topic fascinating and important. And he would keep his own counsel on it, even if it set him apart. At a time, for instance, when President Reagan was touting an all-encompassing, space-based scheme, Cohen favored a limited system and broke with leaders of his own party over the extent of antimissile protection the United States should pursue.

"It was a part of the Senate world that caught his attention," said Bob Tyrer, Cohen's longtime chief aide. "I always saw it as something that he did as a stay against the deadening daily fare that one can have in the job. He obviously made good use of the Armed Services Committee in ways that were important to Maine, regarding bases and what not. But really, if you look at where he actually spent his time, it was on missile defense and other strategic issues."

Cohen's senior colleagues in the Clinton administration credited his enthusiasm with nudging them and the president as far down the path toward developing a system as they had come. One senior official speculated that if William Perry, Cohen's predecessor, had remained as secretary of Defense, the administration would not have gone as far, given Perry's skeptical view of the technology. But Cohen had managed to find an approach that, at least for a time, overcame some of the resistance of the arms control crowd at the State Department and NSC while mollifying missile defense hawks on Capitol Hill. He had sought to focus all parties on a common concern about a growing missile threat, obtain funding for a potential deployment, and define a substantive if limited system architecture.

Now Cohen and the rest of the administration found themselves at the moment of decision. If Clinton was going to take the next step toward deployment, it would be up to Cohen to prod him. Knowing that Sandy Berger, Madeleine Albright, and the others were against proceeding, Cohen felt a particular responsibility to balance their views.

But things were not breaking his way.

The failed July test was not the only setback casting doubt on the technical feasibility of the National Missile Defense plan. Bad news had emerged in another area of the program as well, one that administration officials had expected to be largely trouble-free: a new booster rocket for shooting the kill vehicle into space.

If there was anything that Pentagon authorities figured they already knew how to do, it was build a booster rocket. After all, the United States was a world leader in sending payloads into space atop powerful, reliable engines with names like Atlas and Titan. And yet work on the booster kept encountering one difficulty after another, to the point where by the summer of 2000 it had fallen about a year behind schedule.

This was all the more upsetting for Pentagon officials, since they had chosen to rely on commercially proven motors, thinking this would facilitate development. In July 1998, Jack Gansler, the Pentagon's acquisition chief, had rejected an Air Force proposal to use a converted Minuteman III rocket made by Lockheed Martin and instead opted for a three-stage booster that would be assembled largely from off-the-shelf components. Alliant Techsystems would build the first stage, and United Technologies' Chemical Systems Division received the contract for the second and third stages. The move, which promised cost savings and operational efficiency—and also avoided potential arms control treaty complications over conversion of military missiles—reflected a general Pentagon shift away from ordering weapons built only to military specifications when an existing commercial item would do.

Because several years were required to engineer and assemble the new booster, the plan was to use a surrogate—a two-stage Minuteman II rocket—in the first few intercept tests extending into 2001. But the surrogate was only about half as fast as the real booster would be, so program officials were eager to get the actual version into an intercept test to ensure that the kill vehicle could withstand the added vibration of a faster ride into space.

Confident that the booster was among the least of their worries, program officials at the Pentagon and at Boeing focused most of their attention on developing the kill vehicle and preparing to start construction of the radar on Shemya. These were the areas that officials called the "long poles in the tent": that is, they posed the program's biggest challenges and had the greatest potential to delay everything else. But signs that the new booster might end up a long pole as well began to surface within a year of awarding the contract. Program officials realized by the summer of 1999 that the booster's nozzle-control mechanism, which directs the rocket's thrust, was proving more challenging to design than expected. Most missiles have hydraulic systems, but since the interceptors for the missile defense system were to be placed in canisters and left in the ground potentially for years, hydraulic mechanisms could not be used because they might leak and malfunction. So the new booster needed a battery-powered system to control the nozzles. No one had ever produced such a system for as powerful a booster as the Pentagon had ordered.

Then, in the spring of 2000, program officials discovered a weakness in the system for controlling the roll, pitch, and yaw of the booster. An ignition test revealed heat damage in a metal line that piped hot gases. It had to be redesigned, forcing further delays. Still another problem arose when the writing of tens of thousands of lines of computer code, needed by the kill vehicle to guide the booster during ascent, fell weeks behind schedule. The surrogate booster had not required software, since it guided itself, carrying the kill vehicle as a passenger before releasing it. But in the actual system, the kill vehicle would handle the guidance for both itself and the booster—a change that some program officials had argued was unnecessary and unwise yet nevertheless remained a requirement.

Altogether, these glitches dashed Pentagon plans to conduct a booster demonstration flight by the summer, ahead of Clinton's deployment decision. When Bill Nance learned of the latest snag involving the lag in software code production, he erupted in anger. For months, John Peller and other Boeing officials had assured him that the booster's problems were coming under control. Just a few days earlier, Nance had accepted from Boeing a revised test schedule for the booster. Nance had asked Boeing about reports that concerns about the booster program among some lower-level employees were being suppressed, but he was told that was not the case. Now all of a sudden, during a briefing in Huntsville, Alabama, the Boeing official responsible for overseeing the booster was acknowledging a major software problem. Nance was incensed. Pounding a table, he told the Boeing members present that they had deceived him and the government. Why hadn't they been more forthcoming about the software difficulty? How could they have proposed a new test schedule only days before while aware that this kind of problem would force further delay? Nance was practically shouting. He was red in the face. He called Boeing's performance unacceptable, and in a private conservation afterward with Peller—who himself professed to having been surprised by the booster disclosure—he made clear that he expected to see some management changes.

For Nance and Ron Kadish, the booster episode amounted to a final erosion of their confidence in Boeing's management team. In addition to the booster, development of a critical simulation system known as LIDS (LSI Integration Distributed Simulation), used to generate a broad set of threat scenarios, also had fallen months behind schedule. And Boeing's program costs were running 14 percent above the contracted price of $2.2 billion for the first three years of development. All this contributed to the view held strongly by both BMDO generals overseeing the program that Boeing had not put its A team on the job.

To drive home these concerns, Nance and Kadish decided on a dramatic cut in the bonus fee for the giant aerospace firm. The bonuses, paid every six months, were the company's only source of profit on the contract. In three previous bonus cycles, Boeing had received between 70 and 80 percent of the eligible amount. But in June, shortly before the July test, the company was informed that it would receive only 50 percent of a potential $42 million bonus. Kadish had been inclined initially to pay no bonus at all but eventually concluded that fairness dictated at least some fee.

The extent of the cut shook Boeing and triggered its search for a new management team. Word of the reduced bonus was kept out of the news for more than two months, masking the extent of the Pentagon's displeasure with Boeing, but by early August signs of the company's distress emerged as Peller and a number of lower-ranking managers were taken off the program. These changes, though welcomed by Kadish and Nance, did not entirely resolve their concerns about the program's management. The issue was not simply personnel, it was also conceptual. Boeing, they believed, should not have been acting merely as an integrator focused on fitting subcomponents together, but rather as a traditional prime contractor responsible for delivering the sub-elements as well as integrating them. That way, Boeing could exercise more of the discipline and authority that had been lacking in the program.

"We had come a long way in two years, transitioning from a government-managed hodgepodge of programs to a more coherent national missile defense program," Kadish said. "But unfortunately, we still were operating as if we were in a prototype program, with less than stellar oversight and management of some elements."

BMDO authorities accepted that Boeing alone was hardly to blame for the management troubles. After all, the government had handed the company a virtually impossible task of building a missile defense system in record time using components ordered years earlier before a firm concept had been devised. This naturally drove corporate managers to cut corners and adopt concurrent procedures for producing parts even before they had completed standard qualification testing. "I've been through several meetings where Boeing will come in and say, 'It'll be nine months,' and they were told, 'No, that's not a good answer, come back and do it again,'" said Colonel Mary Kaura, BMDO's program manager for the interceptor. "There's a government here that wants this system as soon as it can get it."

Redefining Boeing's role became a focus of the contract talks between the company and the Pentagon that were renewed in the summer of 2000 and continued through the autumn.

Against this unsettled backdrop, senior Pentagon officials gathered on August 3 for a long-anticipated program review, the aim and structure of which had strangely become less clear as the date for it neared. Just what form the review would take, how many meetings it would entail, and what decisions, if any, it would be asked to make—all had been left open to some speculation. Even the official name of the process—the Deployment Readiness Review (DRR)—had ceased to reflect its actual purpose. The name was a carryover from the three-plus-three plan introduced in 1996, which had scheduled a thumbs-up or -down decision by 2000 on whether to deploy the system by 2003. But in 1999, when the deployment deadline shifted to 2005, so did the schedule for approving some key components. The upgrade of the early-warning radars and the production of communications facilities, for instance, now were not due until 2001, and the green light to produce the interceptors was on the calendar for 2003. In fact, the only real procedural decision necessary in the summer of 2000 was whether to start construction of the Shemya radar given the length of the building time required.

Only belatedly did Pentagon officials try to move away from the idea that conclusions could be drawn about the system's deployment readiness. A year earlier, the Welch panel had recommended calling the review simply a "feasibility assessment" and avoiding the emphasis on deployment. Phil Coyle too had pushed within the Defense Department to change the DRR name. He blamed the reference to "deployment readiness" for exaggerating the review's significance. It complicated prospects for getting a go-ahead to build the radar, which Coyle thought might have come more easily if the review had been portrayed as merely a periodic and entirely normal technical progress assessment and the radar decision a logical next step. At their June news conference, Kadish and Gansler sought to play down the 2000 review by stressing that the real commitment to deployment would not come until the interceptor decision in 2003. They also dropped the term "deployment readiness review" from the title of briefing charts being prepared for the process and renamed it the "NMD status review." But these measures did little to stamp out DRR references. In internal memos and public statements, defense officials continued to use the term.

As the review approached, it became shrouded in great secrecy. Despite the frequent public references, no public notice was given of the date of the meeting. Afterward, no official account was ever offered publicly about what had transpired. Such secrecy reflected a concern by Cohen and his senior aides that the Pentagon not appear on the record formally recommending a course on missile defense that would accentuate differences with the State Department and the White House.

But on August 3, the meeting did occur on the third floor of the Pentagon, down the D corridor, in a conference room encased in thick metal walls and equipped with devices for blocking any eavesdropping efforts. Known officially as the Executive Conference Facility, room 3D1019 is more commonly referred to as the "meat locker" because the temperature there is often cold. Essentially, it is a large vault dressed up as a meeting place with a rectangular table that seats twelve and chairs for another thirty-three people arranged in rows that fill up the rest of the room. It is there that the Pentagon's most sensitive weapons acquisition decisions are taken. The civilians who participated in the August 3 review included an array of undersecretaries and assistant secretaries from the Pentagon branches that usually have a say in procuring major weapons—acquisition, budget, policy, testing, communications, intelligence, and so on—as well as a senior NSC representative. Three generals also were present—Kadish and Nance from BMDO, and General Richard Myers, vice chairman of the Joint Chiefs of Staff.

After a few opening remarks by Gansler, who chaired the session, and an intelligence briefing on the missile threat, Nance delivered an extensive report on the program's technical achievements and shortcomings. He presented a series of charts, the basic message of which was that despite two out of three intercept failures, the program had fulfilled many of its objectives to date. He said ground and flight tests had demonstrated about 93 percent of the system's "critical engagement functions," including such categories as sensor operations and intercept planning. He also said the program had met or exceeded expectations in nineteen out of twenty-three "critical performance areas," covering such aspects as tracking, target acquisition, and discrimination. In yet another self-assessment—this one applying a readiness scale of one to nine used by NASA, with nine being "mission proven"—Nance reported that major components of the national missile defense system warranted either a six or a seven, with one exception. The exception was the software upgrade of the five early-warning radars, which drew a rating of four because of a late start in composing new computer codes.

All in all, the flurry of statistics suggested substantial technical progress even though two out of three intercept attempts had gone awry. Nance and other program officials genuinely believed they had legitimate grounds for portraying the development effort as largely on track. In their view, they had demonstrated the basic concept of the system and shown that the kill vehicle, radars, and battle management systems could function without requiring any more inventions. During the ensuing discussion, the department's leading skeptics—Phil Coyle and Bob Soule—reiterated concerns about the need for a more realistic testing program and the system's ability to deal with the kinds of counter-measures it was likely to encounter. But this technical critique had become familiar after months of public discussion. So the argument soon moved past capabilities and focused more on new worries about scheduling and cost.

With the testing schedule falling further behind as a result of the failed July test and the new booster problems, the question of whether it was realistic to expect to field the system by 2005 provoked the liveliest discussion. Gansler and Walt Slocombe favored sticking with 2005, even if the risk of achieving it had risen considerably. For one thing, they argued, the United States might yet find that it needed an antimissile system sooner rather than later to defend against a rogue state. For another, the administration would have a hard time politically explaining a shift to 2006 or 2007 without appearing to be backing away from the program. Kadish said any chance of meeting a 2005 deployment would require increasing the number of flight tests each year from three to four and "flying through failure"—that is, not postponing a test if the one before it missed.

But even with such intensified testing, Coyle was pessimistic about ever deploying by 2005, as was Soule, who argued for greater "realism" in the program. The viability of the 2005 date was important because, given the potential diplomatic consequences of a deployment, it would be even harder to persuade reluctant administration officials to press ahead with preliminary construction on Shemya if chances that an effective system could be up and working by 2005 were slim. On the question of whether to proceed with Shemya, some officials argued during the meeting that building the radar would make sense even if the deployment date slid by a year or two, since construction might take longer than expected or two, since in the event it did not, the testing program would benefit from having the radar available for use earlier.

The program's ever-rising costs then drew attention when the BMDO program office and the comptroller's office presented conflicting revised estimates through fiscal 2007. The program office said that acquisition costs had jumped from $20.3 billion to $23.9 billion; the comptroller reported an even higher increase to $24.4 billion. Neither total included other big bills that seemed likely, notably about $1.7 billion to implement the Welch panel's recommendations for alternative flight test geometries and expanded work on countermeasures, another $1 billion associated with new Joint Staff requirements to make the system interoperable with other Pentagon weapons, and $1 billion to study alternative booster approaches.

BMDO officials regarded such cost creep as inevitable in a program with new technologies. They offered assurances that the estimates were becoming more reliable as the program matured. But reflecting the exasperation of many others in the room, David Oliver, Gansler's deputy, remarked that by this point in the program he had expected the size of the increases to level off; instead, they appeared to escalate with each new decision point.

Slocombe and his policy staff had raised suspicions that some of the high cost was being driven by measures for combating not just rogue missiles but accidental or unauthorized Russian or Chinese launches. They had pressed for details on what could be deleted if the AUL requirement were dropped. The issue arose again during the review, and the program office reported that the only exclusively AUL-related elements were two of the five early-warning radars due for upgrades—the ones at Clear, Alaska, and Cape Cod, Massachusetts. Each upgrade was projected to cost about $50 million. But if not improved for national missile defense, officials said, each radar still would require modernizing at a cost of about $30 million. So the net savings would amount to about $20 million per radar, or $40 million total. Hearing these figures, Slocombe concluded at the meeting that they were a small price to pay for the added capability sought by the Joint Staff. He declined to press the case for deleting the AUL requirement, saying, "There is no issue."

After four hours of discussion, as the meeting neared its end, one final sharp exchange between Gansler and Coyle seemed to capture the strains within the Pentagon over the National Missile Defense program. Coyle remarked that added tests and hardware might prove difficult to afford, given that the program had been "resource-constrained."

Gansler took this as a challenge to repeated testimony by him and other senior administration officials that money had not been the real limiting factor in getting missile defense to work, technology was. He defended the administration's record, insisting that the program had not been shortchanged and noting that billions of dollars were added in recent years. Coyle's frequent harping on the inadequacies of the program irritated the acquisition chief. He viewed Coyle as inclined all too often to take a glass-half-empty view. It was not that Gansler disagreed with the concerns that Coyle had voiced; he also favored more realistic test scenarios and stepped-up work on countermeasures. But he chose to see the glass as half full.

The meeting ended with no votes or formal decisions being taken. Although the participants had been told to stand by for a possible follow-up session or two, none was ever held. A smaller group of about a dozen senior officials gathered in Cohen's office five days later to go over much the same ground with the secretary, although in a condensed format. Neither Coyle nor Soule was included in the group.

"We said to the secretary, 'You could make the 2005 deployment date, it's still possible,'" Gansler recalled. "We told him: 'Technically, it's high-risk. We'd have to add some additional tests, and this will cost some more money, and we should probably add more discrimination capability. But it's not impossible.'"

Cohen himself accepted 2005 as still achievable and believed that Clinton should not do anything to foreclose on the possibility. Reinforcing this view was the latest report from Larry Welch and his review team. Citing "important program progress" in the previous year, the Welch team members had expressed greater confidence about the Pentagon's technical ability to field a national antimissile system. They did join the chorus of critics urging expanded test challenges and greater attention to dealing with countermeasures, and they voiced doubts about the likelihood of meeting the 2005 deployment date. But they also said they saw "no technical reason to change the schedule at present," noting that if the system's performance faltered in future tests, "the schedule will be self-adjusting."

Cohen also was as convinced as ever of the need for a national antimissile system, given the potential rogue-missile threat. Many others at the time were calling into question the imminence of the North Korean threat, in particular, as a result of a sudden diplomatic opening by the long reclusive Stalinist state. Most dramatically, the North Korean leader, Kim Jong Il, smiling and looking relaxed, had warmly welcomed South Korean President Kim Dae Jung in Pyongyang in the first-ever North-South summit, raising hopes of progress toward reducing a half-century of military hostilities and fierce ideological competition between the two countries. The North's Kim also had recently visited China, marking his first trip outside the country since succeeding his father. He had invited the Russian leader, Vladimir Putin, for a visit in Pyongyang. And he was busily establishing diplomatic relations with a wide range of nontraditional partners, including Italy and Australia.

But Cohen appeared unimpressed by the North's startling emergence from decades of diplomatic isolation. Even the North-South summit left him unmoved. "One summit doesn't change a tiger into a domestic cat," he told the Senate Armed Services Committee in late July.

Cohen was confident too that the program's technology was coming along well. And he considered the projected price tag affordable within the context of the overall defense budget. The new higher estimates did not surprise him. Musing about them during the meeting with aides, Cohen wondered aloud why, once a defense program was under way, contractors invariably seemed to ask for another $5 billion or so.

Where Cohen parted company with other staunch missile defense advocates was over the ABM Treaty, which he favored preserving. He regarded the accord as a stabilizing factor in relations with Russia and worth trying to amend. Only if the Russians refused to revise it would he recommend the United States withdraw from the treaty. He also had come to appreciate the importance of winning the support of NATO allies, particularly since some of the radars necessary for the system were slated to go on European soil. It was an argument he pressed with his former Republican colleagues in Congress who were inclined to disregard European resistance.

But exactly what Cohen would recommend to Clinton he was keeping to himself and a few close aides. He never asked for recommendations from the review meeting or from the smaller group that met with him. Instead, he had begun to compose in longhand (which was later typed) his own confidential memo to Clinton that would make the case for authorizing initial steps toward construction on Shemya.

Coyle too was putting the finishing touches on a report to be forwarded to the White House. His sixty-seven-page paper, which was not publicly released for ten months, provided a detailed assessment of the program and his own recommendations. He stressed that the tests to date, with their limitations and artificialities, were structured more to generate early design data than to produce operationally realistic test results, and so they could hardly support a deployment decision, even if they had been successful. As for whether to proceed with construction on Shemya, he noted that the Pentagon's own criteria for awarding construction contracts had not been met. And anyway, he said, in light of the delays in flight testing and booster development, there was less urgency to start the radar.

Updated assessments of one aspect or another of the proposed antimissile plan also landed at the White House from other government agencies. Sandy Berger had asked the intelligence community to assess likely foreign responses to a U.S. deployment. Berger was searching for some kind of systematic means of weighing the enhanced security that would come from an effective antimissile system against the potentially degrading effects of responses by Russia, China, NATO allies, and others. In retrospect, Berger thought, he should have attempted such an assessment long before.

But he and other senior administration officials found the results disappointing; the intelligence community's assessment was a watered-down product of committee bickering over terms and excessive political caution. Much of what it produced amounted to a compilation of already-known possibilities: that China might expand an already planned increase in its force of about twenty ICBMs, developing both mobile and multiple-warhead weapons and growing its arsenal to more than two hundred warheads; that India and Pakistan could well respond with their own buildups; that Russia would withdraw from arms control treaties and again deploy shorter-range missiles and resume putting multiple warheads on its missiles (something that it had agreed to stop under START II); and that NATO allies would distance themselves from the U.S. effort. Little attempt was made in the report to balance the various potential occurrences. "It didn't really do much in the way of trade-offs," Berger said. Nor had it assessed the implications of *not* going ahead with a deployment.

Intelligence officials had actually advised the White House when the study was ordered six months earlier not to expect a net impact assessment. They could try to predict foreign responses, they said, but the job of analyzing likely trade-offs rested with policymakers at State, Defense, and the NSC. The State Department did attempt its own assessment, in a study headed by John Holum, the department's chief arms controller, but it came out sounding much the same as the intelligence community's report.

In the meantime, administration officials were confronted in daily news reports with the depth of international resistance. While European allies had sought for the most part to muffle their criticisms and avoid a bruising confrontation with Washington, they had continued to voice misgivings about the real extent of the rogue nation threat and the effect a U.S. deployment would have on relations with Russia. Reservations about going along with the U.S. plan were apparent even in those countries slated to host U.S. missile defense radars. The leader of Greenland's home rule government, for instance, declared that permission might not be granted to upgrade the Thule radar facility "if it resulted in increased tension and world destabilization." Even the British were in a difficult position, stretched between their long-standing predisposition to go along with the United States, on the one hand,

and an interest in not straying too far from dominant European opinion on a critical defensive issue, on the other. The House of Commons Foreign Affairs Committee issued a report cautioning that the United States "cannot necessarily assume unqualified UK cooperation" and urging the British government to "encourage the USA to seek other ways of reducing the threats it perceives."

Sounding as ominous as ever, Russia and China issued a joint statement in mid-July at the end of a one-day summit between Putin and his Chinese counterpart, Jiang Zemin. It said the purpose of the U.S. system was "to seek unilateral military and security advantages" and warned that proceeding with it would have "the most grave adverse consequences." The next day, Putin, visiting North Korea, said that North Korea had offered to abandon its missile program if other nations would provide it with rockets to launch satellites into space.

Clinton met with Putin several days later on July 21 during a break at a summit of Group of Eight leaders in Okinawa, Japan. In a casual setting—a private dining room overlooking a hotel pool—the Russian leader offered some candid impressions of Pyongyang. He likened the scene there to a frozen remnant of the 1950s and took particular note of the many generals that Kim Jong Il, the North Korean leader, had around him. "I think you're right about the need for stricter controls" on the North Korean missile program, Putin confided to Clinton. But during their seventy-five-minute discussion, the two leaders said little about North Korea's proposed missiles-for-satellites exchange. It was left to Berger and Strobe Talbott to probe Putin's aides for details and to gauge whether the North Korean move heralded fundamental change or was merely a tactic to undermine the case for U.S. missile defense. They concluded that North Korea was indeed floating a real idea, however problematic, but they preferred to explore it further in direct talks rather than with the Russians as mediators.

For his part, Clinton made clear to the Russian leader that he had not made a decision yet on deployment of a national missile defense but intended to do so during the summer after consulting with Cohen. Putin did not spend much time repeating Russian objections, noting they were well known, but he remarked that his country and the United States could still find a great deal to do together to deal with the missile threat. To that end, the two leaders issued a statement identifying areas of cooperation. These included: a deadline of one year to set up the Moscow-based early-warning center established at the June summit; revision of the Missile Technology Control Regime to incorporate a Russian proposal for a "global monitoring system" and a U.S. proposal for a "missile code of conduct," with incentives for countries not to proliferate dangerous technology; and the renewal and expansion of joint U.S.-Russian efforts on theater missile defense systems.

Putin was not the only leader making the case against a U.S. missile shield at the Group of Eight meeting. President Jacques Chirac of France, speaking to reporters, said the system might never work but nonetheless would prompt an arms race. "It's of a nature to retrigger a proliferation of weapons, notably nuclear missiles," Chirac said. Chancellor Gerhard Schröder of Germany also sided with critics of the system, saying, "I'm skeptical."

Along with its assessment of likely foreign responses, the U.S. intelligence community had provided an updated missile threat report that largely reiterated earlier predictions. It said that despite a pledge not to test long-range missiles, North Korea remained the nation most likely to pose a new danger of missile attack against the United States and could achieve such capability within a few years, if not sooner. It also said that both Iran and Iraq could test intercontinental missiles by 2010, particularly if they received assistance from Russia, China, or North Korea.

The new round of intelligence reporting did clarify one mistaken notion about the 2005 target deployment date. More than a few people in the administration and in Congress apparently had thought the date was predicated on a U.S. intelligence projection that North Korea would field a missile by then. That is even how some senior administration officials had explained the urgency of the missile defense effort in briefings to Congress and foreign governments. But the intelligence community had never predicted the North Korean missile threat would materialize in 2005. What it had said—and reiterated in the new

National Intelligence Estimate—was that the threat, while likely by 2005, could take shape any day if North Korea proceeded with a test of its Taepodong II missile. What had turned 2005 into a fixed point on the calendar was the Pentagon's declaration that an antimissile system probably could not be deployed any earlier.

Talbott, for one, came to this realization while listening to a CIA briefing on the new NIE. Frustrated with himself for not having understood earlier the genesis of the 2005 benchmark, he raised the matter again hours later in a phone conversation with CIA senior analyst Bob Walpole. "He called me and said, 'I don't see 2005 in your report,'" Walpole recounted. "I said, 'That's because that's not our number. That's the Pentagon's number. It's based more on when they could have a system deployed, not on when we think the threat could happen.' But Talbott was not alone. I was getting the same question enough times from others in the administration and on Capitol Hill to make me concerned that the confusion was pretty widespread."

As Clinton prepared to make a decision in August, Berger outlined five options for him. At the extremes, the president could either authorize deployment or terminate the program. But both options were hard to justify and carried deep downsides.

The real choice facing the president was among three middle alternatives, all involving various ways of deferring a deployment decision. One was to announce a deferral on the overall system but authorize contracts for the Shemya radar. Berger likened this approach to putting a car in neutral but letting it roll. A second option was to defer and do nothing about the Shemya radar. This would be equivalent to going into neutral and keeping the car motionless. A third possibility was what Berger called "defer with an attitude": Clinton could announce deferral and express substantive doubts about the program ever going anywhere.

Cohen was alone among the president's senior aides in favoring some kind of forward movement, if only by authorizing site preparation on Shemya. The others had long concluded that going ahead with missile defense would be a technological leap of faith and a diplomatic disaster. Cohen recognized that a full deployment decision was out of the question—and would not have advised it himself, particularly since the political consensus for a limited system had unraveled. In the wake of the test failures and ongoing Russian resistance to amending the ABM Treaty, several leading Democratic lawmakers, including Senators Biden and Levin, who had reluctantly supported the Missile Defense Act in 1999, had moved away from the idea. In the Republican camp, presidential nominee George W. Bush and some influential GOP lawmakers were renouncing the limited approach and declaring that it would be better to make no decision than to decide to proceed with an inadequate system and an unsatisfactory deal with Russia. In view of such political fragmentation, Cohen worried that a decision to deploy would risk deeper political division and would undercut the prospects of ever realizing an antimissile system.

But at the same time, Cohen felt it was important to maintain a sense of momentum in the program, for two reasons. First, it would give Clinton's successor the option of trying to field a system by 2005. Second, it would keep the pressure on the Russians to make a deal on amending the ABM Treaty.

Rudy de Leon, who had taken over as deputy secretary of Defense in the spring, had conceived of a kind of half-step measure that fell short of actually awarding any contracts to private firms. He suggested that the Army Corps of Engineers start buying some of the materials necessary for construction, such as gravel and concrete. Then, in the spring of 2001, two things could happen: the new president could decide to proceed with construction, in which case the Corps could turn the material over to private contractors and little time would have been lost; or the new president could delay or drop the project, in which case the Corps could apply the building materials it had purchased to other projects, and the government would incur relatively minor cancellation costs totaling about $28 million. Either way, the Clinton administration would have avoided a public bidding process, allowing much of the preparation to be handled internally.

This "deferral-plus" approach, as some administration officials called it, provoked fierce objections from the State Department and the NSC staff. The idea that Clinton could decide to take any steps toward building the Shemya radar while declaring he had not decided to deploy the system would fool no one and upset everyone, they said. Abroad, it would be perceived by the Russians, the Chinese, and the NATO allies as tantamount to a deployment decision. At home, it would satisfy neither the missile defense lobby nor the arms control crowd. It also would put undue and unwelcome pressure on the next president, who would inherit a program that was already moving forward. Much better, State and NSC officials argued, for the president to take a crisp, clean deferral decision.

But Clinton's top aides never actually gathered in one room at one time to hash out their differences. Rather, Berger assumed the role of go-between and consensus-builder, as he had done the previous summer with disagreements over the structure of the system and the negotiating approach with the Russians. He was not a neutral player himself, since he was clearly opposed to starting any work on Shemya. But he insisted that Cohen be heard.

Cohen's holdout position created a striking schism among Clinton's national security advisers, who had prided themselves on forging public unity and keeping disagreements to themselves. The "ABC team"— Albright, Berger, and Cohen—had previously never failed to reach consensus on a significant issue before submitting it to Clinton. And only once before had Cohen been overruled, when he argued against imposing sanctions on Burma.

This time Berger could see little room for compromise. Cohen was so invested in missile defense that he could not drop his argument now. Albright was busy with other matters and so unenthusiastic about missile defense that she had largely left the issue to Talbott. With the national security team divided, Berger was careful during much of the summer not to advocate, either privately or publicly, a specific process or timetable for a presidential decision. At the same time, he did not want the decision to drag on past Labor Day, when the president would be joining other world leaders at the United Nations for the Millennium Summit and the presidential campaign would be in full swing. Having waited through July for the Pentagon to hold its review and for Cohen to submit a written recommendation, the president's aides figured their next and probably final window of opportunity for a decision would come in late August, after the Democratic and Republican national conventions.

Following some anticipation earlier in the year that missile defense would become a campaign issue, Bush and Gore had spent little time arguing about it, and surveys showed the public to be largely uninterested. Bush did often talk about the U.S. military while on the stump, but his focus was on modernizing forces and improving readiness. Gore, for his part, saw little advantage in calling attention to the president's impending missile defense decision. With such prominent Republican members of the Foreign Relations Committee as Chuck Hagel of Nebraska and Gordon H. Smith of Oregon calling on Clinton to defer a deployment decision, Gore had become shielded to some extent from any fallout if Clinton chose to leave the issue to the next president.

Acting on Gore's behalf, Leon Fuerth, the vice president's national security adviser, sent Berger a secret one-page memo in July expressing Gore's support for whatever the president decided about deploying phase one, which envisioned one hundred interceptors in Alaska, one X-band radar on Shemya, and five early-warning radars in locations from Alaska to Britain. But Fuerth added a twist. He urged Clinton to consider a different approach to phase two, which had provided simply for an extension of the same midcourse intercept approach to missile defense as phase one, with a second interceptor site and more X-band and early-warning radars. Fuerth argued for pursuing a boost-phase system instead, in cooperation with Russia. He worried that if phase two simply meant a more intensive and more sophisticated defense under U.S. control, Russian resistance would intensify proportionately. Fuerth also believed that the ultimate confidence-building measures with Russia would involve creating cooperative programs. Fuerth regarded the proposal as highly experimental and was not even sure his senior administration colleagues would react favorably to it. But he wanted to foreshadow a different course that Gore himself might take if elected president.

On a trip to Nigeria, Tanzania, and Egypt in mid-August, Clinton read Cohen's paper and an NSC memo analyzing the four criteria that had been set more than a year earlier for deployment—threat, cost, technical feasibility, and impact on international relations and arms control. Berger and John Podesta, Clinton's chief of staff, urged the president to meet directly with Cohen and hear him out. Returning to Washington on August 29, Clinton, visibly tired after an eleven-hour flight from Cairo and still focused on the Middle East, spent an hour with Cohen in the Oval Office. Also present were Berger and Podesta.

Clinton was leaning heavily toward straight deferral but Cohen still hoped to persuade him to go for deferral-plus. The president had always been willing to listen to Cohen and had tended to side with him in major controversies. Cohen attributed this, at least in part, to the lengths he had gone as Defense secretary to play everything straight with Clinton. He had avoided sniping, had often run political interference, and had always coordinated his decisions and even his congressional contacts closely with the White House.

This time, Cohen realized going into the meeting with Clinton that his own case had been undercut by the flight test problems and other program setbacks. But he argued forcefully that the National Missile Defense system was on track technologically and still might be ready to go by 2005. Clinton said the technology had not proven itself. Cohen concurred but noted that the Pentagon's plan called for another sixteen tests in the next few years and emphasized the value of retaining the option of a 2005 deployment date for the next president. Much of the ensuing discussion involved a back-and-forth over the repercussions of deferral-plus. Was construction of the radar still really the long pole for the program? What exactly would be involved in proceeding? How could it be explained to the rest of the world?

Cohen gave the argument his best shot. "He made a helluva case for it, and he nearly talked me into it," Clinton said months later. "He was intense about it. He kept saying we didn't have to do this but we wouldn't even have the option to do it at all by 2005 unless we prepared to pour this concrete. It's not that much money, he said. It's only one site. It's clearly only against North Korea. He knew I was concerned about China's reaction, so he said this first phase wouldn't be enough to threaten the Chinese deterrent. He kept saying this is just about Korea, and this is just the first phase, and we had to do this to preserve the five-year option."

But the president still was not convinced. He had become intrigued with the prospect of a possible deal with North Korea that would diminish the imminence of any missile threat to the United States. He also had doubts about the workability of the proposed antimissile system and the impact a U.S. deployment would have on international relations. "I thought that the tests were, to put it mildly, inconclusive, and I thought that the politics were a long way from being resolved, and I didn't want to see America just sort of slip into something that would wind up making us less secure," Clinton said.

The president left the meeting without declaring his intentions. Later that day, Berger asked the NSC specialist on defense policy, Steve Andreasen, to draft a paper summarizing the downsides of Cohen's deferral-plus proposal. The paper was given to Clinton the next day, a Wednesday, as he flew to Colombia for a one-day visit. It argued that authorizing site preparation on Shemya would be widely perceived not as a deferral but as a step toward deployment and also would leave the next president facing an immediate decision on whether to stop the construction process or withdraw from the ABM Treaty.

On the plane ride back, Clinton announced to Berger and Podesta that he was ready to go public Friday with a decision simply to defer deployment. There would be no start of construction on Shemya. The president just was not comfortable with Cohen's idea of a partial step. "I respected his view, but I also thought that we simply had not advanced the technology, or the diplomacy, to the point that we should start the clock ticking on abrogation of the ABM Treaty and proceed with the Shemya radar," Clinton recounted.

Cohen was informed of the decision on Thursday. "He was disappointed," said a senior Clinton aide. "He really had felt very strongly about this. I think he thought that in the final analysis the president would accept his recommendation." For his part, Clinton felt his senior aides had done a commendable job managing their internal differences. In the weeks before announcing his decision, Clinton had been reading a book by the journalist Frances FitzGerald, *Way Out There in the Blue,* which chronicled the wrenching internal debates that went on in the 1980s among the State Department, the Pentagon, and the NSC over Reagan's missile defense plan. Clinton thought his own administration's debate had been "pretty orderly" by contrast.

To lay out his reasoning, Clinton wanted to give a thoughtful speech rather than simply make an announcement and take questions in the White House briefing room. Talbott and Berger had started composing the speech in June, passing drafts back and forth in the weeks since, incorporating material from the NSC along the way, then handing it off in the final days to a staff speechwriter. Their working assumption had been that the president would end up where he did, deferring deployment. One venue the president's aides had seriously considered for the September 1 speech was the Center for Strategic and International Studies, a Washington think tank, but they hastily settled on Georgetown University, Clinton's alma mater, figuring it afforded a better chance of preserving an element of surprise. Indeed, the timing of the speech caught many journalists off-guard. The university's Gaston Hall, a large lecture hall with stained-glass windows and vaulting supports, also provided a dignified setting as well as a very familiar and comfortable one for Clinton; he had spoken there a dozen times before.

This time his remarks were thorough and tightly woven. He offered a clear and clean-cut rationale for passing the decision to move ahead with the politically charged and diplomatically delicate project to his successor. He said that the threat of future missile attacks from such states as North Korea and Iraq "is real and growing" and that the United States had to search for new defenses. He characterized his proposed land-based system as having made "substantial progress" and said it "could be deployed sooner than any of the proposed alternatives" involving sea- or space-based interceptors. "Still, though the technology for NMD is promising, the system as a whole is not yet proven," the president said. He cited the two recent flight test failures, the booster delays, and the arguments over the system's ability to deal with countermeasures.

> There is a reasonable chance that all these challenges can be met in time. But I simply cannot conclude with the information I have today that we have enough confidence in the technology, and the operational effectiveness of the entire NMD system, to move forward to deployment. Therefore, I have decided not to authorize deployment of a national missile defense at this time. Instead, I have asked Secretary Cohen to continue a robust program of development and testing. That effort still is at an early stage. Only three of the nineteen planned intercept tests have been held so far. We need more tests against more challenging targets, and more simulations before we can responsibly commit our nation's resources to deployment.

Clinton's decision effectively nullified the Pentagon's 2005 target date. Seizing on the Welch panel's analysis that a later deployment date was more likely anyway, Clinton said that his decision "will not have a significant impact on the date the overall system could be deployed in the next administration if the next president decides to go forward." But in reality, Clinton's action delayed the start of work on the Shemya radar until 2002 at the earliest, ensuring that the kind of national missile defense system he had envisioned could not be completed before 2006.

Clinton's decision to defer deployment marked another halt in the decades-long quest to provide the United States with a shield against intercontinental ballistic missiles. As with previous efforts during the cold war, technological shortfalls and political resistance had once again dashed the dream of antimissile protection, at least for the time being. But Clinton stopped well short of dismissing the idea of missile defense. To the contrary, his speech, read closely, amounted to an argument for it, provided it could be shown to work. Significantly too, his embrace of the idea appeared rooted less in a vision of shielding America against a dangerous world than in establishing a new global strategic order. Rather than turning inward and merely seeking its own security, Clinton argued, the United States should pursue missile defense as an opportunity for broader engagement with Russia, the allies, and other nations on arms control, combating terrorism, and preventing the spread of weapons technology.

Clinton affirmed the principle of deterrence, saying it had "served us very well in the cold war" and had even kept Iraqi President Saddam Hussein from unleashing chemical or biological weapons during the Persian Gulf War. He said the notion of deterrence "remains imperative," noting that U.S. forces in South Korea continued to deter aggression by North Korea. But he also questioned whether deterrence was still enough on which to rest U.S. security: "The question is, can deterrence protect us against all those who might wish us harm in the future? Can we make America even more secure? The effort to answer these questions is the impetus behind the search for NMD. The issue is whether we can do more, not to meet today's threat, but to meet tomorrow's threat to our security." Missile defense, he added, would not substitute for diplomacy or deterrence but could provide "an extra dimension of insurance in a world where proliferation has complicated the task of preserving the peace."

It would be far better to move forward in the context of the ABM Treaty and allied support. Our efforts to make that possible have not been completed. For me, the bottom line on this decision is this: because the emerging missile threat is real, we have an obligation to pursue a missile defense system that could enhance our security. We have made progress, but we should not move forward until we have absolute confidence that the system will work, and until we have made every reasonable diplomatic effort to minimize the cost of deployment, and maximize the benefit, as I said, not only to America's security, but to the security of law-abiding nations everywhere subject to the same threat.

Reaction to the speech was favorable among Democratic lawmakers, scores of whom had urged Clinton to defer all decisions on the system. "The president's decision gives us time to perfect our political approach to the ballistic missile threat, as well as our technology," Senator Biden said. Republican reaction was mixed, with some of the program's biggest backers decrying another delay. "Because of the president's misguided decision, our children will remain completely unprotected against the weapon of choice for rogue nations and terrorist groups—the missile," said Representative Curt Weldon of Pennsylvania. Senator Trent Lott, the majority leader, said, "This is yet another example of the Clinton-Gore administration's legacy of missed opportunities." And Senator Ted Stevens of Alaska, who felt a strong proprietary interest in the planned system since it was sited in his home state, grumped angrily at Albright when she phoned with news of Clinton's decision hours before the speech. But Senators Hagel and Smith—the first prominent congressional Republicans to call for deferral after the test failure in January—praised Clinton's choice. "The president made the right decision today," Hagel said. "There will be dangerous consequences for America and the world if we rush to meet arbitrary decision deadlines."

George W. Bush, who during the presidential campaign had expressed support for a decision to defer if it meant avoiding unsatisfactory alternatives, adopted a sharper tone after Clinton's speech and blasted the administration's handling of the missile defense program. "I welcome the opportunity to act where they have failed to lead by developing and deploying effective missiles to protect all fifty states and our friends and allies," he said.

Overseas, Clinton's decision was received with relief and approval. Leaders throughout Europe, who had been caught between America's ambitions for missile defenses and the strong opposition from Russia and China, applauded the president's action. Russian officials expressed satisfaction that the diplomatic campaign by Putin and others to galvanize opposition against the American program had apparently helped persuade Clinton to abandon it. Privately, they had expected Clinton to proceed with work on Shemya. Hours before the speech, when Talbott had phoned Mamedov with news of Clinton's decision, the Russian diplomat was surprised, though pleased.

As for Cohen, he had very little to say publicly. He made no appearance after the speech, issuing only a brief statement that did little to mask his disappointment over the rare rejection of a recommendation from him to the president. "I have noted on many occasions that several emerging threats warrant the deployment of an effective missile defense program as soon as technologically feasible, and I will work closely with my successor on providing all appropriate information," the statement said. "In the meantime, we will aggressively proceed with the developmental testing program and also continue our consultations with the Congress, our allies, and with Russia."

A week after his decision, Clinton met Putin on the sidelines of the United Nations Millennium Summit, a gathering of about 160 presidents, kings, and prime ministers. The suspense that had surrounded Clinton's missile defense decision was gone, and the two leaders saw little purpose in hashing out their differences on the issue one more time. But Clinton did tell Putin that the cuts in nuclear weapons being sought by Russia would continue to depend on progress toward missile defense. Keeping open the prospect of U.S.-Russian cooperation in the area, the two presidents recommitted their countries to sharing early-warning information about missile launches, strengthening joint controls on the proliferation of missile technology, and expanding cooperation on theater missile defenses.

By then, the future of missile defense clearly rested on the outcome of the U.S. presidential election. A Gore victory would probably lead to a renewed attempt to amend the ABM Treaty and new deadlines for deployment of the phase-one system. A President Gore also would probably put greater emphasis on exploring a boost-phase system in cooperation with Russia, given the interest in such an approach that Fuerth, Gore's national security adviser, had expressed in his confidential memo to Clinton and Berger. A Bush victory, on the other hand, promised an even more radical departure from the past, marked by either an extensive revision of the ABM Treaty or the treaty's termination altogether, plus pursuit of sea-, air-, and space-based missile defense elements.

As the campaign entered its final weeks, top Pentagon officials did what they could to avoid making news about missile defense. They chose, for instance, not to finalize a draft report summarizing the Deployment Readiness Review. The report had put a generally positive gloss on the program, asserting that the testing had demonstrated the "feasibility and effectiveness of hit-to-kill technology" and declaring that "no technical reason" existed that would prevent the system from meeting its requirements. In this sense, the report was more a reflection of BMDO's briefings at the review than of the skeptical discussion that followed, during which much concern had been expressed about slipping schedules and undemonstrated performance. Both Coyle and Soule had objected to the draft; Coyle called it misleading, Soule considered releasing it politically inappropriate at that point, since its thrust was inconsistent with the president's decision.

But even though they shelved the generally upbeat review report, Cohen's aides also resisted appeals from a House panel to release Coyle's critical assessment of the program. The report, dated August 11, recommended increasing the planned number of tests, using more complicated targets, improving the ground simulation effort, and adopting alternative flight test scenarios that would better represent the real-life conditions that interceptors might encounter.

BMDO officials had actually started acting on many of the concerns raised by Coyle and other critics. With further testing in limbo as a result of the failed July flight and the election, Kadish and Nance huddled with staff to reexamine options and produce a revised plan. They had received yet another critical jolt that autumn from Larry Welch and several of his review team members who had examined the countermeasures controversy. Unlike the Union of Concerned Scientists, Welch's group had access to the Pentagon's classified information. Their report did support a couple of arguments that defense officials had been making—namely, that countermeasures were indeed difficult to produce and that the planned antimissile system did have capabilities inherent in its sensors and radars to combat countermeasures. But the report also confirmed what many critics had suspected about the Pentagon's program: no clear plan existed for getting from the initial C-1 phase to the later C-2 and C-3 phases. The report urged that the issue be given much greater attention, both in determining the kinds of countermeasures a missile defense system might encounter and in figuring out how to overcome them.

Kadish himself readily admitted that there had been shortcomings in the program that were not evident to him and his aides at the start. "Truth is a function of time," he would often say when discussing the benefits of hindsight. But Kadish also remained confident about the program's essential direction, convinced that a ground-based system provided the quickest, most reliable route toward a national missile defense. He saw no major technical obstacles in the way of realizing such a system. "In general, there are basically two ways to look at the program to date, and they could be termed the 'glass-half-full' and the 'glass-half-empty' views," Kadish said a week after Clinton's decision in testimony before the national security subcommittee of the House Committee on Government Reform. "My assessment at the moment is that it is half full. I say this because we have made remarkable and substantial technical progress despite two high-profile test failures."

If the program now looked inadequate, he believed, it was largely because the requirements had shifted over the previous year, placing added demands on him and the other senior planners. "We put together a pretty good program and started executing it and made remarkable progress," Kadish said in an interview. "But somewhere along the line, all the external expectations increased, and we moved from doing just basic technology against a very well-defined threat of a few, simple countermeasures to worrying about confronting a lot of other counter-measures and doing other things. The system we started out to build turned out not to be the one that people ended up wanting to build."

While looking at ways to revamp the testing plan, Kadish and his deputies also sought to bear down on nagging quality control problems and what they perceived as a general lack of rigor and attention to detail by some of the contractors. They were troubled by what they considered inadequate discipline on the part of major contractors for such a high-priority, high-profile Pentagon program, something they attributed in large part to doubts about the strength of the government's own commitment. "While we believed we had a national priority program that we'd been tasked to do, we didn't find that level of commitment to the program at the contractor or subcontractor level," Nance said. "In many cases, we found it was being treated as just another program. We ought to have had the A team at every level of the program. But it wasn't there."

Nance spent the autumn drafting a new contract with Boeing that would make the aerospace giant more responsible for the performance of individual subcontractors, redefining Boeing's role from one of integrator to prime contractor. Boeing shook up its own management team, replacing John Peller with Jim Evatt, a retired two-star Air Force general with extensive corporate management experience. In fact, the company installed new senior managers in just about every segment of the program. All in all, Boeing made more than forty personnel changes. It also altered the structure of its relationships with subcontractors, taking a more hands-on approach.

In addition, Kadish and Nance delivered stern messages to some of the major vendors for the system, urging them to review all product designs and production procedures to avoid quality problems. In a December visit to Raytheon's Tucson facility, where the kill vehicle was being produced, Kadish told company managers that the government was going to start getting "ruthless" in insisting on quality workmanship. The next flight test was delayed for months as teams of outside experts were called in to comb through both the Raytheon kill vehicle and the Lockheed booster. Much of the equipment was disassembled and rebuilt, while the design histories and production records of many parts were scrutinized. At one point, one senior Raytheon manager counted as many as nine audit groups examining the management, engineering, and manufacturing of the kill vehicle alone.

With the missile defense project stalled, the Clinton administration in its final few months pursued a diplomatic opening with North Korea that had the potential of neutralizing the very missile threat that had been a central justification for the Pentagon program. North Korea's second-highest-ranking military officer, Cho Myong Rok, came to Washington in September bearing an invitation to Clinton to visit Pyongyang. Cho indicated that Kim Jong Il was ready to strike a deal and reaffirmed an earlier proposal that Kim had raised with Putin: that North Korea would forgo its long-range missiles if the West agreed to launch civilian satellites for Pyongyang.

The North Korean invitation prompted a vigorous debate within the administration. State Department officials were eager to send the president on the trip if an accord curtailing North Korean missile production and exports could be struck—and Clinton himself wanted to go. Having backed South Korean leader Kim Dae Jung's forward-leaning push for links with the North, Albright and other senior U.S. diplomats believed the United States now had its own chance to facilitate a thaw on the Korean peninsula and also reduce the missile threat. But a visit by a U.S. president would mark an enormous political concession to North Korea, and Pentagon officials worried that Clinton might be driven to accept an inadequate deal. They wanted North Korea to give up not only its long-range missile development but also shorter-range missiles that posed a threat to Japan. And they wanted North Korea not only to stop production of new missiles but also eliminate existing stocks of weapons.

Tensions over the issue highlighted some of the same underlying considerations that had generated differences between the State and Defense Departments over missile defense. Albright regarded the North Korean opening as just the kind of diplomatic alternative to building antimissile weapons that the United States should be pursuing—and probably should have started months before. Seeing parallels between the new Korean détente and the crumbling of the Soviet Union a decade earlier, she believed that totalitarian regimes were often more prone to influence through outside contacts than they appeared. Cohen, however, opposed granting North Korea a presidential visit until the United States was sure of gaining most, if not all, of what it wanted in return. He fretted that Albright and her senior policy coordinator on North Korea, Wendy Sherman, regarded the very act of getting a president into North Korea as an important U.S. objective in itself.

"State had it backwards, claiming a presidential trip would be an achievement for the United States," a senior Cohen aide said. "For years, North Korea had been seeking high-level contacts in order to bypass the South Koreans. A Clinton visit would help North Korea in several big ways: it would contribute to the regime's internal legitimacy; it had potential for creating fissures between the United States and South Korea; and it would open the door for the Europeans and others to have relations with North Korea."

State officials disputed this portrayal of their position as a sellout and charged the Pentagon with missing a golden diplomatic opportunity. "There are some at the Defense Department who don't want to do anything unless they can do everything," a senior Albright aide said. "That's just the way they are."

In any case, Clinton decided to send Albright to Pyongyang in late October to explore the terms of a possible presidential trip. Before she left, State and Defense officials clashed further over what should constitute America's bottom line with the North Koreans. The lack of resolution concerned Berger, who himself wavered over the wisdom of a presidential visit. At a meeting among top national security advisers just ahead of Albright's trip, Berger wanted to ensure that Albright did not commit to a presidential summit while she was in North Korea. "You're going on a listening mission to hear what they're proposing, not to negotiate," he said to her. Peeved at the remark, Albright shot back, "God forbid the secretary of State should engage in diplomacy."

The mood at the Pyongyang meetings was good. Albright's visit narrowed the gap with the North Koreans and challenged the Western image of Kim Jong Il as an irrational leader. Until his emergence earlier in the year at the summit with the South Korean president, Kim had been largely a mystery to the world. Accounts swirled of a hermit given to excessive drinking and movie binges in his private theater. But several foreign visitors who had seen Kim in recent months had described him as a polite host, articulate and knowledgeable.

That is how Albright found him as well. And in their two days of talks, Kim was firmly in command. On the second day, Albright alluded to a list of questions that her staff had handed to Kim's aides hoping to get clarification on a number of points. Kim asked for a copy of the list and, in front of Albright, proceeded to run through each item. By the end of the talks, Kim had made several important concessions. On exports, he offered to halt all missile sales, including missile components, technical advice, and brokering services—which was more than U.S. officials had expected. The ban would even cover systems that North Korea had already contracted to provide to Third World countries, although which countries were never specified. Kim also offered not to produce, test, or deploy missiles with a range of more than 500 kilometers. U.S. officials were not sure whether he intended to include Scud C missiles, which were capable of reaching Japan and which the Pentagon wanted covered in any deal, but Kim's statements appeared promising nonetheless. In return, the North Koreans indicated that they would like foreign help in launching civilian satellites as well as nonmonetary assistance, such as food, coal, and other commodities. The Americans did not quantify how much aid they would be willing to provide North Korea.

Several important issues remained unresolved, including whether North Korea was willing to destroy missiles it had already produced and how the agreements would be verified. Kim made it clear that he would not welcome inspections, stating that Washington had adequate means to monitor compliance through satellites and other technical assets. The Americans wanted at least a declaration by the North Koreans of the numbers and types of missiles in their arsenal and stipulation of what facilities would be destroyed and how. Hoping to resolve lingering differences, U.S. and North Korean representatives met several days later in the Malaysian capital of Kuala Lumpur. But when the Americans presented draft documents that put forward precise terms for a summit agreement, they were berated by the North Koreans for trying to expand the scope of the deal sketched by Kim. "We said we weren't negotiating but just trying to understand their position," recalled a defense official who attended the talks. The North Koreans continued to insist simply that if Clinton came to Pyongyang, Kim would make it worth his while.

As U.S. officials conferred on their next move, a deal appeared within reach but still hinged on getting the North Koreans to be more forthcoming about final terms. Tensions between State and Defense persisted, with Cohen insisting that any deal address the elimination of those missiles already fielded, while Albright and Berger were inclined to leave the ultimate resolution of this matter to a follow-on negotiation. Looming over the debate were State Department concerns that a successor Republican administration might close the door on any North Korean deal in order to preserve the rationale for building a missile defense system. At a meeting of top advisers at the White House on November 21—with the outcome of the U.S. presidential vote still in dispute—discussion turned to how a Bush administration and a Republican-led Congress might view a deal. "A lot of those people, they're for missile defense," Albright remarked, suggesting that they therefore would oppose a diplomatic solution. At which point, the Pentagon's Walt Slocombe interjected, "Some of us are for missile defense, too." This prompted Berger to step in with a reminder that missile defense was not the subject on the table.

The meeting ended with tentative plans to send Wendy Sherman and a team of Pentagon, State Department, and NSC officials to Pyongyang to give the North Koreans a date for a Clinton visit in return for more concessions on missiles. Any deal, it was decided, would have to include a ban on new deployment of Scud C missiles, a definite framework for addressing the removal of missiles already deployed, and some provisions for verifying compliance with the accord. But the Sherman group never went. Until the election was settled and the potential for a constitutional crisis was removed, Clinton did not want to commit to a trip. After Bush emerged the winner in mid-December, two of his top advisers—Condoleezza Rice and Colin Powell—were briefed on the potential deal. They made clear that Bush would not undercut a Clinton initiative, but neither would he endorse a deal. "We felt very strongly that Clinton was still the president and we weren't supposed to have an opinion on what they ought to do," Rice said later.

In late December, Clinton decided reluctantly to drop the whole idea. Some of his advisers had worried that a last-minute trip to North Korea would have run the risk of looking opportunistic and inadequate, not to mention the prospect that the new administration might have rejected out-of-hand any deal that Clinton concluded. But the main reason that Clinton did not go appears to have been his involvement at the time in a final, frantic effort to strike a peace deal between the Israelis and the Palestinians. Traveling to North Korea would have required a week, factoring in requisite stops in South Korea, Japan, and possibly China. "I thought we were close enough to a Middle East peace that I couldn't be gone for a week in Asia when I might be needed in the Middle East," Clinton said. He also expected that Bush would quickly resume where his own administration had left off with the North Koreans and conclude an agreement. "I figured it was a bird's nest on the ground that Bush could easily pick up," Clinton said.

Albright was disappointed at missing the chance for a deal with North Korea. Pentagon officials were relieved that Clinton had chosen not to go.

Chapter 16

ONCE MORE, WITH FEELING

George W. Bush's interest in missile defense, by his own account, dates back to the 1980s, although before his campaign for the presidency in 2000 he had not spent much time touting the issue—or most other foreign policy issues, for that matter. As the governor of Texas, he often said, the only foreign country he had to worry about was Mexico.

But missile defense had captured his imagination, as it had the imaginations of many Republicans. "I don't think I'd really focused on missile defense until Ronald Reagan brought up Star Wars," Bush said when asked during an Oval Office interview in the summer of 2001 to recount how he had become such an advocate. "His idea was, let's focus on defenses as opposed to offenses. He was the first public leader that brought what I would call a paradigm shift to the public arena.

"Of course, it's a different strategy than the one I'm going to employ because the world changed," Bush went on. "He viewed the Soviets as still the enemy; I don't because they're not. But he did talk about thinking differently about missiles and defenses. As I recall, he even talked about sharing technologies, if I'm not mistaken. I always thought it was a fascinating idea, that here the United States says we'll develop technologies necessary to make the world more peaceful and share them with people."

Bush could recall no moments of epiphany similar to Reagan's visit to Cheyenne Mountain that contributed to his own belief in the need for missile defense. He named no particular mentor or adviser who had helped shape his early views on the subject, although once his presidential campaign got underway he would be surrounded by a number of advisers deeply committed to the cause. He did recall going to the Middle East shortly after winning re-election as governor of Texas in 1998 and seeing a film of Israel's antimissile system, the Arrow, designed for use against medium-range targets. "To me, it spoke about the potential," he said. During the visit, he asked an Israeli general about the accuracy and feasibility of the system and recalled being assured it worked. "Of course their line is it does, because the minute somebody thinks it doesn't, then the country is much more vulnerable. I found that to be an interesting concept unto itself. Deterrence, real or not real, is still deterrence."

In any case, missile defense was on Bush's mind in the summer of 1998 when he interviewed Condoleezza Rice about becoming his foreign policy adviser for the run for the presidency. At the time, Rice was a political science professor and provost at Stanford University. She had served as a Soviet specialist on the NSC staff under Bush's father, and it was the senior Bush who had set up the meeting between his son and Rice at the Bush family's summer residence in Kennebunkport, Maine. By that time, the Rumsfeld commission had issued its report on the growing rogue nation missile threat, but North Korea had yet to launch the Taepodong I missile.

Bush focused on only a few issues with Rice during their talk, and missile defense was one of them. He was curious about the revival of debate between the Clinton administration and Congress over a nationwide antimissile system. He was particularly interested in why previous U.S. administrations had chosen a doctrine of nuclear vulnerability. Was it a technical problem, or a kind of policy choice? "I think he was just fascinated with the fact of this vulnerability, and appalled by it," Rice said. "In that sense, he was like Reagan, who also just couldn't believe we would choose to blow each other up rather than try and defend ourselves. I think his interest in the subject came out of that sense of disbelief."

Soon afterward, Bush began a series of briefings with Rice and other national security experts on a range of foreign and defense policy issues. One session at the Texas governor's mansion in Austin in February 1999 focused on U.S. military readiness and modernization requirements but also touched on missile defense. Addressing the controversial nature of such topics, Bush told the gathering that he viewed elections as a means for getting a mandate for leadership and political capital for enacting programs. "You have to raise the issues," one participant quoted him as saying, "you have to lay them out there, because you don't have the legitimacy of the process behind you if you suddenly show up in January and say, 'Here's my issue.'" Pointing up toward a spot on the wall of the dining room where the group was meeting, Bush said it would be his job as president to set a high mark because compromise was inevitable in a democracy, and if he did not start out high, he would not get anywhere near where he wanted to go.

At another meeting with his advisers in May, Bush received a briefing on missile defense that included an assessment from Don Rumsfeld of the global threat. Bush found the briefing encouraging. "I thought it made eminent sense; it kind of confirmed my philosophy," Bush recalled two years later. His own fixation on missile defense appeared rooted less in ideology than in a straightforward concern that the United States had no defense against missile attack. Such vulnerability, he worried, left the United States open not so much to actually being hit but to being blackmailed by the threat of a missile strike. And this could clearly influence U.S. diplomatic and military action.

"See, there's a threshold question for a person who wants to be president," Bush said in the interview. "And that is: How internationalist are you? If the answer is not very, then the nuclear shield we've got now is perfect. Don't dare tread on us, we'll blow you up. If you want to be more aggressive, then you have to identify the true threats. My instincts and my logic—at least I thought it was logical—led me to believe that."

Bush listened at the May meeting as the discussion turned to what to do about the ABM Treaty; some advocated abrupt withdrawal, while others favored at least some attempt at diplomacy with the Russians. Bush was inclined to try the diplomatic path first, but he made his ultimate objective clear. "A couple of the folks got into the minutiae of the treaty," recalled another participant, "and the governor interrupted them and said: 'My concern isn't with the treaty, my concern is to get a missile defense. And if the treaty doesn't interfere with that concern, then fine, we'll modify it. But if it does, then I don't care about the treaty."

Immersed in a tough fight for the Republican nomination with Senator John McCain of Arizona, Bush decided that missile defense was an issue better suited for the general election campaign, when it would help define him against a Democratic opponent. So he barely raised it in the primaries. In a defense policy address delivered in September 1999 at the Citadel, a military college in South Carolina, he did declare his intention to deploy both theater and national antimissile systems "at the earliest possible date"—and to do so even if Russia refused to accept changes in the ABM Treaty that he said his administration would propose. Most of the speech, though, focused on military readiness and transformation.

Bush and his senior advisers had not yet discussed what position to take on Clinton's planned deployment decision when, in December 1999, the *Washington Post* columnist Jim Hoagland asked Bush whether he would criticize Clinton if the president chose to defer a decision to his successor. "No. I might even praise him," Bush said, telling Rice afterwards that he worried about Clinton locking the United States into an inadequate system and an overly restrictive deal with the Russians.

In the interview, Bush also called on Clinton to begin "to increase the latitude of thought and examination and research on the confines of the current ABM Treaty" to allow for development of "effective" antimissile systems. Bush acknowledged that overcoming Russian opposition to treaty changes would be difficult, and that the war in Chechnya was straining U.S.-Russian relations. What would he do as president to create conditions to cause Moscow to change its view? "I'm not sure where the leverage points are going to be then," he said. "But my job would be to convince them it is in their interest and in our interest to prevent the rogue launch."

Working with Rice to advise Bush on foreign and defense issues were seven people who had held national security posts in previous Republican administrations: Richard Armitage, an international consultant; Robert Blackwill, a Harvard University professor; Stephen Hadley, an international lawyer; Richard Perle, a fellow at the American Enterprise Institute; Paul Wolfowitz, dean of the Johns Hopkins School of Advanced International Studies; Dov Zakheim, a defense consultant; and Robert Zoellick, a Harvard research scholar. The group called itself the "Vulcans," taking their name from the ancient god of the forge, whose statue is a symbol of Birmingham, Alabama, Rice's hometown.

With Clinton due to see Putin in early June 2000, the Vulcans gathered in Washington in April to outline a Bush speech that would put the governor's views on missile defense and nuclear weapons on the record ahead of the summit. In conference calls and e-mails preceding the meeting, the advisers had searched for some way to break the conventional logic that portrayed antimissile systems as leading inevitably to a destabilizing arms race. They were eager to frame missile defense in the larger context of a new strategic doctrine that would link defensive systems with smaller, not larger, arsenals of offensive weapons.

Richard Perle, who had served as an assistant secretary of Defense under President Reagan, led off the meeting with a forceful argument for moving away from formal arms control negotiations toward unilateral weapons cuts, based not on notions of U.S.-Russian nuclear parity but on realistic assessments of post–cold war targeting requirements. Perle himself had once been one of the nation's most prominent cold warriors, with a reputation as a shrewd analyst and tough political tactician. As far back as the 1970s, when he worked as a Senate aide, he had fought against arms control deals. In those days, he worried that such agreements would erode U.S. strategic advantages over the Soviet Union, which he viewed as a barbarian regime. Now, he believed the United States had many more nuclear weapons than it needed. With the fear of a preemptive Russian attack all but gone, Perle argued, there was little sense in tightly linking the shape and size of American nuclear forces to Russian developments. It was time, he said, for a more flexible approach, one likely to result in lower numbers of U.S. nuclear weapons and a missile defense system.

Perle went on to propose more than just a restructuring of the U.S.-Russian nuclear relationship. He suggested an even broader transformation of relations with Moscow that would focus less on negotiating arms control treaties and more on ways the United States could assist in reshaping the Russian economy and civil society. But Rice thought that trying to deal with the whole picture of U.S.-Russian relations was too ambitious for the upcoming Bush speech. So the ensuing discussion focused on revamping nuclear strategy.

Here, agreement on general principles came quickly. The group had no problem accepting the notion that international stability should no longer depend on massive mutual vulnerability. The concept of mutual assured destruction, they agreed, had been meant to fit a world of Soviet troops poised to invade Europe and nuclear weapons on hair-trigger alert, but that world had ceased to exist. They regarded traditional arms talks as protracted and inefficient, inclined to produce excessively rigid accords. In their view, the old arms control framework actually had the perverse effect of keeping the United States and Russia locked in a cold war mind-set and armed with unnecessarily large nuclear arsenals kept on high alert.

"We all kind of picked up on each other, fed on each other," one participant said. By the end of the meeting, there was what another participant called "almost a revival rally atmosphere, a sense of, 'Yes, let's get on with it, the sooner the better.'"

In Bush they saw a presidential candidate whose own lack of experience in nuclear affairs could be turned to some advantage. "Here was a possible president who wasn't mired in the cold war, who didn't have to carry the baggage that so many people—including some of us in that meeting—had carried during the cold war," Perle observed later. "He could think in fresh ways about it."

Asked about this depiction of him, Bush concurred. "We're beyond cold war. I'm 55 years old, I'm not a cold warrior," he said. "My instincts were that we needed to think differently, that the cold war is over." He recalled growing up with a fear of possible nuclear attack. "I remember as a kid, during the Cuban crisis, I was frightened. It was a period of time—I don't know how much it went into my thinking—but the concept of being so armed up that we can obliterate each other fifty times over—whatever it is—it's a notion that does not reflect where we sit today."

Steve Hadley was assigned to draft a concept paper. A onetime assistant secretary of Defense under Bush's father, Hadley had a knack for presenting complicated issues clearly and was one of the fastest writers in the group. His paper outlined choices in three general areas: How low should the United States go in reducing nuclear weapons? How should the cuts be achieved? And what should the new relationship be between deterrence and defense?

Rice spent hours with Bush going over the concept paper, reviewing how nuclear targeting plans were generated and what the new approaches to arms cuts might be. In drafting the speech, Bush decided not to endorse any specific lower number, just the idea of relatively rapid reductions possibly executed unilaterally. But that in itself pointed to a dramatic break from traditional nuclear planning and approaches to arms control. Just what would replace the old framework of arms control treaties, with its rigid rules, stringent verification requirements, and firm ceilings, was not clear. Nor was a replacement doctrine for MAD articulated. Still, Bush appeared resolved to commit himself to a new course.

His advisers were excited, although they worried that such a break from the past ran the risk of drawing fire from Republican conserva-

tives for weakening U.S. defenses. They ran the plan past several prominent national security experts who had served under Bush's father—
among them, Dick Cheney, Colin Powell, and Brent Scowcroft. In
approaching this earlier group of leaders, Bush's advisers had a sense of
the younger generation questioning the conventions of the old. Rice
later likened the scene to the deacons of a church challenging the senior
clergymen. And the reception they got was initially skeptical. The old
guard expressed some skepticism and urged that Bush at least make
clear in the speech that he was not abandoning nuclear deterrence and
would ensure that the military chiefs had a say in any future nuclear
reductions. Finally, however, they not only endorsed the speech but
agreed to join Bush on the podium when he delivered it at the National
Press Club on May 23—the same day the military chiefs went to testify
on Capitol Hill expressing reservations about deeper cuts in nuclear
arsenals. Surrounding Bush as he spoke were five men who had held
high-level national security posts in previous Republican administrations—Rumsfeld, Powell, Scowcroft, Henry A. Kissinger, and George P.
Shultz—several of whom had been instrumental in negotiating the
nation's cold war arms control treaties.

Declaring that "Russia itself is no longer our enemy," Bush said U.S.
security "need no longer depend on a nuclear balance of terror." He recognized deterrence as still "the first line of defense against nuclear
attack," but he said the requirements of deterrence had changed. He
called for reductions in American nuclear weapons to a level "significantly" below what had been negotiated under START II and the
removal of "as many weapons as possible from high-alert, hair-trigger
status." He added that the cuts "should not require years and years of
detailed arms control negotiations," suggesting they might be taken
unilaterally.

"The premises of cold war nuclear targeting should no longer dic-
tate the size of our arsenal," he said. "We should not keep weapons that
our military planners do not need."

Bush coupled this appeal for reduced nuclear arsenals with a call for
a much larger missile defense program than the "flawed" system pro-

posed by the Clinton administration. He characterized Clinton as "driving toward a hasty decision, on a political timetable," and urged deferral rather than approval of a plan "that ties the hands of the next president and prevents America from defending itself." As president, he said, he would build defenses "at the earliest possible date" to protect not only the United States but also its allies and interests overseas. He steered clear, however, of proposing any specific architecture, saying he wanted to examine all options for a global system, including "whether or not a space-based system can work."

By making his pitch for a more ambitious antimissile system part of a larger plan for overhauling U.S. nuclear strategy, Bush not only joined the missile defense debate but started to recast it. He did not want to argue over how to fit the new system into an existing nuclear order; he wanted a new paradigm altogether, one based on defense as well as deterrence and on flexible, unilateral arms control moves rather than rigid treaties. Previous calls for unilateral reductions had carried with them the notion that one side would reduce first in order to encourage the other side to follow. Indeed, that was much of the motivation behind the 1991 initiative by Bush's father removing tactical nuclear weapons from Europe. But the younger Bush was going for something else this time. He was pushing a revolutionary change in doctrine.

The Gore campaign and the Clinton administration belittled much of what Bush proposed as impractical or reckless. Bush had been vague on how far he would cut nuclear forces; he had not addressed the inherent loss of verification and inspection procedures when treaties are discarded; he had offered no specific alternative missile defense system, and no price tag for one. Nonetheless, the speech received favorable editorial reviews for its bold effort to break out of the cold war mold. And even Clinton confided to aides that he found the speech clever and well crafted. It had appealed, he said, to both the right, in its call for a more robust missile defense, and the left, in its backing for deeper nuclear weapons cuts.

Just what kind of missile defense system did Bush have in mind? He seemed genuinely to have no plan. He and his advisers believed that the

ABM Treaty had so constrained development of sea- or space-based alternatives that there was no way of judging the reasonableness of various options until those constraints were lifted and more aggressive development pursued. Pressed on ABC's *This Week* two months after the speech to detail his thinking, Bush spoke only generally of expanding research and conducting a "full-scale effort" to develop a system that worked. He remarked that many people thought the boost-phase approach would provide the "most effective" system, but his comment was hardly an endorsement of the idea.

Although Bush had spelled out his position in broad terms on both national missile defense and the future of nuclear arms, neither issue factored much in the race against Gore. Bush sought for months in stump speeches to play up several military-related themes, including the need to improve the readiness of U.S. troops and transform them into a more modern force. But polls suggested that the public was much less interested in national security than in such domestic themes as education, Medicare, and Social Security. Still, the military messages served some purpose for Bush. They shored up his conservative base and countered his perceived lack of experience in military and foreign affairs. And they outlined an agenda that, if he were elected, he could point back to later as a mandate on which to act.

After the election was decided and while preparing to take office, Bush and his top cabinet appointees made clear that deploying missile defenses would rank among the new administration's top priorities. Bush spoke in terms of an "obligation" to do all he could to protect the United States and its allies from missile attack. His appointment of Don Rumsfeld as secretary of Defense, while not the initial person he had in mind for the job, nonetheless appeared to reinforce this commitment to going forward. Although Rumsfeld in his previous service as Defense secretary during the Ford administration had presided over the dismantling of Safeguard, he was known as a forceful advocate of building another antimissile system. The commission on the missile threat that he had chaired in 1998 had never addressed the actual question of missile defense, but it had provided a major impetus for stepping up work on a national antimissile system. More recently, Rumsfeld had chaired a commission on U.S. space capabilities that argued the deployment of weapons in space was inevitable, and he supported the aggressive exploitation of space for missile defense and perhaps eventually for space-based lasers.

At his confirmation hearing in January, Rumsfeld presented the case for deploying a nationwide missile defense, saying that the United States needed to develop a new kind of deterrence against emerging missile threats. He said it would help persuade potential adversaries "that they're not going to be able to blackmail and intimidate the United States and its friends and allies." National missile defense also would give a U.S. president more options in a crisis besides a preemptive strike, he said. As for whether the technology was ready, Rumsfeld sought to play down the significance of the early failures. "I was reading the book *Eye in the Sky,* about the Corona program and the first overhead satellite, and recalling that it failed something like eleven, twelve, or thirteen times during the Eisenhower administration and the Kennedy administration," he said. "And they stuck with it, and it worked, and it ended up saving billions of dollars . . . because of the better knowledge we achieved."

But for all the new administration's emphasis on missile defense, the Bush team appeared inclined not to rush ahead with a specific architecture or to withdraw abruptly from the ABM Treaty. To be sure, Bush's advisers had little good to say about the treaty. Rumsfeld derided it as "ancient history" and a "straitjacket," noting it was so old that it preceded even his previous tenure as Defense secretary. And more than a few influential Republicans urged Bush to set in motion a plan to scrap the treaty altogether. Senator Thad Cochran, for instance, who had sponsored the legislation in 1999 that called for deployment as soon as technologically possible, recommended either giving the Russians six months to accept modifications or withdrawing from the treaty.

But lacking a clear notion of what would replace the treaty, and given the risk of international uproar over a U.S. withdrawal, the Bush group signaled that there would be no sudden upheaval. In this case, it was Colin Powell, the new secretary of State, who emerged early as a voice of caution and deliberation in the administration. He asserted that there would be "a long way to go" and "a lot of conversations" with the Russians before the United States would walk away from the treaty.

Had Bush himself wanted to start the clock on a countdown to withdrawal, he easily could have done so during his first few weeks by simply approving a plan to begin X-band radar construction on Shemya. Clinton had decided to skip a December deadline to award contracts that would have initiated construction in the summer of 2001. Under a timetable drafted by BMDO for the incoming administration, some work could still have started in 2001 if Bush approved contracts by March. Several of Rumsfeld's advisers—including Steve Cambone, Bill Schneider and Chris Williams—argued for pushing ahead with the project. Rumsfeld himself appeared uncertain, showing interest in an alternative plan to float the radar on a movable oil rig platform, which would increase the range of uses for it. That plan, though, was undercut when Schneider presented Rumsfeld with a photo of an oil rig that had capsized off the coast of Brazil.

Ultimately, Bush decided to let the March deadline pass, thus ensuring at least a year's delay, since the harsh weather in the Aleutian Islands meant that work could not start until the spring of 2002.

"We wanted more time to think about what to do, about what kind of system ultimately to build," Rice explained. "We felt that an early Shemya decision might have seemed to prejudice it. Also, Shemya belonged to a configuration conceived by the Clinton administration that was perceived as decoupling the United States from the allies, while our very strong message has always been that we would not decouple. We didn't want the first thing out of the box to be seen as decoupling."

The new administration also decided to take its time in deciding what to do about a potential deal with North Korea. A visit in March by the South Korean president, Kim Dae Jung, confronted Bush with an early decision on whether to pick up where Clinton had left off—as the

South Koreans were urging—or to back off. Two days before Kim's arrival, Bush's top aides met and agreed to wait. They would appear supportive of South Korea's "sunshine policy" of encouraging economic cooperation, family reunification, and other exchanges but would make clear they were not ready to resume their own dialogue with Pyongyang and would take time to review U.S. policy toward North Korea. Some confusion about the administration's position then emerged when Powell, a day later, appeared to signal a different course. "We do plan to engage with North Korea to pick up where President Clinton and his administration left off," Powell said at a news conference. "Some promising elements were left on the table, and we'll be examining those elements."

But Bush announced a different position the next day, telling Kim Dae Jung that the United States would not resume missile talks anytime soon. Powell portrayed his own remarks as a case of having gotten "a little too far forward on my skis"—out ahead of the president and the administration. The U.S. decision came as a clear rebuff to the South Korean leader, who had told American officials that he believed only a narrow window of opportunity existed for seizing on North Korea's recent willingness to emerge from its diplomatic seclusion. But the Bush team seemed eager to slow the pace of the Clinton rapprochement with North Korea. Some of the president's advisers were concerned that verification procedures for any missile deal had not been nailed down. They also had questions about what to do about the 1994 agreement that had frozen North Korean production of nuclear weapons material at Yongbyon. And they argued that the détente between the two Koreas had not yet led to any change in the North's massive military deployments along the border, or to any apparent relaxation in Pyongyang's repressive rule of its population. Some Republican lawmakers also were urging the new administration to reconsider the 1994 agreement and shift toward a more guarded North Korea policy. They complained that the Clinton administration had rushed to embrace the Kim Jong Il regime without getting sufficient returns.

In light of the cold shoulder from the new U.S. administration,

North Korea had warned that it might resume missile testing. But in early May, Kim Jong Il told a visiting delegation of European Union (EU) leaders that he would extend the test moratorium to 2003. The very fact that the EU had sent its own envoys, thus departing from its past deference to U.S. leadership on Korea, indicated Europe's rising concern about the future course of U.S. policy. In June, Bush announced a willingness to resume talks with North Korea, but he indicated that the agenda would have to widen to cover not just North Korea's production and export of missiles but also the deployment of soldiers on the South Korean border, with the aim being to bring about a "less threatening conventional military posture."

Beyond the issues of North Korea and missile defense, the future direction of U.S. strategic doctrine in general remained a source of speculation and unease throughout the spring and summer of 2001. Particularly disturbing to traditional arms controllers were Bush's additional appointments at the Defense and State departments. In filling key jobs at the undersecretary and assistant secretary levels, Bush drew heavily from a crop of intellectual talent deeply rooted in skepticism toward the existing arms control framework. This group included Douglas Feith, the new undersecretary of Defense for policy; Jack Crouch, assistant secretary of Defense for international security policy; and John Bolton, undersecretary of State for arms control and international security. An early indication of the administration's thinking was revealed in a report by the National Institute for Public Policy, a Washington research group headed by Keith Payne, a Republican strategist and longtime critic of deterrence. The study, which had started in 1999, was not intended to become a road map for a strategic revolution, according to Payne and other participants. But at its release at the start of the new administration, it took on just that reputation, largely because many of its key participants were assuming senior posts in the new administration, including Steve Hadley, who had been named Bush's deputy national security adviser; Robert Joseph, the NSC aide for counterproliferation; Steve Cambone, a special assistant to

Rumsfeld; and Bill Schneider who was advising Rumsfeld. It was thus perceived as the rationale for Bush's new course.

The report was based on the premise that the spread of nuclear and missile technology had made it impossible to predict deterrence requirements with the kind of certainty that had prevailed during the cold war. Consequently, the report argued, arms control had to take a different tack than the rigid, legalistic approach it had followed for decades, with long-term warhead ceilings codified, based on principles of parity and mutual vulnerability. A more flexible framework was needed, the report concluded, one that would allow for defensive as well as offensive systems and possibly involve unilateral adjustments as well.

The new administration embraced this upending of the old order but had few specifics to substitute for it—no new arms levels, no replacement for the ABM Treaty, no blueprint for a national missile defense system. Bush's advisers did not believe, however, that they needed such specifics at the start. Their plan was to try first to sell the rest of the world on the idea of a need for change while consulting with the allies and the Russians about the details. They wanted to make the case for a strategic shift, building a logic that allies and potential adversaries would find compelling. "We set out to try to win the intellectual argument first—that there did need to be a new way of thinking about nuclear weapons, the offense-defense link, about deterrence," Rice said.

In early February, at the annual Munich Conference on Security Policy, which brought together top security officials and defense experts from Europe, the United States, and Asia, Rumsfeld underscored the administration's determination to proceed with an antimissile system even if it could not overcome the objections from the Russians, the Chinese, and the Europeans. He described missile defense as nothing less than a moral imperative. But he also sought to defuse opposition to the administration's plans by assuring allies that the United States would consult with them and by offering to help European nations and other allies to deploy missile defenses—something the Clinton administration had been reluctant to do.

The new Washington team also started using meetings with key allies to try to move Bush's agenda forward. Joint statements with the leaders of Britain, South Korea, Japan, and Germany included language that spoke of recognizing new missile threats and bringing defensive as well as offensive weapons to bear in addressing them. Such repeated references to missile defense suggested the administration was intent on dealing with international resistance partly by simply impressing the rest of the world early on that the United States now considered missile defense a reality, not a possibility. It was as if enveloping missile defense in an aura of inevitability would somehow facilitate not just European support but Russian acquiescence. Powell even drew a parallel between the resistance to missile defense and European opposition in the early 1980s to Washington's decision to deploy Pershing and cruise missiles in Europe. "Our European allies at that time were going nuts," he told senators at his confirmation hearing. "And it took quite a selling story. . . . But lo and behold, we were able to do it by convincing our friends that this made sense."

And indeed, there seemed reason at first to think the approach might have some effect. European governments appeared increasingly resigned to the idea and wary of expending political capital by directly opposing the United States. The overall tone of the European response was polite, if restrained, with allied officials seeming relieved that Bush's advisers at least wanted to consult with them. But the Europeans continued to stress that however determined Bush was to move ahead, some way needed to be found to preserve arms control. And as commentators noted, the situation was different from the mid-1980s. Reagan's missile deployment had always been backed by the various European governments; the resistance at the time had come from public opinion. On missile defense now, the European governments were deeply skeptical of Bush's initiative—and considerably more likely to be influenced by appeals from Moscow than they had been in 1983.

Bush figured that consultations with European allies would get him only so far, that the key to overcoming the opposition both abroad and at home would be to work out an understanding with Russia. And here, several top Bush advisers had as a frame of reference the missile defense talks begun in the final year of the administration of Bush's father. At the time, Russian President Boris Yeltsin and his reformist government had indicated a willingness to develop a global missile defense system jointly with the United States. Just how this would have worked and what it would have meant for the ABM Treaty had never been decided, and the talks were halted after Clinton took office in 1993. But the discussions had put a number of ideas for technology cooperation on the table and presented an intellectual framework that Bush's team was hoping to revive at least in part.

The Russians, for their part, had shown signs of sharing the Bush team's desire to move beyond old antagonisms to a new strategic dialogue. While still wary of national missile defense and any precipitous discarding of the ABM Treaty, they had made statements indicating a desire for dialogue. They also had presented an alternative proposal for a mobile antimissile defense system for Europe. A nine-page document, delivered to NATO Secretary General George Robertson in February, detailed Putin's concept of a limited, theater-based system. The plan provided a largely theoretical framework for how a European-based system might be developed using Russian technology, rather than the more elaborate network of defenses targeting intercontinental missiles that Bush envisioned. It was a plan long on generalities and short on specifics, offering little technical evaluation and no cost estimates, development timetables, or organizational structures. Absent from the new proposal was any mention of Russia's earlier interest in pursuing the more ambitious approach of shooting down missiles immediately after launch in their boost phase. On its face, the proposal did little to address Washington's strategic concerns, because it would not defend American territory and would be intended to counter short- and medium-range missiles. But it was taken as a sign that Russia was interested in pursuing something—a recognition that missile defenses were an appropriate response to the growing threat.

Facing an intensive round of European consultations in May and a first meeting with Putin in June, Bush decided to provide a sense of where he wanted to go with missile defense and arms control and scheduled a speech at the National Defense University on May 1. Although the major themes of the speech had already been forecast in Bush's National Press Club address a year earlier, the president assigned a certain historic importance to the event. Before leaving for the university that afternoon, he was in the Oval Office with a few close aides. When the time came to go, he announced, "Well, let's go transform the world." There was some banter among the aides, who were accustomed to the president joking and were not quite sure whether he was serious. Someone said, "Okay, that's a good thing to do this afternoon." But Bush then stopped and looked very solemn. "No, this is really important," he said. "This is an important day."

He began the speech recalling the tensions and distrust of the hostile rivalry with the Soviet Union, with thousands of nuclear weapons on hair-trigger alert, a U.S. strategic command post in the air twenty-four hours a day, seven days a week, and one and a half million Soviet troops stationed in Eastern Europe. Today, he noted, the Iron Curtain had vanished, and the United States faced a Russia with a democratically elected president. But the world was less certain and more dangerous now, he said, since more countries possessed nuclear weapons and were pursuing long-range missiles. In such a world, he said, cold war deterrence, which relied on nuclear retaliation, was no longer enough. A new concept of deterrence that relied on defensive systems as well as offensive missiles was needed. To get there, he added, the United States and Russia had to "move beyond the constraints of the thirty-year-old ABM Treaty." His characterization of the treaty was harsh. While the Clinton administration had referred to the accord repeatedly as "a cornerstone of international stability," Bush lambasted it for hamstringing U.S. pursuit of promising antimissile technologies and perpetuating a U.S.-Russian relationship "based on distrust and mutual vulnerability."

Still, Bush stopped short of declaring that the United States would withdraw from the accord. Nor did he spell out what he and his advisers had in mind to take its place. He spoke only vaguely of replacing the treaty "with a new framework that reflects a clear and clean break from the past, and especially from the adversarial legacy of the cold war." He depicted the treaty not as an inherently bad agreement but rather as one whose time had come and gone. Whatever its value had been in 1972, it was no longer the right foundation for security in 2001. "What we'd like to do is to be able to lay it aside or bury it with honors," a senior White House aide explained.

In fact, what the president and his advisers envisioned was establishing a sort of loose set of understandings with the Russians. And much of the speech was intended as an appeal to the Russians to help work this out.

> This new cooperative relationship should look to the future, not to the past. It should be reassuring rather than threatening. It should be premised on openness, mutual confidence, and real opportunities for cooperation, including the area of missile defense. It should allow us to share information so that each nation can improve its early-warning capability and its capability to defend its people and territory. And perhaps one day we can even cooperate in a joint defense. I want to complete the work of changing our relationship from one based on a nuclear balance of terror to one based on common responsibilities and common interests. We may have areas of difference with Russia, but we are not and must not be strategic adversaries.

In effect, Bush was offering an implicit deal to Russia to cooperate with his plan and in return share in some kind of broader "joint defense," although what form this might take or how it might come to pass remained vague even in the minds of Bush's top advisers. "We're all kind of struggling with it together," Rice said afterward. "No one believes that there ought to be a new treaty, where it takes you ten years to negotiate on a new offense-defense mix or something like that. But there is a general belief in drafting a set of common understandings— maybe some new truly cooperative security arrangements, perhaps around warning or around information-sharing. And by the way, we think it would be more representative of the kind of relationship we'd like to have with Russia generally on other kinds of issues as well. We don't think arms control and nuclear issues ought to be the centerpiece of the U.S.-Russian relationship; that was abnormal."

Conspicuously, Bush did not make the same offer to China. In fact, he included only a glancing reference to China in his speech. Relations had been badly frayed several weeks earlier by the April 1 collision of a U.S. Navy surveillance plane and a Chinese jet fighter, whose pilot was killed, and by China's 11-day detention of the U.S. crew. Bush had phoned Putin ahead of the speech but had made no overture toward the leaders in Beijing.

As for what a U.S. missile defense system might look like, Bush sounded rather general about that as well. He referred to some "near-term" options that Rumsfeld had identified that would allow for an initial capability. He indicated that work would proceed on a variety of both midcourse and boost-phase systems. He suggested that a period of sorting out lay ahead. "We will evaluate what works and what does not," he declared. But there was no mention of the four criteria that had guided the Clinton administration's selection. Indeed, no criteria were mentioned at all. Bush simply pledged to undertake "real consultations" with allies and acknowledged a need to "reach out" to Russia and China.

In fact, Rumsfeld had spent his first weeks in office intensively reviewing options with Ron Kadish. The new Defense secretary immersed himself in the details of various approaches, studying the state of existing research into possible sea-, air-, and space-based alternatives to the land-based plan pursued by his predecessor. Kadish initially advised Rumsfeld that the ground-based system still provided the earliest opportunity and probably the only option that could be ready this decade, although he made clear that even this option faced substantial technical hurdles and a funding shortfall of several billions of dollars over the next few years. Rumsfeld recognized that basic elements of the Clinton administration's proposed architecture might be the most promising. "We have to deal with where we are, not where we wish to be," he told Kadish at an early meeting. But the Pentagon leader also pushed the BMDO director to look at all options anew and come up with alternatives. For instance, did the X-band radar really need to go in Shemya and risk the inhospitable construction conditions there? How soon could even a skeletal system of ship-based interceptors be ready? Could an airborne laser program still in early development and focused on combating shorter-range missiles be expanded to shoot down ICBMs?

In a major shift from the Clinton administration's approach, Rumsfeld instructed Kadish and the BMDO staff to think outside the constraints of the ABM Treaty. He told them to design the kind of system they thought the United States needed regardless of the treaty's provisions. He also directed them to conceive a system or set of systems that would extend protection to European and Asian allies and that could be completed as soon as practical.

The emphasis on defending allies abroad contrasted with the Clinton administration's desire to duck the question of extended protection and concentrate, at least initially, on fielding a U.S. territory–only system. Even the official name of the Clinton plan—National Missile Defense—had underscored its national focus. It also set the plan apart from theater missile defense systems intended to guard troops in the field. Signaling both the global reach of the new approach and an interest in extending the capabilities of some previous theater systems, Rumsfeld told reporters at a March 8 news conference, "I've concluded that 'national' and 'theater' are words that aren't useful." Lord Robertson, the secretary general of NATO, was in Washington at the time and confirmed that dropping "national" as a modifier was important in getting European allies to take a fresh look at missile defense. "I think taking the 'N' out of 'NMD' has changed perceptions on that and encouraged a more rational debate," he said.

With the new administration looking at alternatives to a land-based scheme, Navy officials saw a fresh opportunity to push for the sea-based option. They called Rumsfeld's attention to a joint Navy-BMDO report on the subject. The report was a revised version of the study that Pentagon officials had sent back for redrafting in the spring of 2000. Although it was completed in December 2000, the Clinton administration had continued to sit on it during its final weeks in office. Rumsfeld promptly signed it and sent it to Congress. The report was not altogether favorable to the Navy. Reflecting the skepticism toward sea-basing among Pentagon policy analysts in the Clinton administration and in BMDO, it said land-based sites could probably be replicated more cheaply than ship-based interceptors could be deployed, and it suggested that any Navy option was at least a decade away. It also noted that the Navy still faced many of the same technical challenges in getting a ship-based system to work as others were having fielding a land-based system. But the report also concluded that a ship-based system could significantly enhance U.S. missile defenses by extending coverage and providing greater flexibility in positioning radars and interceptors. Navy planners had gone on to analyze four detailed options involving modifications of existing Aegis ships and missile interceptor upgrades and presented them to administration officials. One concept described equipping two Aegis cruisers with improved radars and at least fifty interceptors as early as 2005 at a cost of between $1.4 billion and $1.8 billion.

The sea-based approach was popular among some influential Bush aides such as deputy national security adviser Steve Hadley; it also had a strong following among some Republican lawmakers and had been heavily promoted by the Heritage Foundation, a leading conservative think tank. Much of its attraction derived from the promise of greater mobility and versatility, and it also carried the possibility of providing an emergency quick response capability within a few years.

So did an airborne laser weapon under development by the Air Force. The project, which involved putting a chemical laser in the nose of a modified Boeing 747–400 jumbo jet, had been designed to shoot down short- and medium-range missiles like Iraqi Scuds to protect troops in the field. If freed from ABM Treaty considerations, the weapon might be adapted for use against intercontinental ballistic missiles, although questions remained about the accuracy of the laser beams amid atmospheric distortions and the lumbering aircraft's ability to reach more than a few hundred miles inside an enemy's border.

In reviewing options with Rumsfeld, Kadish took pains to play down exaggerated claims by proponents. "It took a while for people to understand the art of the possible," Kadish said. "It's not as easy to do some things as people had postulated." Nonetheless, the general and the BMDO staff favored the idea of an interlocking, multi-layered system of land-, sea-, and air-based weapons rather than a single architecture. This way, enemy missiles could be shot at several times: in boost phase, while rising from their launchpads; in midcourse, while high in space; and again if necessary in terminal phase, while reentering the atmosphere heading toward their targets. Developing these alternatives to the single ground-based setup was no sure thing; it would require upgrades of defense systems designed for other purposes and further development of radars and sensors and rockets. But even the ground-based system, while most advanced, was in doubt given the earlier test failures and fresh delays.

By April, BMDO and Boeing officials had come up with a list of twenty-nine alternative architectures. The number was narrowed to twenty-one and then to about a dozen, as Bush and his advisers settled on the notion of a broad research and development effort that would pursue multiple weapons for targeting missiles in all three stages of flight. In the boost phase, BMDO would experiment with the airborne laser and also with interceptors fired from Aegis ships; in the midcourse phase, it would continue with the land-based system favored by Clinton but also explore a ship-based option; and in the terminal phase, it would look at extending the Army's THAAD system to counter longer-range missiles. The idea was that while some systems would succeed and get accelerated, others would prove less promising and be dropped.

Central to the plan was establishment of a new test site in Alaska that would include a command center and five missile silos at Fort Greely, near Fairbanks, and about five more silos on Kodiak Island off the state's southern coast. It also called for upgrading the Cobra Dane surveillance radar on Shemya Island, with no decision still about the X-band radar's future. The Kodiak facility, together with existing launch capabilities at California's Vandenberg Air Force Base and in the Kwajalein Atoll, would provide a triangle of Pacific sites enabling testers to pursue more challenging interception geometries. Instead of the forty-year-old practice of firing target missiles away from the continental United States and interceptors toward it, the pattern could be reversed by shooting interceptors out of Alaska.

While the Clinton administration had envisioned using Fort Greely as the interceptor site for a fully operational system as early as 2005, the new Bush approach limited the base to serving as simply a storage site for interceptors that would be launched from the more remote Kodiak facility largely for safety reasons. But the administration would have the option of declaring the Fort Greely facility a working missile defense system as early as 2004, just as Bush neared the end of his term. In fact, two of the other new experimental programs—using missiles from Aegis ships to do boost-phase intercepts and flying the airborne laser against long-range missiles—also could conceivably provide rudimentary emergency response capabilities by the middle of the decade, according to BMDO's scheme.

The absence of a specific architecture—or even a firmly established deadline for drafting one—was in one sense a step back from the very defined Clinton approach. In its breadth of experimentation, the Bush plan effectively acknowledged that military contractors had yet to figure out how best to mount a national missile defense. In this way, it resembled Reagan's early scattershot search for the most promising technologies. The Bush program even included new emphasis on some of the more exotic weapons pursued in the 1980s, including accelerated development of chemical lasers that would fly in space and fresh research on the old Brilliant Pebbles idea of stationing small interceptors in orbit around Earth. But Bush's objective of a limited defense was much more modest than Reagan's goal of an all-encompassing shield, and his development effort, at least initially, was focused less on far-out space-based weapons than on ground-, sea-, and air-based ones. If some of the envisioned systems worked, Bush's notion of multiple layers of protection carried a greater chance of actually providing a leak-proof defense than what Clinton had proposed.

It also threatened to bring the United States into conflict with the Russians over the future of the ABM Treaty sooner rather than later. Just how soon remained a subject of some conjecture and tension between administration officials and congressional Democrats. On a number of fronts—construction of the test facility in Alaska, initiation of testing of sea-based interceptors and radars, use of the airborne laser against long-range missiles—the new plan appeared headed toward clear violations of the treaty. But at what point precisely the treaty limits would be crossed was subject to various legal interpretations, as was the question of whether the treaty could survive at all, even in amended form. In the case of the test bed, for instance, some influential arms controllers saw it as potentially a clever means of deploying some system while keeping the arms control framework largely intact. The test bed operations would be so limited in scope, the argument went, that the Russians were not likely to worry about any risk to their nuclear force. But others attacked the plan as a devious and destabilizing effort to field a missile defense quickly under the guise of improving testing. Democratic lawmakers, who had urged the Pentagon to conduct more realistic tests on antimissile technology, accused the administration of attempting to trap them in their own rhetoric. And a *New York Times* editorial warned that the arrangement "could dangerously blur the distinction between testing and the fielding of an operational system."

From a developer's point of view, the new let's-see-what-works approach was a dream, and BMDO officials were very excited about it. "Previously, we were building architectures before we had confidence in the stuff; now, the idea is to wait, gain some confidence and then devise the system structure," Kadish said. He and his staff also presented a schedule of intensified testing that included ten flights of the ground-based system and seven of the not-yet-tested sea-launched program through 2002, with many more to follow. No longer would a test failure lead inevitably to months of delay before the next flight; barring the discovering of some systemic flaw, every effort would be made to continue testing even while failure reviews were underway.

As for the countermeasures challenge that had undercut confidence in Clinton's program, BMDO adopted a recommendation by the Welch panel to set up three teams—dubbed red, white, and blue. The red team would gather whatever information it could about the kinds of countermeasures that potential adversaries might muster. The blue team would determine how best to combat such measures. And the white team would referee. Larry Welch accepted a Pentagon offer to lead the white team.

This open-ended approach to development of a major weapons program marked a sharp departure from traditional practice. Gone were the detailed operational requirements and specific milestones normally applied by Pentagon officials. Gone, too, was the previous attempt to build the system to a specific projected threat. Instead, the idea was to pursue a more general set of capabilities and try to reach them in phases or developmental "blocks" spaced in two-year intervals. This shift made several of the military chiefs nervous, and they voiced their concern at a meeting with Kadish and Rumsfeld in April. The chiefs worried that BMDO, after fielding some initial capabilities, might stop short of developing an optimal system and settle for something of insufficient quality and performance. They also were wary of handing BMDO billions of dollars more in research and development money without some way of controlling how the funds would be spent, since the military services had their own equities in certain programs.

Some board-of-directors-like arrangement was promised to ensure the services a say in BMDO's decision-making. But in general, the idea of relaxing the traditional requirements process appealed to Rumsfeld and other top administration officials eager to move ahead with whatever could be made to work. This kind of measured development process, they argued, would ensure a more cost-effective and reliable system in the long run. Rumsfeld also made clear publicly that the Bush administration was prepared to field a system even if it had proven considerably less than 100 percent effective. In the secretary's view, the striving for near-perfection had been a barrier to getting into the missile defense business at all. Critics quickly accused the administration of proposing "a scarecrow defense." *New York Times* columnist Tom Friedman ridiculed the notion that a rogue state—said to be crazy enough not to be deterred in the first place by the threat of U.S. nuclear retaliation—would be rational enough not to launch a missile against a U.S. defense that did not fully work.

One advantage of the multi-layered system envisioned by Bush was that no one layer needed to be perfect. Even so, contained in Rumsfeld's lowering of performance standards appeared to be the recognition that perfection in missile defense may simply be unattainable.

As Bush and his team pushed ahead with plans for an expansive missile defense system, Democratic lawmakers were drafting their own strategy of resistance, hoping at least to slow the renewed Republican drive. The Democratic leadership in Congress appeared no less determined to make missile defense as defining a point of contention between the two parties under Bush as Republicans had sought to make it under Clinton. For one thing, the Democrats had been freed from having to follow policies adopted under Clinton. For another, missile defense seemed to provide a natural wedge issue. Just as it had energized the Republicans' conservative base, it promised to rally large numbers of Democratic loyalists and primary voters who were strongly opposed to an expensive and unproven missile defense system.

Leading Democrats wasted no time after Bush delivered his May 1 speech to warn the president against abandoning the ABM Treaty and pursuing a missile defense system that, they said, could leave the nation less secure by sparking a new arms race. Standing alongside the ranking Democrats on the Armed Services and Foreign Relations Committees, Senator Tom Daschle of South Dakota, the minority leader, declared that Bush had begun "one of the most important and consequential debates we will see in our lifetime." Daschle questioned whether a missile defense was technologically feasible and added, "A missile defense system that undermines our nation politically, economically, and strategi-

cally—without providing any real security—is no defense at all."
Senator Levin, the ranking Democrat on the Armed Services Committee,
said his party would "try in some way to stop the expenditure of funds
for a system that would abrogate the ABM Treaty."

The power of Democrats to make good on such threats increased
significantly in June when the defection of Senator James Jeffords of
Vermont from the Republican Party suddenly shifted control of the
Senate to the Democrats. The move put two of the most outspoken critics of missile defense at the head of critical committees—Levin at
Armed Services Committee and Joe Biden at Foreign Relations. Just
how significantly this would change the dynamic became evident quickly as Levin began a series of hearings in which he pressed senior
Pentagon officials on their missile defense plans, making clear they
could not expect support for added funds if it meant violating the ABM
Treaty. Similarly, Biden announced his own set of hearings to explore
what he said what his central concern: the net impact that a U.S. missile
defense deployment would have on the global balance of forces.

Overseas, the missile defense issue was enmeshed in a wider set of
concerns—Bush's repudiation, for instance, of Kyoto Treaty restrictions
on gas emissions to control global warming, and his administration's
talk of a U.S. retreat from security commitments in places like the
Balkans—all of which contributed to a general perception of the new
Washington team as holding themselves arrogantly above majority
international opinion. European governments continued to remain
troubled by Bush's proposal to push aside the ABM Treaty. They argued
that if past arms control accords had become outdated, they should be
replaced by new treaties, not a kind of amorphous set of understandings
or arrangements.

Bush concentrated his efforts on striking a deal with Putin. At their
first meeting in June at a sixteenth-century castle nestled in the Alps
near the Slovenian capital of Ljubljana, the two leaders talked hopefully
about bridging the differences between them on missile defense. The
Russian leader cautioned Bush about developing a missile defense shield
without Moscow's consent, saying such an action could seriously strain

relations between the two countries. He continued to refer to the ABM Treaty as "the cornerstone of the modern architecture of international security." After the summit, Putin gathered with a group of American reporters and said that a U.S. withdrawal from the ABM Treaty would prompt Russia to keep its missile forces equipped with multiple warheads, reversing one of the great achievements of the START II accord negotiated by Bush's father during his presidency. But the summit had been striking for its warmth and for the relative ease with which the two leaders seemed to have related. Their upbeat statements at a joint press conference suggested a potential for cooperation on the security issues that had divided Washington and Moscow in the past. And in the interview with reporters, Putin acknowledged that he did not see an immediate threat from U.S. missile shield plans—a significant change from his comments a year earlier to Clinton. "I am confident that at least for the coming twenty-five years," he said, American missile defenses "will not cause any substantial damage to the national security of Russia."

Bush was optimistic about his ability to bring Putin around. Asked after his meeting in Slovenia whether he intended to force the ABM Treaty issue within a year or two, Bush said, "Less than two years, for certain," indicating he expected to have a better sense of the timing by the autumn. "I think we need to give my relationship with Putin time to grow," he said. But he also declared that he would press ahead if the Russians tried to drag out discussions. "As I told Ivanov, the foreign minister—they were all in here—and I said, 'Look, you can [try to] talk me to death, but I'm not going to let you talk me to death.'"

Critics faulted Bush for overestimating the potential of U.S.-Russian relations to support an end to formal arms control treaties and a move toward unilateral actions. While Russia was no longer an enemy, they said, it remained a competitor, vying for economic and political influence in Eastern Europe and in the oil-producing states of the Caucusus, and aiding several U.S. enemies, including Iran and Iraq. Even with the end of the cold war, the United States and Russia harbored distrust and uncertainty about each other's intentions. "I believe that we are better off trying to have clear-cut understandings with the Russians," said

Leon Fuerth, who was Al Gore's national security adviser. "Loose, totally flexible arrangements—where neither side knows what the other can
do and has no dibs on what the other does—can help undermine today's
trust and turn it into tomorrow's paranoia."

Moreover, as emphatic as Bush sounded about the need for a new
strategy with Russia, he had considerably less to say during his first
months in office about how he planned to deal with China. And yet
China remained more strident than Russia in opposing a U.S. missile
defense and warning of grave international consequences if the United
States proceeded with deployment. "I would love for the Chinese to participate," he said in the interview, while acknowledging that was not
likely to happen. "They're not threatened like Russia or the United
States or allies." Pressed on how China fit into his strategic picture, he
said: "First of all, we don't have the same type of relationship that we
had with China, the history is different. My problem with Russia is,
we've got to quit that history. With China, it's an evolving relationship."

Clinton, looking back on his own experience, remarked that concerns
about how China might react had been one of the principal factors
behind his reluctance to proceed with deployment. "I always thought we
could work it through with Russia eventually," he said. "The real big
question was: What would China do? And that's what always bothered
me about getting into the deployment—before we had an understanding
about what was going to happen to China, and what chain reaction
might be triggered, most importantly with India and then with Pakistan."

In July, the Bush administration detailed its missile defense plan in
public for the first time, laying out costs and rough timetables. The proposed 2002 budget sought an increase in missile defense spending of 57
percent, from $5.3 billion to $8.3 billion. The request was actually
about $1.1 billion less than BMDO had requested, but it was still a hefty
boost, particularly for a program that had yet to demonstrate substantial success. At the same time, the administration issued a policy statement saying it had informed allies and Russia that the testing program
"will come into conflict with the ABM Treaty in months, not years."
This provided the clearest signal to date of the administration's intentions. The statement said that the administration did not intend to con-

duct its antimissile tests "solely" to exceed treaty constraints. But unlike the Clinton administration, there was no intent to "design tests to conform to, or stay within, the confines of the treaty," the statement said.

In testimony on the budget request, Democrats accused the administration of being intentionally vague about the new plan to lure Congress into financing work that might violate the ABM Treaty. And they threatened to block any spending that seemed likely to breach the treaty. "The administration's plans for missile defense for fiscal year 2002 have been harder to zero in on than a target in a missile defense test," Levin declared. Deputy Defense Secretary Paul Wolfowitz told the committee: "We are on a collision course. No one is pretending that what we're doing is consistent with that treaty. We have either got to withdraw from it or replace it."

The budget roll-out came in a week that culminated with another test in the Pacific of the ground-based, mid-course system—newly named the Midcourse Defense Segment. It was the first in a year since the embarrassing failure of the kill vehicle to separate from the booster. Nagging quality control concerns about the booster and kill vehicle had repeatedly postponed tentative test dates. Both had undergone extensive reviews, but problems kept cropping up through the winter and spring. Operation of a navigation device in the kill vehicle, for instance, was found to be impeded by natural impurities in the device's components. The kill vehicle's cooling system, which failed in intercept test two and leaked in test three, was again troubled by contamination and flow issues. Sent to California for a ship readiness review in May, the kill vehicle was promptly returned to Arizona until all its software programs could be completed and fully integrated at the factory rather than in the field.

The caution paid off. By mid-summer, as final preparations were under way in the Marshall Islands for a launch on July 15 (July 14 in the continental United States), the process was going much smoother than ever before. Absent were many of the mysterious snafus—what the testers called "unverified anomalies"—that had bedeviled previous runups. A few glitches did arise. Raytheon discovered yet another nitrogen leak in the kill vehicle's coolant system, although this one was even less than the last and deemed no threat to the test. A self-destruct unit on the booster was found to be defective and replaced. Then, during a simulation twenty-five days before the launch, a range safety operator inadvertently pressed the button that, in an actual test, would have destroyed the interceptor in flight.

Two days before launch, the test team faced a brief scare when Lockheed employees preparing the booster reported ten tiny pieces of metal missing from access panel fasteners. Since the booster is what had failed on the previous flight—most likely as a result of some kind of particle contamination—alarm bells went off at the prospect that the little metal chips had fallen inside the booster. An intensive search found most of the pieces lying outside the missile, and the booster's interior was checked to ensure it had been adequately insulated against the possibility of a missing chip causing an electrical short.

The most significant difference about the test set-up this time was the extensive security on and around Kwajalein. Mindful of how easy it had been for two Greenpeace protesters to enter the launch area last time, the Pentagon deployed a platoon of soldiers from the Army's 25th Infantry Division in Hawaii to guard the site. Marshalese authorities declared a twelve-mile maritime exclusion zone around Kwajalein, and patrol boats and helicopters swept the lagoon. The "threatcon" level rose a notch to "Bravo," mandating identity checks and searches. And several days before the test, a one-star general from the Army's Space and Missile Defense Command arrived to help coordinate the security operation. The only sign of any protester activity in the area was a lone Greenpeace activist who had camped out on a nearby island and kept a regular diary of his experiences on the Internet. Two days before the launch, Marshallese officials ordered him removed and taken to Majuro, the capital, where he remained for several weeks until his visa expired. A businessman described as sympathizing with the activist then promptly asked U.S. authorities for help getting the next flight out of Kwajalein as well.

As part of the added security, an S-band surveillance radar was installed on Meck Island to scan for boats. But its location a few hundred yards from the launch pad raised concerns among the testers that the energy emitted by the radar might somehow damage the interceptor's electronics. A decision was made to station a barge off the island and use its radar instead. But three days before launch, a technician who had not gotten word to leave the S-band off turned it on for a half hour to do a routine maintenance check. This sent the test team scurrying to assure themselves no harm had been done to the kill vehicle or booster.

No protest activity emerged in Kwajalein on launch day. But about ten minutes before the scheduled liftoff of the target missile in Vandenberg, the countdown was halted when three small rubber boats suddenly zipped into the California range's exclusion zone and two Greenpeace activists swam onto a nearby beach. They were quickly cleared, and the count resumed.

The test scenario was to be exactly the same as before. And many of the fifty or so people gathered in the control room on Meck to monitor the event were also the same, with the notable exception of some senior Boeing managers. The Boeing team shirt also had changed; red was out, thin Navy blue stripes were in.

At 2:40 P.M. Kwajalein time (7:40 P.M. PDT), the target missile soared skyward, triggering cheers and applause in the Meck control room. Deployment of the mock warhead and decoy went smoothly, and tracking radars detected the objects. Right on schedule, the interceptor launched at 3:01 P.M., again to applause. "Go get 'em," someone shouted. As many in the control room edged forward in their seats, anxiously watching performance indicators on a giant video screen, the kill vehicle ran through all its operations—separating from the booster, orienting itself with a couple of star checks and receiving two guidance updates from the IFICS ground transmitter. With two minutes to go before anticipated intercept, the kill vehicle started homing in on its target. A hush settled over the room. A mission control voice began calling out the time until intercept: one minute, thirty seconds, ten, nine, eight. . . . And then, just when expected, a large burst of light filled the screen—the blinding moment of collision captured by a long-range camera outside.

All in the room leapt to their feet—clapping, hugging, exchanging high-fives. Champagne flowed. In that split-second flash, the future of the land-based interceptor concept had gone from cloudy to bright. At a late night news conference in Washington, where Kadish had watched the test, the BMDO director tried to keep things in perspective. He knew as well as anyone that the test merely represented a demonstration of the principle of hit-to-kill; it was far from proof that the United States could build a reliable, effective system capable of operating under combat conditions. "We have a long road to go," Kadish cautioned reporters. "So this test is just one on a journey, one stop on a journey."

But for Nance and his testers, there was something enormously satisfying about the test. The euphoria of the first intercept success nearly two years earlier had lasted for months. But the hit this time was somehow sweeter because of all the technical scrutiny that had gone into it and the enormous sense of pressure that had surrounded it. In a brief speech to the control room crowd after the intercept, Nance commended all for their commitment and their perseverance. "To hell with the critics," he declared. "A great job." The room broke into much laughter and applause.

Afterword
to the Paperback Edition

Just as a new set of political and diplomatic showdowns over missile defense appeared to be taking shape under the Bush administration, terrorists seized four commercial airliners on September 11, 2001. They flew two into the World Trade Center towers in New York and another into the Pentagon; the fourth jet crashed in a Pennsylvania forest. The astonishing assault on the nation's financial and military centers, which resulted in more than three thousand deaths, marked the worst terrorist attack in U.S. history. It jolted the United States into declaring an all-out war against terrorism, raising what had been a comparatively low-grade counterterrorist effort by military and law enforcement agencies into a high-profile national campaign. Greater attention to securing the American homeland against attack promised to be as much a part of the new war as rooting out terrorists abroad and eliminating their sources of support. The intensified focus on homeland defense would have a profound effect on the course of national missile defense.

Critics of missile defense tried arguing that the terrorist strikes in New York and Washington proved that Bush had been concentrating on the wrong threat. Since the United States had much more to fear from low-tech terrorism than from a high-tech missile attack, they said, the money the administration had planned to spend on missile defense should be diverted to counterterrorism activities. But supporters of missile defense seized on the attacks as added justification for their cause. They said the events of September 11 reinforced the case for protecting American cities by all available means, including antimissile systems. "Can anyone doubt that if the terrorists behind Tuesday's attacks had had access to a ballistic missile, they would have used it?" the *Wall Street Journal* asked in an editorial. "Why settle for toppling the World Trade Center if you can destroy all of New York in an instant, without having to go to the trouble of sneaking a crew over the border and arranging for pilot training in Florida?"

Funding both a crackdown on terrorism and a buildup in missile defenses would have been problematic before the September 11 attacks. But in the aftermath, the financial constraints receded. Democrats who had been preparing to cut the administration's $8.3 billion request for missile defense and attach conditions to ensure that testing remained within the terms of the ABM Treaty sought to avoid a partisan wrangle and let Bush's initiative proceed, recognizing that they lacked the votes to mount a challenge amid a new surge of patriotic feeling and military spending. Forced to share the stage with the new war on terrorism, missile defense seemed destined to lose some of its prominence as the administration's dominant military project. But the administration showed little inclination to scale back the program.

To the contrary, Bush brought to a head an argument that had simmered among his top national security advisers over how to remove the ABM Treaty as an impediment to testing and development. For months there had been little doubt that Bush would get rid of the treaty's constraints. The question had hinged on how he would do it. Would it be with the cooperation of the Russians and Europeans or over their objections? Would it be by amending the treaty or by withdrawing from it altogether?

The Russians, who had been unwilling to enter talks with Clinton about revising the treaty, had indicated a willingness to do so with Bush in the context of deeper reductions in nuclear weapons. But Bush had raised the ante. He was proposing a more ambitious missile defense system than Clinton had in mind and was showing little desire to enter into the kind of detailed, rigid accords that had marked past arms control negotiations between Washington and Moscow. Clinton's stated purpose in seeking negotiations with Russia had been to strengthen the ABM Treaty. Bush's desire was to do away not only with constraints on missile defense development but with the kind of set-piece arms control process practiced by Republicans and Democrats alike for forty years.

Not all the president's top advisers were ready to dump the treaty. Powell had emerged as the most forceful voice for compromise, arguing that a changed treaty was better than none at all. Although he regarded research on antimissile systems as worth pursuing, he was wary of letting the effort disrupt relations with the Russians, Chinese, and Europeans. A concession to the Russians on the formalities of the ABM Treaty, he figured, would be more than repaid in other ways—their acquiescence in the expansion of NATO, the continued reform of their economy, and their cooperation in the war against terrorism. He thought a way could be found to conduct broader testing within the confines of an amended treaty. Failing that, he favored a less formal understanding that would permit Russia to blink at certain specific tests.

But senior officials at the Pentagon and White House wanted to drop the ABM Treaty altogether as a demonstration that America would accept no artificial limits on its national interest. They contended that the new administration had to remain aggressive if it stood any chance of overcoming the hindrances and inertia of the past. In their view, simply a modified ABM Treaty would not be enough to encourage the kind of investment in alternative antimissile systems they envisioned. They argued that amending the treaty would entail a time-consuming ratification battle in the Senate and that even something less than a formal amendment, such as a Russian side letter promising not to challenge certain tests, would give the Russians too much leverage. With the Pentagon planning to start building missile silos for a new test bed in Alaska and expand experimentation with sea-based weapons and airborne lasers, the administration's research into missile defense systems soon risked breaching limits imposed by the treaty.

Bush's own political standing had been enhanced in the months following the September 11 attacks as he rallied the nation behind a campaign against al Qaeda's terrorist network and routed the terrorists and their Taliban backers from Afghanistan. With national security still very much on the minds of Americans, Bush gave Russia formal notice on December 13, 2001, that the United States would be withdrawing from the ABM Treaty. "I have concluded the ABM Treaty hinders our government's ability to develop ways to protect our people from future terrorist or rogue state missile attacks," the president said. "Defending the American people is my highest priority as Commander in Chief, and I cannot and will not allow the United States to remain in a treaty that prevents us from developing effective defenses." At the same time, he noted the development with Russia of "a new, much more hopeful and constructive relationship" and reiterated his interest in forging a new strategic accord.

Under the terms of the treaty, the termination would not take effect for six months. But the deed was done, the historic threshold crossed. And yet, the reaction both at home and abroad was considerably more muted than many had anticipated. None of the dire predictions of ruptured relations, new armament plans, or unraveling of other arms control accords occurred. In fact, the Russians responded more with regret than outrage. Putin called the move "mistaken" and said it could create a "legal vacuum" in arms control at a time when the world faced new threats. Many Russian analysts and politicians were more blunt, saying Bush's decision had gratuitously humiliated Moscow after Putin stood with the United States in its campaign against terrorism. But the Moscow government's more restrained official line was what counted most and served to deflate the potential for fierce European protests that the American move would undermine strategic stability. As for China, when a high-level U.S. delegation went to brief Beijing authorities on the new plan, they found Chinese officials more agitated about U.S. arms sales to Taiwan than about missile defense.

Such measured responses partly reflected a wait-and-see attitude on the part of the Russians and Chinese. It was clear to them that the technical complexity of missile defense would keep the United States from fielding anything in the near term that could threaten their security. But some credit must also go to the groundwork that administration officials had laid. They had signaled their intention to jump beyond the treaty and presented a reasoned case for a new strategic framework. Additionally, Bush had sought to develop a personal relationship with Putin, underscoring his central belief that the United States and Russia had entered a new era of friendlier ties. Domestically, too, Bush faced an American public newly sensitized to matters of homeland security and generally sympathetic to the idea of doing everything possible to ensure against potential threats.

"Arms control agreements are not forever," observed James Schlesinger, a former defense secretary, in an opinion piece in the *Washington Post*. "Strategic conditions change. The bipolar world of the '70s and '80s is gone—thus the feared two-sided competition to deploy additional offensive vehicles did not reappear. Rather than being 'the cornerstone of strategic stability,' the treaty turned out to be more like the cornerstone of arms control theology. As the treaty over time became less relevant, it was defended with increasing passion."

By the time the treaty formally died, on June 13, 2003, it had lasted thirty years and eighteen days. The White House issued a four-paragraph statement saying simply that the demise of the treaty was well deserved and that both the United States and Russia should look forward to a new era of missile defense. It even dangled the prospect of U.S.-Russian cooperation on antimissile systems, citing the possibility of expanding military exercises, sharing early warning data, and exploring potential joint research and development of missile defense technologies. Russia declared it would no longer be bound by the START II accord, but that action was more symbolic than substantive. The accord had not entered into force and was effectively superseded by a new treaty on nuclear arms cuts.

Indeed, the two countries had managed a month earlier to complete work on an accord codifying a reduction in their nuclear arsenals by roughly two-thirds over the course of a decade. Since the presidential campaign, Bush had coupled the development of missile defenses with deep cuts in nuclear weapons. What had been only a vague notion shared by him and a number of senior advisers had taken shape in the administration's first year as Pentagon officials undertook a broad review of the country's nuclear requirements. In the autumn of 2001 they had recommended to Bush a fundamental turn in nuclear war planning. Instead of sizing the nuclear arsenal on the prospect of war with Russia, they proposed a more flexible strategy reflecting the passing of the cold war and contemporary concerns about hostile states possessing chemical, biological, or nuclear weapons. If U.S. war planners focused more on the potential for nuclear strikes against Third World adversaries and less on holding targets in Russia at risk, a case could be made for shrinking nuclear inventories substantially—say, to the neighborhood of 2,000.

Clinton had favored a similar cut, but he had met stiff resistance from the military chiefs, who insisted that any reduction come only in the context of a change in nuclear doctrine. Bush, by proposing just such a shift in doctrine, was able to bring around the uniformed leadership, including the hawkish Admiral Richard W. Mies, who was still in charge of the Strategic Command. Bush's plan also included some hedging and accounting provisions that made the reductions less deep and permanent than they seemed—and more acceptable to the chiefs. Instead of destroying many of the warheads due for removal, for instance, the plan called for putting them in storage for possible reactivation in the event of conflict with a nuclear-armed adversary. It also altered how nuclear weapons were counted, excluding those on submarines and bombers that were being overhauled.

Meeting with Putin in November 2002 in Washington, Bush had told the Russian leader that the United States would unilaterally reduce its nuclear arsenal to between 1,700 and 2,200 warheads. Administration officials insisted that their move was the natural outgrowth of a hardheaded look at nuclear requirements, but some foreign policy benefits were evident as well. By pledging to slash nuclear arms, Bush and his aides reinforced their case to the Russians and Europeans that the U.S. missile defense program should not be seen as posing a bid for strategic dominance. In fact, U.S. officials said, the decision to reduce America's dependence on nuclear weapons was made possible in part by the assumption that the country would be erecting some sort of missile defense system.

The Russians welcomed the move, having expressed a strong interest for some time in cutting their own arsenal. Putin reciprocated by announcing his intention to cut the Russian arsenal to between 1,500 and 2,200 warheads. But he also insisted that the arms cuts be codified in a formal arms control treaty. Bush had opposed etching the commitment into a treaty that he feared would take years to negotiate and ratify. Others at the White House and the Pentagon also wanted to avoid getting drawn into negotiating a legally binding document that could tie U.S. hands and open the process to a Senate debate and vote. They argued that with the prospect of a U.S.-Russian nuclear conflict now far less likely than during the cold war, detailed treaties stipulating the exact balance of warheads between the two powers were no longer central to global stability. Instead, they envisioned a process of parallel cuts based essentially on trust.

But Putin wanted something that would obligate future U.S. presidents. Within the U.S. administration, Powell argued for giving the Russian leader a formal document. He believed that a binding deal would help Putin at home, reassure the Europeans, and install some limits on Russian behavior. He also thought a written promise might ease the sting that many Russians felt at the U.S. withdrawal from the ABM Treaty.

Eventually Bush sided with Powell and agreed to pursue a new treaty. Putin, in turn, yielded to almost all of Bush's demands. The result was a three-page agreement, announced on May 13, 2003, that was historic for its brevity and the speed with which it was negotiated. It was the shortest yet one of the most far-reaching treaties in four decades of arms accords with Russia. True to Bush's minimalist vision of nuclear arms control, it had none of the lengthy rules, extensive verification requirements, or vast appendixes typical of previous nuclear deals.

Called the Strategic Offensive Reductions Treaty, the accord, which was signed during Bush's first trip to Russia in late May, committed the United States and Russia to reduce nuclear arsenals to between 1,700 and 2,200 warheads each by the end of 2012. But it did not require destruction of the warheads, essentially giving both sides the right to keep the weapons in storage and available for reinstallation on missiles or aboard nuclear-armed submarines on relatively short notice if either country faced new threats. The Russians had wanted both sides to eliminate delivery platforms—missiles, long-range bombers, and submarines. They had reasoned that if the platforms were taken out of service, then the warheads would follow. But the Bush administration had focused on reducing just the number of deployed warheads. Unlike the past treaties, the new document allowed both nations freedom to shape their own mix of land-based, bomber-based, and submarine-based weapons however they pleased.

Some arms control advocates faulted the accord for not making deeper, more permanent cuts to reduce the threat of nuclear proliferation. But many Senate Democrats, relieved that the administration had agreed to any accord, joined Republicans in praising the new document for affirming sizable reductions in nuclear stockpiles. Even those who disagreed with the administration's nuclear strategy predicted the new treaty would face no significant obstacles to ratification.

One other great advantage of the new accord, in the administration's view, was that it placed no prohibition on building missile defense systems. The Russians had tried for months to include limits on the U.S. plan, first seeking a pledge in the treaty that any U.S. defensive system would not threaten Russian strategic forces. When the administration rejected that, the Russians pressed for a firm statement in the treaty's preamble, which U.S. officials also rejected.

The Pentagon, meanwhile, was taking full advantage of the new strategic framework and the opening it provided for missile defense. Two days after the end of the ABM Treaty in June, defense officials broke ground at Fort Greely in Alaska for construction of six interceptor missile silos and assorted military facilities. At the time the project was said to be intended primarily for ground testing, under cold-weather conditions, of actual site elements—the silos, the communications networks, the maintenance buildings, and other support functions. But defense officials made no secret of their intention to be able to use the site as an operational antimissile system should the need arise, particularly in the event of a North Korean threat.

The facility was scheduled for completion by September 30, 2004, just as the next presidential election campaign—and Bush's expected run for a second term—would be peaking. Longtime critics of missile defense saw politics as the driving force behind this schedule. But administration officials insisted the motivation was military, not political, with the urgency justified by concern that North Korea might already have developed a missile capable of reaching the western United States.

Even as the first shovels of dirt were being turned at Fort Greely, Pentagon officials were looking at broadening the purpose of the project. Their eagerness seemed somewhat at odds with the administration's initial stated objection that time should be allowed to pursue various technological approaches before locking the nation into a specific missile defense architecture. Defense officials had pointedly avoided presenting a plan for fitting any of the experimental systems together, saying years would be needed to test and affirm which weapons would work best. Early in the summer of 2002, Rumsfeld himself, at least in public, was still sounding disinclined to set a specific deployment date. "Our position on this is that we're not going to set artificial deadlines," he told the *Washington Times* in July. "We're basically in a research and development mode, and to put target dates out there in an R and D—in an uncertain world like that—is not useful."

But behind the scenes the administration was moving toward a deployment decision, pushed by a number of Rumsfeld's senior policy aides who were bent on getting something built sooner rather than later. No new intelligence about North Korean missile plans explained the urgency. Although the North Koreans were continuing to run ground tests on elements of a long-range missile, they were abiding by their moratorium on flight testing. Nonetheless, Kadish and his team were directed to study what could be done to accelerate deployment of an antimissile weapon even while preserving the basic experimental nature of the Pentagon's program.

A number of people outside the administration had been critical of the program's open-ended nature. Congressional Democrats accused the administration of being fiscally irresponsible in not putting forward a specific architecture that then could be priced. Republican missile defense advocates also expressed frustration, arguing that greater definition in the program and an emphasis on some approaches over others would speed deployment and secure those elements considered most essential. Defense contractors griped about the absence of a clearer, long-range program.

The Defense Science Board, an influential Pentagon advisory group, also urged in a summer study that the administration start narrowing the focus of its missile defense program and opt to proceed with both ground-based and sea-launched interceptors. The panel concluded that enough was already known to warrant some choices, which in turn would increase the prospects for a timely deployment of a workable system. This group's recommendations carried particular weight because it included some of the nation's most respected authorities on missile defense and was headed by retired General Larry Welch, who had chaired three independent missile defense studies over the previous four years, and William Graham, a former science adviser to President Reagan and a onetime NASA deputy administrator. Both men—along with the chairman of the Defense Science Board, William Schneider Jr.—had served on Rumsfeld's 1998 commission on missile threats.

The panel's recommendation reflected the thinking of Kadish and the Pentagon's chief acquisition official, Edward C. "Pete" Aldridge Jr., who had replaced Jacques Gansler with the change in administrations. They were interested in shedding the program of its image as little more than an expensive technology incubator. But Kadish in particular also recognized the risks of attempting too much too fast. With the treaty only just buried and many of the experimental options not fully explored, he felt that not all the potential options were known. He firmly opposed making a commitment to a specific architecture or spelling out some fixed, grand design for missile defense. He agreed that the ground-based midcourse system and the sea-based boost-phase option were the most promising approaches and should be emphasized. Flight tests, while relatively few, had persuaded him of the technical viability of both. But he continued to favor an evolutionary approach to development, telling listeners that the "most commonsense" course was to build a missile defense network "a piece at a time when it's ready."

Over the summer a plan emerged to transform the Alaska site into an operational facility. Instead of functioning as a test bed set up to handle occasional exercises, the site would have to be staffed and equipped to remain on alert around the clock, seven days a week. And it would need as many interceptors as could be produced, as fast as possible. Kadish and his team figured that only six interceptors could be built and put in place by the 2004 deadline. But they identified an opportunity to add four more in the near term using silos at Vandenberg Air Force Base.

The addition of Vandenberg offered several advantages. Not only would use of the base provide more interceptors, but it ensured uninterrupted protection should something put Fort Greely out of action. Additionally, the existence of a second site would enable Pentagon authorities to begin to evaluate the battle management and communication links that come from trying to operate a multi-site system. And Vandenberg's location in the continental United States would afford better coverage of the far reaches of the southeastern United States, which were a stretch for interceptors fired from Fort Greely.

To help track enemy missiles, the Pentagon also counted on upgrading an early-warning radar at Beale Air Force Base in California and the Cobra Dane radar on the island of Shemya. Under Clinton, officials had also planned on building a high-resolution X-band radar on the island. In fact, it was construction of the radar that was to be the limiting factor in how quickly the whole system could be assembled. The Bush administration had spent months considering whether to float the X-band radar at sea on a giant platform. The technical community was skeptical that the platform could support the radar's five-million-pound weight.

Political resistance to the idea also came from Senator Ted Stevens, the powerful Republican senator from Alaska. But ultimately, Kadish recommended the platform plan. Two foreign-based early-warning radars—one at Fylingdales in Britain, the other at Thule in Greenland—also were considered critical for tracking and targeting missiles that might be launched at the United States from the Middle East. The U.S. plan called for upgrading these radars by 2005.

The Navy, too, entered the picture in a significant way, marking a major change in its relationship with the national missile defense program. Kadish was interested in incorporating the Aegis system's SPY-1 radar to track longer-range missiles, but he wanted to be sure that the Navy was willing to allocate enough ships to ensure coverage when needed. He also wanted a dedicated Aegis vessel for testing. Navy leaders recognized that if their service was ever going to break into a missile defense business long dominated by land-based options, the time was upon them to make the commitment. They agreed to assign the *Lake Erie* full-time as a test ship. They also pledged to work out a concept of operations that would make surveillance ships available for missile defense missions and ensure use of the Navy's new Standard Missile-3 interceptors to counter short- and medium-range missiles. The SM-3 had undergone its first intercept tests in 2002, scoring hits in each of three attempts.

Traveling to Bush's ranch in Crawford, Texas, in mid-August, Rumsfeld and Kadish briefed the president on the deployment plan, which was being held in great secrecy. Pentagon officials were not looking for a decision at that point, just an indication of the president's general readiness to support a deployment. "It makes sense to me," the president said at the end of the hourlong session.

Through the autumn, Pentagon officials refined the plan, pinning down production schedules and costs and pursuing discussions with European officials about upgrading the radars in Britain and Greenland. Kadish wanted to be sure that if he told Rumsfeld and Bush that something could be built by 2004, it really could be. "One of the caveats that General Kadish had given us all was, don't overpromise what you can deliver in this time frame," recalled Keith Englander, the general's technical director. "He did not want to end up going back to the president and saying, 'I can't deliver because we overpromised.'"

The construction timetables at Fort Greely left little slack if the 2004 deadline was to be met. But the biggest schedule risk involved development of the interceptors, especially the boosters used to lift the kill vehicles into space. An initial plan overseen by Boeing to design a single type of booster had suffered months of delay and given way in early 2002 to two competing models. Lockheed had taken over the Boeing design, and Orbital Sciences Corporation was working on an alternate version derived from the company's space launch vehicles. The two models were each to be flight-tested twice in the summer of 2003. Depending on their performance, the Pentagon could choose one or keep both for eventual deployment. Kadish expected that both designs would ultimately be used, enabling the production of more interceptors in the short run and providing various performance options for the system, since each version had certain advantages over the other. Orbital's model, for instance, was slightly faster, but Lockheed's version came in a canister and so could be shipped and handled more easily and required less on-site assembly.

On December 17, 2002, Bush announced his decision to begin fielding a rudimentary system for defending the United States against missile attack with an initial ten land-based interceptors to go in Alaska and California in 2004 and ten more at Fort Greely in 2005. The plan also envisioned employing as many as fifteen Aegis ships equipped with SPY-1 radars. To defend against shorter-range missiles, Bush also said that up to twenty Standard Missile-3 interceptors would be placed on three Navy ships and that production of the new Patriot system interceptor, the PAC-3, would be increased.

In National Security Presidential Directive 23 ordering the deployment, Bush set out the broad outlines of the missile defense policy that had taken shape over the previous two years. Describing the need for antimissile systems, the directive asserted that a growing missile threat from "hostile states" represented a "fundamentally different" set of circumstances than the United States faced during the cold war standoff with the Soviet Union and required "a different approach to deterrence and new tools for defense." It described missile defenses not as "a replacement for an offensive response capability" but as "an added and critical dimension of contemporary deterrence." The directive also stressed that the nationwide system set for deployment in 2004 was only "a starting point" of what would be "an evolutionary approach" to better defenses over time. "The United States will not have a final, fixed missile defense architecture," the directive said. "Rather, we will deploy an initial set of capabilities that will evolve to meet the changing threat and to take advantage of technological developments."

Special emphasis also was placed on using antimissile systems to defend "allies and friends." Reaffirming a statement made by Rumsfeld in the early months of the administration, the directive formally eliminated the Clinton-era distinction between "national" and "theater" missile defenses. To counter any impression of a burgeoning Fortress America, the directive included a section promoting the idea of cooperation with Russia and other countries in developing and fielding missile defenses. But just what form this cooperation would take was not clear. While Britain and France definitely had a role by granting permission to use the early warning radars on their territory, and Canada was likely to play a central part in operating the system as a longtime partner in NORAD, opportunities for other foreign participation appeared limited, at least in the near term. U.S. officials spoke of direct industry-to-industry deals as the fastest and most promising means for international cooperation, but what this might entail remained vague.

The deployment announcement put Bush on record for the first time committing to a specific date and defining an initial operational system for defending U.S. territory from missile attack. Pentagon officials stressed that the system would have very limited capability at first, aimed largely at knocking down North Korean missiles—and only a handful of them at most. Many more years of development and testing would be necessary, they said, to provide a truly comprehensive U.S. antimissile shield. Rumsfeld described the initial plan as "better than nothing," saying at a news conference: "The reason I think it's important to start is because you have to put something in place and get knowledge about it and experience with it." In a statement, Bush called the new initiative "modest" in scope, but said it "will add to America's security and serve as a starting point for improved and expanded capabilities later, as further progress is made in researching and developing missile defense technologies and in light of changes in the threat."

Critics in Congress and in scientific and advocacy groups opposed to missile defense slammed the move as premature and likely to result in a waste of billions of dollars on inadequate technology. Even some die-hard advocates of missile defense were frustrated, preferring to have seen greater emphasis placed on such alternatives as long-range sea-based interceptors or space-based weapons. Having won the battle over the ABM Treaty, some members of the pro-missile defense crowd fell to vying among themselves over what kind of system to concentrate on building first.

The Pentagon had every intention of pursuing other options. It was exploring boost-phase interceptors and "multi-use" interceptors. It was evaluating designs for satellites armed with interceptors. It also had started looking into setting up "forward-based" radars in such places as South Korea and Japan to help identify warheads as they took off and before they had time to release any decoys. And it was studying conceptual designs for miniature kill vehicles that would permit several to launch atop a single booster.

But the most mature option available to the Pentagon remained the use of land-based interceptors in what defense officials had labeled the Ground-Based Midcourse Defense Segment. Other approaches in which administration officials had earlier placed some hope were faltering. The airborne laser, which defense officials had initially projected might provide a rudimentary capability by 2004, was struggling with weight and other technical issues. Existing sea-based interceptors, while showing promise against short- and medium-range targets, were regarded, after further review, as simply too slow to fire against long-range missiles. And the space-based laser program was scaled back in the face of determined congressional resistance.

As evidence that the ground-based system was the best bet, at least for the near term, defense officials pointed to a series of successful flight intercept tests. Indeed, the testing had gone nearly as well as could have been expected. The pace of flight trials had picked up, with one run every few months, and the first four under Bush had scored hits.

A fifth test, which had taken place a week before the deployment announcement, flopped when the kill vehicle failed to separate from its booster shortly after launch in the Marshall Islands. A similar mishap had led to the failure of the July 2000 test, although for a different reason. This time, officials blamed a tiny broken metal pin used to connect a computer chip to a circuit board, which in turn was responsible for activating a miniature laser that was supposed to sever the booster's restraining bolts. The pin broke because a small piece of insulation foam intended to cushion it against vibrations had been removed as part of a design change by a subcontractor. Neither Raytheon nor Boeing had known of the change—a fresh example of the kind of gap in management oversight and quality control that senior program officials had tried hard to close. Everyone again vowed to do better.

Taking advantage of the end of the ABM Treaty, defense officials also had begun incorporating a ship-based SPY-1 radar and a land-based radar from the Army's THAAD system to observe targets. The hope was to use these additional sensors eventually to help discriminate between actual warheads and decoys. All in all, counting the three flight tests during Clinton's term, the Pentagon could claim five intercepts in eight attempts. Along with other tests involving shorter-range interceptors—the Navy's Standard Missile-3 and the Army's PAC-3 system—Pentagon officials declared that they had demonstrated the feasibility of the hit-to-kill concept. "That's why we have gained the confidence that we could take this next, modest step," Kadish said at a Pentagon news conference in December.

But the tests had gone only so far. They were still being run under highly scripted conditions far from those of an actual missile attack, using the same limited Vandenberg-to-Kwajalein course and employing surrogates or prototypes for key components of the system, including the booster, the kill vehicle, and the X-band radar. Assessing the results of the tests also had become more challenging for those outside the program as a result of new limits imposed by Kadish on the release of information. Stunned by a confidential study showing that many details about the system were readily available from public sources, Kadish had clamped down in the spring of 2002, putting a stop to the disclosure of such specifics as the types and numbers of decoys in the flight tests.

Opponents of missile defense protested. They accused the Pentagon of stifling constructive debate on the program by denying the kind of standard reports and disclosures used in the past to hold the administration accountable for the billions of dollars it was investing. They dismissed the national security argument, noting that the program was too early in its development for basic test flight information to be useful to an enemy. Some lawmakers also complained about a dearth of detailed cost estimates and timetables for the antimissile programs. Kadish assured them that his office would provide whatever information it had on hand. But such leading critics as Senator Jack Reed, a Rhode Island Democrat who was heading the Armed Services Subcommittee on Strategic Forces, suspected that the administration wanted to reduce oversight to prevent technological flaws and cost overruns from garnering public attention.

The argument over secrecy was part of a larger debate about the extent of oversight that the program should be receiving. The tighter hold on information followed an order from Rumsfeld at the start of 2002 establishing new authority for the missile defense team and granting it an extraordinary exemption from the planning and reporting requirements normally applied to major acquisition programs. At that time, too, Rumsfeld elevated the team to full agency rank, changing its name from the Ballistic Missile Defense Organization to the Missile Defense Agency. "The special nature of missile defense development, operations, and support calls for nonstandard approaches to both acquisition and requirements generation," Rumsfeld asserted.

The wide latitude being given missile defense planners to skirt traditional Pentagon accountability and oversight rules drew warnings from watchdog groups and some members of Congress concerned that Rumsfeld was handing missile defense officials what amounted to a blank check. But Pentagon officials insisted they were not trying to cover up anything. They just could not provide cost estimates or production schedules, they said, because the program's experimental character made any such projections meaningless.

In congressional testimony and interviews, Kadish promised that the Missile Defense Agency would police itself even more rigorously than before and receive oversight within the Pentagon from a panel of top-level civilians called the Senior Executive Council, consisting of the service secretaries as well as the deputy secretary of defense and the undersecretary for acquisition. But Democrats countered that the new liberties afforded the agency—and the increased silence around it—circumvented the checks and balances designed to give Congress and others sufficient information to form technical and budgetary assessments. While acknowledging the experimental nature of much of the missile defense work, critics argued that the intent of the law was nonetheless to compel the Pentagon to come up with plans and projected milestones—and not seem to be spending money blindly.

As a political reality, Democratic lawmakers recognized that they were powerless to halt Bush's plan or even cut back funding for it. The 2002 elections had left Republicans in control of both houses of Congress again, ensuring the necessary legislative support for the administration's missile defense drive. North Korea was looking as threatening as ever, having admitted in late 2002 to a secret program to enrich uranium, then moving in early 2003 to expel international monitors, withdraw from the Nuclear Non-Proliferation Treaty, and restart the shuttered nuclear reactor at Yongbyon. By early 2003, the leading Democratic critic of missile defense, Carl Levin of Michigan, conceded that the debate about whether to deploy a nationwide antimissile system was over. Bush had decided the matter.

But that did not mean there was no longer anything to debate. Instead, the focus of controversy shifted. It was no longer an ideological battle over arms control. The arguments had become centered on such practical considerations as the proposed system's likely performance, the adequacy of the test program, and the projected long-term costs of deployment.

Reflecting this shift, the strategy of Democratic opponents was also adjusted. Instead of trying to force a vote on the scope and expense of the program—and suffer likely defeat—they concentrated on writing language into defense authorization bills requiring the Pentagon to provide cost estimates, measurable performance criteria, and operational test schedules. They also sought to mandate operational assessments from the Pentagon's chief test evaluator. The outspoken Philip Coyle no longer held that job, but his successor, Thomas Christie, had a similar reputation for honest, informed judgments. Christie had no desire to become a public critic, as Coyle had, and he assured Rumsfeld when appointed that he would take any concerns to the secretary directly. But Christie also had an obligation to Congress, which had confirmed him, to judge the operational readiness of major weapons systems before they were deployed. In the case of missile defense, Christie and his office were being put in the unusual position of being asked to assess the operational readiness of a system that would still be in its infancy at the time of its scheduled initial fielding. Just what criteria would be used to judge its readiness was not yet clear. The administration had discarded the traditional practice of stipulating specific requirements. What capabilities should be considered adequate, and against what threat? The answers were still to come.

One factor complicating any assessment would be the relatively low number of flight tests by the time of deployment. For even as the administration moved ahead with its deployment plan, it cut back on flight trials. Beset by delays in developing the new three-stage boosters, Major

General John Holly, who had succeeded Nance as program manager, recommended in November 2002 holding off on more intercept attempts. Running more flight tests using slower, two-stage surrogate boosters struck Holly and other program officials as unlikely to yield enough new information to justify the expense. So Kadish declared that 2003 would be the "year of the booster" and agreed to cancel three intercept tests that had been planned, hoping to be able to resume flight trials with a new booster in the autumn of 2003. It was a frustrating setback, since only a year earlier Kadish had told Congress and others of his desire to get the program on an aggressive schedule of roughly one test every quarter. He had actually achieved that goal in 2002, only to have to draw back again. Even if testing did restart late in 2003, there would be time enough for only two or three more intercept attempts before the scheduled deployment date.

The ambitious scope and fast pace of the administration's approach reflected Bush's determined embrace of missile defense. Indeed, given how frustratingly elusive the dream of a nationwide antimissile system had proven for his predecessors, the confidence and boldness that Bush and his advisers continued to bring to the endeavor were all the more impressive—or foolhardy, depending on which side of the argument one was on. At a minimum, the fresh investment of billions of dollars—not to mention a major chunk of presidential political capital—had raised the stakes: either Bush and his fellow missile defense enthusiasts would win big, or they would fail spectacularly.

Bill Clinton had sought to limit the cost of his missile defense venture by proposing to keep it imbedded in a revised ABM Treaty. Militarily, his strategy was intended to be just enough to deal with an emerging Third World threat without upsetting the strategic balance with Russia. Politically, it was an effort to neutralize whatever partisan advantage the Republicans might derive from the missile defense issue, without incurring too much expense or inciting too much international opposition. Some in the Bush administration, including the president himself, were convinced that Clinton structured his approach in a way that was calculated to fail—a kind of poison pill for national missile defense. The proposed land-based system for midcourse intercepts would never have provided an adequate defense by itself, they said. The question of whether to go forward with it was constrained by too many criteria. And U.S. diplomacy was conducted in a way destined to array Russia, China, and the NATO governments against the program.

This was too cynical a view. Although most members of Clinton's national security team never had their hearts in the program, they genuinely thought for a time that they had a shot at a complex three-part deal—what Sandy Berger labeled the trifecta—involving Russian agreement to amend the ABM Treaty, U.S. deployment of a modest missile defense system, and a U.S.-Russian pact on deeper cuts in offensive nuclear weapons. The political backing for such an approach was manifested in the 1999 Missile Defense Act's endorsement of a limited system if built in the context of new arms reduction talks with Russia. The overwhelming vote for that measure marked only the third time, following legislation in 1969 and 1991, that Democratic lawmakers had gone on record with Republicans favoring a national antimissile system.

But as in the past, the fragile consensus quickly unraveled in the face of fresh doubts about the technical feasibility and costs of the envisioned system, as well as second thoughts about the net impact of a deployment on foreign relations and global security. For the Clinton administration, missile defense turned into a strenuous exercise in damage control and deployment avoidance.

If Clinton's handling of the issue did anything to advance the cause of missile defense, it was at least to alert the rest of the world that the United States was again serious about the issue. Nearly a decade after the demise of the Soviet Union, Washington was coming around to questioning a cold war nuclear deterrence doctrine based on the notion that the only defense was a strong, stable offense. By stirring the rest of the world to start focusing on new missile threats and old treaties, Clinton helped set the stage for Bush, even if the two men had sharply different views about how to proceed.

As this book goes to press in the summer of 2003, the future of missile defense remains difficult to project. Bush's plan to test a broad range of possible systems, while deploying a rudimentary one in the near term, offers greater promise than Clinton's single-solution architecture of eventually finding an optimum design—if one is to be found. But the cost of all this deploying and experimenting is sure to be substantial, jeopardizing the financing of other military projects more popular with senior military commanders. Moreover, concerns about the rigor and realism of the testing continue to hound the program, with critics charging that the decoys and countermeasures used in the tests are overly simplified. It is not inconceivable that opposition could emerge if further testing leads to a string of failures and ever-rising costs.

The history of missile defense is littered with exaggerated claims of progress, coupled with assurances that a workable system requires just a little more testing and engineering. While some significant advances have occurred in recent years, particularly in hit-to-kill technology, going from demonstrations to a combat-ready operational system has proven much tougher than expected, even for the land-based, midcourse intercept approach on which the Pentagon is concentrating.

Opponents see the competition for resources in the Pentagon as possibly working in their favor: the military chiefs' fight to preserve other programs could rein in missile defense. Some also predict that a combination of budget pressures and technical limitations will cause the sites in Alaska and California to suffer the same fate as the Safeguard antimissile system fielded by President Nixon but later abandoned. And even though the rest of the world appears to have grudgingly accepted U.S. abandonment of the ABM Treaty, some still fear that once an American missile defense system is deployed, a new arms race will be sparked. Russia could halt or reverse cuts in its nuclear forces, and China could respond by adding to plans to build up its offensive weapons, which in turn could spur India and then Pakistan to follow suit.

Still, more favorable conditions have been set for the realization of a missile defense system than have ever existed in the long argumentative history of whether to attempt such a feat. Treaty constraints that advocates had blamed for blocking the pursuit of optimal technologies have been removed. Spending has been increased, surpassing $9 billion in

2004. No country is known to have launched or expanded a weapons buildup in response to the U.S. initiative. And the Bush administration has lowered the bar for what will be considered acceptable to deploy, discarding earlier requirements for a near-perfect solution.

Under the circumstances, program officials figure they either will have to manage to build some kind of antimissile system or acknowledge that missile defense simply is not meant to be. "If we can't succeed this time," Kadish said as the administration embarked on its quest, "we might as well hang it up."

Washington, D.C.
August 2003

ABM Treaty: Anti-Ballistic Missile Treaty, signed in 1972 by Richard Nixon and Leonid Brezhnev. It prohibits the development, testing, and deployment of sea-based, air-based, space-based, and mobile land-based systems that would provide a complete territorial missile defense.

Aegis: A computerized combat system used on U.S. Navy ships capable of simultaneous operation against surface, underwater, and air threats.

Air-based system: An antimissile weapon fired from an aircraft.

Airborne laser: A proposed system using a modified Boeing 747–400 jumbo jet equipped with a chemical laser designed to shoot down short- and medium-range missiles.

AUL: Accidental and unauthorized launch.

Ballistic missile: A missile that after an initial burst of power coasts toward its target without any significant lift from its surface to alter the course of flight.

BMC³: The battle management, command, control, and communications network that processes data from satellites and tracking radars into a targeting plan for the missile interceptor.

BMDO: Ballistic Missile Defense Organization, established in 1993 to succeed SDIO.

Booster: The rocket that carries a payload into space.

Boost phase: The first phase of a ballistic missile's trajectory as the missile rises through the atmosphere with its booster still burning. The typical total burn time for an ICBM is three to six minutes.

Brilliant Pebbles: A concept for a space-based antimissile system composed of thousands of tiny, independently targeted interceptors; under study in the late 1980s and early 1990s.

C-1, C-2, and C-3: Capability ratings devised by BMDO to describe three stages of system development. C-1, the initial system architecture, would cope with no more than five "simple" warheads—that is, warheads with decoys no more advanced than balloons. C-2 would deal with a maximum of five "complex" warheads that involve complicated decoys, radar jammers, and other deception measures. The C-3 system would defeat up to twenty warheads loaded with complex deception devices.

Countermeasures: Measures such as decoys, jammers, and chaff taken by an attacker to deceive a missile defense system.

Defense Support Program: First launched in 1971, a constellation of early-warning satellites, now numbering four, that scan for the intense infrared light of missiles taking off or nuclear detonations.

Demarcation protocols: Unratified clarifications to the ABM Treaty—announced by U.S. and Russian officials in 1997 after four years of negotiation—that attempt to define the line under the ABM Treaty between "theater" and "strategic" defenses.

Discrimination: The ability to distinguish real warheads from decoys.

DOT&E: Pentagon's Director of Operational Test and Evaluation.

Early-warning radar: A surveillance radar that provides detection and tracking of approaching missiles or aircraft.

ERIS: Exoatmospheric Reentry-Vehicle Interceptor Subsystem, an experimental antimissile weapon first tested in 1991.

Exoatmospheric: Sixty or more miles above the earth's surface.

GPALS: Global Protection Against Limited Strikes, a missile defense architecture proposed by Bush in 1991.

Helsinki Accord: A 1997 agreement between Bill Clinton and Boris Yeltsin affirming a readiness in future negotiations on START III to accept reductions in nuclear arsenals to 2,000 to 2,500 warheads.

Hit-to-kill: A missile defense approach in which an interceptor rams the target, destroying it by force of impact rather than by explosives.

HOE: Homing Overlay Experiment, a prototype antimissile weapon first tested in 1982.

ICBM: Intercontinental ballistic missile, a land-based missile with a range of more than 5,500 kilometers.

IFICS: In-flight Interceptor Communications System, a ground-based facility for sending target information to the interceptor after launch.

Infrared: Invisible light waves composed of wavelengths slightly longer than those forming the color red. Every type of object radiates a unique infrared signature, which can be identified by measuring the curve of the received energy.

Interceptor: A kill vehicle joined with a booster that together are launched against an offensive missile.

Kill vehicle: A self-contained package of sensors, thrusters, and navigation gear that, once separated from its booster, can identify a target warhead and maneuver into a collision with it. Sometimes referred to as an EKV, or exoatmospheric kill vehicle.

Land-based system: An antimissile system that uses locations on land to shoot interceptors at incoming warheads.

Layered defense: A network of overlapping antimissile systems providing more than one opportunity to engage an incoming missile.

MAD: Mutual assured destruction.

Midcourse phase: Coming between the boost and terminal phases, it is the longest period of flight and covers the time during which a missile rises to its apogee and begins to arc back toward earth.

Multilateralization agreement: An unratified U.S.-Russian memorandum of understanding, announced in 1997, stipulating that the Soviet Union has been succeeded by four countries—Russia, Belarus, Kazakhstan, and Ukraine—as signatories of the ABM Treaty.

NIE: National Intelligence Estimate.

NMD: National Missile Defense, a Clinton administration program intended to provide nationwide protection against ballistic missile attack.

Nodong: A North Korean missile, first tested in 1993, with a range of 1,000 to 1,300 kilometers.

NORAD: North American Aerospace Defense Command.

NSC: National Security Council.

PA&E: Pentagon's Office of Program Analysis and Evaluation.

Rogue nation: Originally applied in the 1980s to countries whose internal policies were oppressive, the term was being used by the mid-1990s for such countries as North Korea, Iran, and Iraq to denote a refusal to abide by international norms and a potential for irrational action.

SALT I: Emerging from the Strategic Arms Limitation Talks, a set of agreements that included the ABM Treaty and an interim accord freezing the number of U.S. and Soviet launchers. Both agreements were signed by Nixon and Brezhnev in 1972.

SALT II: An accord signed by Jimmy Carter and Brezhnev in 1979 that set an initial overall ceiling of 2,400 ICBM launchers, submarine-based missiles, heavy bombers, and air-to-surface missiles, with the ceiling to be lowered in later stages and with subceilings for various forms of launchers and their warheads. SALT II was not ratified by the U.S. Senate, but both countries abided by its provisions until 1987, when the United States exceeded its limits. The treaty had been due to expire in December 1985.

SBIRS-High: The six-satellite Space-Based Infrared System-High system designed to replace the thirty-year-old DSP system by 2006. It promises faster scanning rates and a new "stare" capability, and thus more accurate data for launching antimissile interceptors. Still greater tracking features, plus the ability to discriminate warheads from decoys, is to come with SBIRS-Low, a constellation of two dozen or more satellites planned for 2011.

Scud: A missile first designed by the Soviet Union with a range of 300 to 600 kilometers.

SDIO: Strategic Defense Initiative Organization.

Sea-based system: An antimissile system that operates from floating platforms, whether Navy ships or specially outfitted barges.

SIOP: Single Integrated Operational Plan, the nuclear targeting list.

Space-based laser: An experimental project aimed at equipping satellites with lasers for zapping enemy missiles.

START I: The first Strategic Arms Reduction Treaty, in negotiation through most of the Reagan administration and signed in July 1990 by George Bush and Mikhail Gorbachev. It reduced the number of Soviet long-range nuclear warheads from about 11,000 to 6,600 and U.S. warheads from about 12,600 to 8,500.

START II: The second Strategic Arms Reduction Treaty, signed by George Bush and Boris Yeltsin in January 1993 but still not in force. It would reduce U.S. and Russian strategic arsenals to 3,000 to 3,500 warheads.

Star Wars: The nickname given by critics to Reagan's notion of a system for shielding the United States from massive missile attack.

STRATCOM: U.S. Strategic Command.

Taepodong I: A North Korean missile, first tested in 1998, with a range of 2,000 kilometers in its two-stage configuration and up to 5,500 kilometers in a three-stage configuration. Its follow-on version, the Taepodong II, has not been flight-tested, but U.S. intelligence analysts say its design should enable it to deliver a nuclear warhead more than 6,000 kilometers.

Terminal phase: The final phase of flight after a missile reenters Earth's atmosphere.

THAAD: Theater High-Altitude Area Defense, a U.S. intermediate-range interceptor developed by the Army.

Three-plus-three: The Clinton administration's proposal in 1996 for three years of additional research on national missile defense to be followed, if warranted, by three years of building an initial system.

TMD: Theater missile defense, a system to guard against short- and medium-range missiles.

UCS: Union of Concerned Scientists.

Welch reports: A series of reviews of the missile defense effort in 1998, 1999, and 2000 by a panel headed by Larry Welch, a retired Air Force chief of staff and now president of the Institute for Defense Analyses.

X-band radar: A high-frequency radar capable of precision tracking and of discriminating between warheads and benign objects.

Notes

This book is based on more than three hundred interviews, as well as notes taken in meetings, memoranda of conversations, talking points, reporting cables, briefing charts, calendars, and other primary documents. In many cases, those interviewed did not want to be cited as sources. But there were numerous opportunities to check, recheck, and triangulate various accounts with different participants. Only when someone spoke on-the-record is the interview noted below. Any direct quotations that appear in the text come only from sources with firsthand, immediate knowledge of what was said. Where material was available in public printed sources to augment confidential interviews, the documented sources are listed.

INTRODUCTION

xxi **Colorado's Cheyenne Mountain:** Tour of the NORAD facility, June 12, 2001.

xxii **Ronald Reagan certainly did:** Martin Anderson, *Revolution* (New York: Harcourt Brace Jovanovich, 1988), pp. 80–86.

xxiii **Clinton himself had been deeply skeptical:** Clinton's written answers to questions from the author, June 5, 2001.

xxiv **the 1972 Anti-Ballistic Missile (ABM) Treaty:** Formally named the Treaty Between the United States of America and the Union of Soviet Socialist Republics on the Limitation of Anti-Ballistic Missile Systems.

xxv **more than $100 billion over fifty years:** Stephen I. Schwartz, ed., *Atomic Audit: The Costs and Consequences of U.S. Nuclear Weapons Since 1940* (Washington, D.C.: Brookings Institution Press, 1998), p. 270.

xxvi Opinion surveys routinely show: Opinion has changed little over the decades. For polls in the 1960s, see Ernest Yanarella, *The Missile Defense Controversy: Strategy, Technology, and Politics, 1955–1972* (Lexington, Ky.: The University Press of Kentucky, 1977), p. 142, 218n; for polls in the 1980s, see Peter Grier, "U.S. Public Opinion Generally Favors 'Star Wars,'" *Christian Science Monitor*, November 21, 1985; "Public Opinion and SDI," *Strategic Defense Initiative Fact Book* (Arlington, Virginia: American Defense Preparedness Association, 1986); for 1995 poll by Luntz Research Company, see "Public Opinion on U.S. Foreign Policy and Defense Issues," *Issues '96: The Candidate's Briefing Book* (Washington, D.C.: The Heritage Foundation, 1996); and see July 1998 poll conducted for the Center for Security Policy, the Claremont Institute, the Free Congress Foundation, and the Heritage Foundation.

xxvi the gruesome consequences: Frank Ervin et al., *The New England Journal of Medicine*, May 31, 1962, pp. 1127–1155.

xxvii at least twenty-five countries: Office of the Secretary of Defense, *Proliferation: Threat and Response* (Washington, D.C.: U.S. Department of Defense, January 2001).

xxviii no so-called rogue state: The term "rogue" was applied by the Clinton administration to a handful of countries, including North Korea, Iran, Iraq, and Libya. Although flawed, it is used in this book as a window onto the way policymakers and others approached the problem of emerging missile states.

xxviii "about half a dozen things": Interview with General (retired) Larry Welch, December 9, 1999.

xxix merely an engineering one: Frederick Seitz, "Missile Defense Isn't Rocket Science" (op-ed article), *Wall Street Journal*, July 7, 2000.

xxix only eleven of twenty-eight flight tests: History provided to the author by BMDO.

xxx all the money will be found: James Kitfield, "The Ultimate Bomb Shelter," *National Journal*, July 8, 2000.

CHAPTER 1: BACK TO THE FUTURE

3 striking the United States: Donald R. Baucom, *The Origins of SDI, 1944–1983* (Lawrence: University Press of Kansas, 1992), p. 4.

3 scope of contemporary technology: Benson D. Adams, *Ballistic Missile Defense* (New York: American Elsevier Publishing Co., 1971), p. 17.

3 "a bullet with a bullet": B. Bruce-Briggs, *The Shield of Faith: The Hidden Struggle for Strategic Defense* (New York: Simon and Schuster, 1988), p. 102.

4 by the end of 1962: Fred Kaplan, *The Wizards of Armageddon* (New York: Simon and Schuster, 1983) p. 161.

4 took on a new urgency: Scott McMahon, *Pursuit of the Shield: The U.S. Quest for Limited Ballistic Missile Defense* (Lanham, Md.: University Press of America, 1997), p. 15.

4 in their boost phase: Adams, *Ballistic Missile Defense*, p. 44.

4 in orbit over ICBM sites: Baucom, *Origins of SDI*, pp. 3–7, 16.

5 a new kind of radar: McMahon, *Pursuit of the Shield*, p. 17.

6 McNamara had initially focused: Kaplan, *Wizards of Armageddon*, pp. 315–17.

6 in the 1968 elections: Morton H. Halperin, *Bureaucratic Politics and Foreign Policy* (Washington, D.C.: Brookings Institution, 1974) pp. 297–310.

7 "not your defensive missiles": Kaplan, *Wizards of Armageddon*, p. 346.

8 strategic, technical, and diplomatic considerations: Adams, *Ballistic Missile Defense,* p. 245.

8 into an arms control deal: McMahon, *Pursuit of the Shield*, p. 46.

8 thus intensify the arms race: Baucom, *Origins of SDI*, p. 22.

9 warheads overwhelming the system: Richard L. Garwin and Hans A. Bethe, "Anti-Ballistic Missile Systems," *Scientific American*, March 1968, pp. 21–23.

9 This was a new phenomenon: Kaplan, *Wizards of Armageddon*, p. 349.

9 saving most of a population: Freeman Dyson, "A Case For Missile Defense," *Bulletin of the Atomic Scientists*, April 1969, pp. 31–33.

9 ready to argue it would: McMahon, *Pursuit of the Shield*, p. 42.

10 "ever developed by the United States": Adams, *Ballistic Missile Defense,* p. 1.

11 coining the enduring acronym MAD: Bruce-Briggs, *Shield of Faith*, pp. 372–73.

12 the program absorbed $5.5 billion: Schwartz, *Atomic Audit*, p. 71–72. Figure is in then-year dollars.

13 the chances of intercepting the target warhead: General Accounting Office, "Ballistic Missile Defense: Records Indicate Deception Program Did Not Affect 1984 Test Results," NSIAD–94–219 (July 1994).

13 the promise of directed-energy weapons: Baucom, *Origins of SDI*, p. 113.

13 none was on the horizon: Frances FitzGerald, *Way Out There in the Blue* (New York: Simon & Schuster, 2000), p. 120.

14 one of a number of options: Janne E. Nolan, *Guardians of the Arsenal* (New York: Basic Books, 1989), p. 164.

15 one count found about one thousand: Robert M. Lawrence, *Strategic*

Defense Initiative: Bibliography and Research Guide (Boulder, Colo.: Westview Press, 1987).

15 **if they put their minds to it:** Frances FitzGerald, *Way Out There in the Blue*, p. 258.

16 **explored the range of technologies:** Ibid., p. 246. Also, for summary of the APS findings, see "Report to the APS of the Study Group on Science and Technology of Directed Energy Weapons," *Physics Today*, May 1987, pp. S–3 to S–16.

16 **split with the administration:** Janne E. Nolan, *Guardians of the Arsenal*, p. 222.

17 **pegged its cost at $146 billion:** David B.H. Denoon, *Ballistic Missile Defense in the Post–Cold War Era* (Boulder, Colo.: Westview Press, 1995), 108.

18 **of up to 200 warheads:** Denoon, *Ballistic Missile Defense in the Post–Cold War Era*, p. 132.

18 **two Republican senators:** McMahon, *Pursuit of the Shield*, p. 98.

19 **moved quickly to ally:** Ibid., p. 99.

20 **Gorbachev provided a new impetus:** Ibid., p. 110.

21 **an attempt at compromise:** Ibid., p. 115.

22 **the Gulf War had faded:** Ibid., p. 117.

23 **when it had not:** R. Jeffrey Smith, "SDI Success Said to Be Overstated," *Washington Post*, September 16, 1992.

23 **"end of the Star Wars era":** R. Jeffrey Smith, "Threat Gone, 'Star Wars' Is Banished," *Washington Post*, May 14, 1993.

25 **make the most of it:** Joseph Cirincione, "Why the Right Lost the Missile Defense Debate," *Foreign Policy* (Spring 1997).

25 **assumed the United States already had a defense:** Bradley Graham, "Missile Defense Failing To Launch As Voting Issue," *Washington Post*, July 28, 1996.

25 **appeal for a strong military:** James Kitfield, "Going Ballistic," *National Journal*, June 6, 1996.

27 **than on any other defense issue:** Cirincione, "Why the Right. . . ."

28 **"rush to failure":** Bradley Graham, "Panel Fires at Antimissile Programs," *Washington Post*, March 22, 1998.

29 **"a great failure on my part":** Interview with General (retired) Larry Welch, December 9, 1999.

CHAPTER 2: THE ROGUES ARE COMING

32 **a National Intelligence Estimate:** "Emerging Missile Threats to North America During the Next 15 Years," PS/NIE 95–19, November 1995.

34 **a dramatic display of congressional ire:** Kitfield, "Going Ballistic."

34 **acknowledged in congressional testimony:** Hearing of the House Committee on National Security, February 28, 1996.

34 **rushed and incomplete:** Independent Panel Review, December 1996.

38 **"because they had not been exposed":** Interview with Barry Blechman, March 3, 2000.

38 **"we're not going to consider it":** Interview with William Graham, April 13, 2000.

42 **"everyone sort of looked":** Interview with Stephen Cambone, March 13, 2000.

44 **released its findings and recommendations:** *Report of the Commission to Assess the Ballistic Missile Threat to the United States: Executive Summary* (Washington, D.C.: U.S. Government Printing Office, July 15, 1998).

44 **Rumsfeld attributed his group's:** Bradley Graham, "Iran, North Korea Missile Gains Spur Warning," *Washington Post*, July 16, 1998.

48 **Gingrich hailed the panel's study:** Ibid.

49 **Brazil, Argentina, Egypt, and South Africa:** Hearing of the Senate's Governmental Affairs subcommittee on International Security, Proliferation and Federal Services, February 9, 2000.

49 **quietly delivered a "side letter":** Unclassified version of letter issued in March 1999.

CHAPTER 3: WAKEUP CALL

52 **U.S. intelligence analysts had characterized:** Steven Lee Myers, "Missile Test by North Korea: Dark Omen for Washington," *New York Times*, September 1, 1998.

52 **North Korean news agency:** Nicholas D. Kristof, "North Koreans Declare They Launched a Satellite," *New York Times*, September 5, 1998.

53–54 **extended more than 4,000 kilometers:** Joseph S. Bermudez, Jr., "A History of Ballistic Missile Development in the DPRK," Occasional Paper 2, Center for Nonproliferation Studies, Monterey Institute of International Studies (November 1999).

57 **"compliance with arms control agreements":** Interview with Robert Walpole, April 18, 2000.

58 **appeared "a bit odd":** Dana Priest, "North Korea May Have Launched Satellite," *Washington Post*, September 5, 1998.

59 **"a friend at the CIA":** Interview with William Graham, April 13, 2000.

61 **General Shelton had written:** Dated August 24, 1998, as reprinted in the *Congressional Record*, 105th Congress, November 9, 1998, p. S10049.

62 "'Ode to Rumsfeld' played backwards": Interview with Richard Haver, October 4, 2000.

63 "a resistance to change": Interview with Lieutenant General (retired) Patrick Hughes, June 7, 2000.

63 issued new assessments: Robert Walpole, speech to the Center for Strategic and International Studies, December 8, 1998.

64 were not really North Korean at all: Robert Schmucker, "Third World Missile Development—A New Assessment Based on UNSCOM Field Experience and Data Evaluation," Twelfth Multinational Conference on Theater Missile Defense, Edinburgh (June 1999).

66 photos of the North Korean launch site: The FAS published the photos and its analysis of the imagery on its web site, www.fas.org.

68 "these are not military capabilities": Interview with William Schneider Jr., May 24, 2000.

CHAPTER 4: DISCONNECTED TIMELINES

75 found its way into the Washington Times: John McCaslin, "Inside the Beltway," Washington Times, March 3, 1998.

77 "focus on the treaty timelines": Interview with Robert Bell, May 12, 2000.

77 "nobody had talked about the timeline": Interview with John Hamre, March 18, 2000.

78 "I didn't get a pushback": Interview with General Joseph Ralston, March 2, 2000.

CHAPTER 5: SHOOTING FOR 2005

83 the other by the General Accounting Office: "Even with Increased Funding, Technical and Schedule Risks Are High," General Accounting Office, GAO/NSIAD-98-153, June 1998.

84 "by the judicial branch": Sheila Foote, "Judge Dismisses Missile Defense Suit," Defense Daily, October 11, 1996.

84 he characterized the schedule: Lieutenant General Lester Lyles, testimony before the Senate Armed Services Committee, March 24, 1998.

84 told a Senate panel: Hearing of the Senate Armed Services Committee, October 2, 1998.

85 gone before Congress that autumn: Bradley Graham, "Senators Scold Military Chiefs," Washington Post, September 30, 1998.

90 **"didn't mean we should start there"**: Interview with Robert Soule, March 31, 2000.

91 **"where we were going with it"**: Interview with Bill Clinton, July 6, 2001.

91 **"networking among these countries"**: Interview with Madeleine Albright, January 3, 2001.

92 **"to take on more challenges"**: Interview with James Steinberg, April 12, 2000.

95 **pledged an additional $112 billion**: Bradley Graham, "Weeks of Give-and-Take Led to Defense Spending Boost," *Washington Post,* January 14, 1999.

99 **which led the front page**: Steven Lee Myers, "U.S. Asking Russia to Ease the Pact on Missile Defense," *New York Times,* January 21, 1999.

99 **administration's position on the treaty**: White House briefing, January 21, 1999.

CHAPTER 6: A POLITICAL TIPPING POINT

103 **"facts about the problem"**: Interview with Mitch Kugler, January 19, 2000.

105 **"Nevada's own senators"**: Carla Anne Robbins, "High-Stakes Debate on Missile Defense Is Here Again," *Wall Street Journal,* August 7, 1998.

109 **"until after the vote"**: Interview with Robert Bell, March 16, 2001.

110 **"brains of the operation over there"**: Interview with Mitch Kugler, March 29, 2000.

114 **"built consistent with its terms"**: Henry Cooper, testimony before the Senate Foreign Relations Committee, September 26, 1996.

117 **"become a higher national priority"**: Henry Kissinger, "India and Pakistan: After the Explosions," *Washington Post,* June 9, 1998.

117 **"a felony against the future"**: Thomas W. Lippman, "Arms Control Foes Faulted by Albright," *Washington Post,* June 11, 1998.

117 **no longer bound by the treaty**: George Miron and Douglas Feith, "Memorandum of Law: Did the ABM Treaty of 1972 Remain in Force After the USSR Ceased to Exist in December 1991 and Did It Become a Treaty Between the United States and the Russian Federation?" prepared for the Center for Security Policy, Washington, D.C., January 22, 1999; and David Rivkin, Jr., Lee Casey, and Darin Bartram, "The Collapse of the Soviet Union and the End of the 1972 Anti-Ballistic Missile Treaty: A Memorandum of Law," prepared for the Heritage Foundation by Hunton & Williams, Washington, D.C., June 15, 1998.

117 **surprised some scholars**: George Bunn quoted in *Inside Missile Defense,*

June 24, 1998; Michael Glennon, "Yes, There Is An ABM Treaty," *Washington Post*, September 9, 2000.

118 **"nuclear threat that Russia poses":** Pat Towell, "Can U.S. Build Missile Shield Without Shredding a Treaty?" *Congressional Quarterly*, December 4, 1999.

118 **"security for the American people":** Senator Jesse Helms, "Amend the ABM Treaty? No, Scrap It," *Wall Street Journal*, January 22, 1999.

119 **"defunct 1972 ABM Treaty":** Thomas W. Lippman, *Madeleine Albright and the New American Diplomacy* (Boulder, Colo.: Westview Press), p. 239.

120 **"would be very opposed to that":** White House news conference, March 5, 1999.

CHAPTER 7: GETTING THE RED OUT

125 **"the North Koreans couldn't hit anyway":** Bell interview, May 12, 2000.

126 **"for getting the Aleutian Islands":** Interview with John Hamre, March 18, 2000.

136 **two books on forgiveness:** Al Kamen, "In the Loop," *Washington Post*, April 11, 1999.

138 **an item about the discussion:** Bill Gertz and Rowan Scarborough, "Inside the Ring," *Washington Times*, July 23, 1999.

148 **"like trying to lasso a train":** Interview with Sandy Berger, March 5, 2001.

148 **"let ourselves be blind-sided":** Interview with Strobe Talbott, September 18, 2000.

CHAPTER 8: AGITATION ABROAD

153 **Senate's refusal to ratify:** Eric Schmitt, "Senate Kills Test Ban Treaty in Crushing Loss for Clinton," *New York Times*, October 14, 1999.

157 **a United Nations resolution:** "Preservation of and Compliance with the Treaty on the Limitation of Anti-Ballistic Missile Systems" UN resolution, December 1, 1999.

158 **hearing talk of new threats:** Camille Grand, "Missile Defense: The View from the Other Side of the Atlantic," *Arms Control Today* (September 2000): 12.

158 **once it got one:** Ibid., p. 13.

159 **the attitude on his side:** Stephen Cambone, Ivo Daalder, Stephen J. Hadley, and Christopher J. Makins, "European Views of National Missile Defense," Atlantic Council of the United States Policy Paper (September 2000), p. 18.

160 **"by domestic political considerations"**: Interview with Ambassador Gebhardt von Moltke, March 16, 2001.

161 **they had long since abandoned**: Cambone et al., "European Views of National Missile Defense," p. 4.

163 **would be downright irresponsible**: Keith Payne, speech to seminar series on missile defense sponsored by the National Defense University Foundation, September 20, 2000.

164 **"important to the United States"**: Walter Slocombe, speech at the Center for Strategic and International Studies, November 5, 1999.

164 **"defense also has a role"**: John Holum, speech to the Conference on International Reactions to U.S. National and Theater Missile Defense Deployments, Stanford University, March 8, 2000.

164 **"not only fatal but futile"**: Walter Slocombe, testimony before the Senate Armed Services Committee, March 23, 2000.

165 **"states of concern"**: Madeleine Albright, *The Diane Rehm Show*, National Public Radio, June 19, 2000.

170 **he would be stepping down**: Celestine Bohlen, "Yeltsin Resigns, Naming Putin as Acting President to Run in March Election," *New York Times*, January 1, 2000.

171 **"their end-game was yet"**: Interview with Strobe Talbott, June 15, 2000.

171 **a protocol that would cover all the changes**: A copy of the protocol was first published by the *Bulletin of the Atomic Scientists* in Russia in April 2000.

173 **"China may be one of the countries"**: Interview with Jon Kyl, March 7, 2001.

173 **on its existing CSS-4 missiles**: "Foreign Missile Developments and the Ballistic Missile Threat to the United States Through 2015," National Intelligence Council, September 1999.

174 **at a rate of fifty missiles a year**: Admiral Dennis Blair, commander of U.S. Forces in the Pacific, cited by Steven Mufson and Thomas Ricks, "Pentagon Won't Back Taiwan Deal," *Washington Post*, April 17, 2000.

175 **"the survivability of their deterrent"**: Interview with Strobe Talbott, February 26, 2001.

175 **"no obligations to them"**: Steinberg interview, April 12, 2000.

176 **"a grand design aimed at China"**: John Holum speech, Stanford University, March 8, 2000.

CHAPTER 9: UP, UP, AND AWAY

179 **Boeing won the LSI contract**: Bradley Graham, "Boeing Wins Big Contract," *Washington Post*, May 1, 1998.

180 "the challenges we had on space shuttle": John Peller, testimony before the Senate Armed Services Committee, February 24, 1999.

180 "we had some quality problems": Interview with John Peller, May 18, 2000.

182 "can't stamp out all the wrong": Interview with Harry Stonecipher, November 9, 2000.

183 "my call in the end": Peller interview, May 18, 2000.

185 "suddenly the issue was resolved": Stonecipher interview, November 9, 2000.

185 "built-in, institutional conflict": Interview with Lieutenant General Ronald Kadish, May 1, 2000.

186 "an awful lot of training to do": Interview with Ron Meyer, May 16, 2000.

188 "on the first flight test": Interview with Major General Willie Nance, Jr., May 4, 2000.192

190 the decoy balloon rather than the warhead: James Glanz, "Antimissile Test Viewed as Flawed by Its Opponents," *New York Times,* January 14, 2000.

192 "might be seen as cheating": Nance interview, May 4, 2000.

192 "a no-win situation": Interview with Keith Englander, October 27, 2000.

193 constituted a stinging critique: The report was reprinted in *Arms Control Today* (November 1999).

194 "accept a high-risk program": Bradley Graham, "Panel Faults Antimissile Program on Many Fronts," *Washington Post,* November 14, 1999.

194 "can be disastrous": Ibid.

195 "you'd rather do things serially": Interview with Lieutenant General Ronald Kadish, December 6, 1999.

CHAPTER 10: MISSED

197 "reached agreement on two": Nance interview, May 4, 2000.

198 "completely realistic testing": Interview with Robert Soule, April 11, 2000.

198 "a virtual repeat of that": Interview with Bill Carpenter, May 15, 2000.

201 "a high probability of failure": Ibid.

202 "the maddening part": Interview with Lieutenant General Ronald Kadish, March 22, 2000.

202 "not good enough in this game": Interview with Major General Willie Nance, Jr., February 4, 2000.

203 **simplistic target scenarios:** Director, Operational Test and Evaluation, FY1999 annual report, February 14, 2000.

203 **personal success and failure:** William J. Broad, "Words of Caution on Missile Defense," *New York Times,* January 16, 2001.

205 **the number of actual flight tests:** "Budgetary and Technical Implications of the Administration's Plan for National Missile Defense," Congressional Budget Office, April 2000, p. 25.

206 **"these guys certainly aren't ready":** Englander interview, October 27, 2000.

207 **"we didn't know that":** Ibid.

208 **up to $60 billion:** "Budgetary Implications of HR3144, The Defend America Act of 1996," Congressional Budget Office, May 15, 1996.

208 **varies by type of system:** David Mosher, "Understanding the Extraordinary Cost Growth of Missile Defense," *Arms Control Today* (December 2000), p. 9.

210 **at least 95 percent effective:** Michael Dornheim, "Missile Defense Design Juggles Complex Factors," *Aviation Week and Space Technology,* February 24, 1997, p. 54.

211 **"from more nuclear-capable states":** William Perry, Defense Department Annual Report to the President and the Congress (March 1996), p. 223.

214 **a confidential memo:** Bradley Graham, "Navy Chief Promotes Missile Defense Role," *Washington Post,* February 28, 2000.

214 **the case for such a system:** Missile Defense Study Team, "Defending America: A Near- and Long-Term Plan to Deploy Missile Defenses" (Washington, D.C.: The Heritage Foundation, 1995).

CHAPTER 11: THE DECOY DILEMMA

222 **overcome the planned antimissile defense system:** Union of Concerned Scientists and the Security Studies Program at the Massachusetts Institute of Technology (UCS/MIT), *Countermeasures* (Cambridge, Mass: Union of Concerned Scientists, April 2000).

223 **new U.S. intelligence assessment:** "Foreign Missile Developments and the Ballistic Missile Threat to the United States through 2015," National Intelligence Council, September 1999.

223 **"to deploy effective countermeasures":** UCS/MIT, *Countermeasures,* p. xxii.

224 **"no defense at all":** Vernon Loeb, "Antimissile System Is Called 'No Defense,'" *Washington Post,* April 12, 2000.

224 reasserted his agency's confidence: Ibid.

224 raised disturbing questions: William J. Broad, "Ex-Employee Says Contractor Faked Results of Missile Tests," *New York Times,* March 7, 2000.

226 laid out his findings: William J. Broad, "Antimissile System's Flaw Was Covered Up, Critic Says," *New York Times,* May 18, 2000.

226 "the subtlety of a Gatling gun": Elaine Sciolino, "Scientist Is Not Subtle in Taking Shots at Missile Shield," *New York Times,* July 10, 2000.

227 moving to classify Postol's letter: William J. Broad, "Pentagon Classifies a Letter Critical of Antimissile Plan," *New York Times,* May 20, 2000.

227 arrived unannounced at Postol's office: David Abel, "Critic Accuses Pentagon of Trying to Silence Him," *Boston Globe,* June 24, 2000.

227 "his conclusions are wrong": William J. Broad, "U.S. Defends Antimissile Plan," *New York Times,* May 26, 2000.

227 revised its target set: William J. Broad, "Pentagon Has Been Rigging Antimissile Tests, Critics Maintain," *New York Times,* June 9, 2000.

228 "edit your language": Elaine Sciolino, "Scientist Is Not Subtle . . . ," July 10, 2000.

228 "no basis in fact": Tony Capaccio, "FBI Clears TRW Inc. of Fraud Charge in Missile Defense Test," Bloomberg.com, May 4, 2001.

229 "like teaching freshmen": Interview with Richard Garwin, February 2, 2001.

229 making the case: Hearing of the Senate Foreign Relations Committee, May 4, 1999.

231 "new interceptors and the radar in Alaska": Garwin interview, February 2, 2001.

232 "its long-range missiles": William J. Broad, "Physicist Group Says Missile Defense Tests Fall 'Far Short,'" *New York Times,* May 11, 2000.

232 any such attempt at defense: William J. Broad, "Nobel Winners Urge Halt to Missile Plan," *New York Times,* July 6, 2000.

233 call for delay: Jane Perlez, "Biden Joins GOP in Call for a Delay in Missile Defense Plan," *New York Times,* March 9, 2000.

233 some Republicans started doing so: Elizabeth Becker and Eric Schmitt, "Delay Sought in Decision on Missile Defense," *New York Times,* January 20, 2000.

233 they suggested an alternative: John Deutch, Harold Brown, and John White, "National Missile Defense: Is There Another Way?" *Foreign Policy* (Summer 2000).

234 "sort of quiet it down": Interview with John Deutch, February 1, 2001.

234 "giving them a way out": Interview with William Perry, April 9, 2001.

235 **defend the system:** Pentagon news conference, June 20, 2000.

239 **decoys and other countermeasures:** DOT&E FY1998 annual report, February 1999.

CHAPTER 12: NUMBERS GAME

249 **Clinton's new guidance:** R. Jeffrey Smith, "Clinton Directive Changes Strategy on Nuclear Arms," *Washington Post,* December 7, 1997.

252 **the SIOP has consisted:** Bruce G. Blair, "Trapped in the Nuclear Math," *New York Times,* June 12, 2000.

258 **to bolster this point:** Eric Schmitt, "Pentagon Feels Pressure to Cut out More Warheads," *New York Times,* May 11, 2000.

259 **would damage national security:** Ibid. See also Bill Gertz, "Joint Chiefs Oppose Russian Plan to Cut 1,000 U.S. Warheads," *Washington Times,* May 11, 2000.

260 **limiting U.S. forces and weapons:** Hearing of the Senate Armed Services Committee, May 23, 2000.

263 **"our nuclear posture and force levels":** Congressional Record, 106th Congress, June 30, 2000, p. S6291.

CHAPTER 13: LOOPHOLES

272 **"says yes when he means no":** Interview with Strobe Talbott, February 3, 2000.

272 **from North Dakota to another location:** Jane Perlez, "Russian Aide Open Doors a Bit to U.S. Bid for Missile Defense," *New York Times,* February 19, 2000.

272 **approval of START II:** Michael Gordon, "Putin Wins Vote in Parliament on Treaty to Cut Nuclear Arms," *New York Times,* April 15, 2000.

276 **more incoming missiles and decoys:** Steven Lee Myers, "Russians Get Briefing on U.S. Defense Plan," *New York Times,* April 29, 2000.

276 **alternative to the administration's plan:** Igor Ivanov, speech at the National Press Club, April 26, 2000.

278 **twenty-five Republican senators:** The letter was dated April 17, 2000.

278 **he wanted no part:** Helen Dewar and John Lancaster, "Helms Vows to Obstruct Arms Pacts," *Washington Post,* April 27, 2000.

279 **contained several phrases:** "Joint Statement by the Presidents of the United States of America and the Russian Federation on Principles of Strategic Stability," June 4, 2000.

279 "put up these umbrellas": Vladimir Putin, interview with Tom Brokaw, June 1, 2000.

280 Putin's plan was aimed: Michael Gordon, "Putin Suggesting Alternative Plan on Missile Defense," *New York Times,* June 3, 2000.280

280 "unethical not to do so": Charles Babington, "U.S. Set to Share Its ABM Research," *Washington Post,* June 1, 2000.

280 outlining his own missile defense vision: George W. Bush, speech at the National Press Club, May 23, 2000.

281 "has not been made": Babington, "U.S. Set to Share Its ABM Research," June 1, 2000.

281 "pushing the edge on this": Interview with Steve Andreasen, April 19, 2001.

283 "either one of these approaches": Vladimir Putin, remarks at a press conference, June 4, 2000.

284 "security of every European country": Alessandra Stanley, "Putin Goes to Rome to Promote Russian Arms Control Alternative," *New York Times,* June 6, 2000.

284 launches around the world: William Drozdiak, "Putin Urges Joint Missile-Warning Center," *Washington Post,* June 17, 2000.

284 use theater antimissile systems: Michael Gordon, "Putin Seeks Allies in Quest to Fight U.S. Missile Plan," *New York Times,* June 11, 2000.

284 announced plans to visit North Korea: Michael Gordon, "Putin to Visit North Korea," *New York Times,* June 9, 2000.

284 "activist Russian foreign policy": Gordon, "Putin Seeks Allies . . . ," June 11, 2000.

285 strikes by long-range rockets: William Drozdiak, "U.S. Rejects Russian Plan for Joint Missile Defense," *Washington Post,* June 10, 2000.

285 getting the interceptor to distinguish: Press conference at NATO headquarters, Brussels, Belgium, June 8, 2000.

286 for a new S-500: Kerry Gildea, "BMDO Reviews Proposals to Work with Russia on NMD," *Defense Daily,* June 23, 2000.

CHAPTER 14: MISSED AGAIN

287 "Missed Again": A version of this chapter appeared in the *Washington Post Sunday Magazine,* December 10, 2000. It is drawn largely from a reporting trip to Kwajalein at the time of the July 2000 launch.

287 band of Micronesian coral: Howard W. French, "U.S. Missile Defense Plans Renew Atoll's Role," *New York Times,* June 11, 2001.

294 **cost about $90 million:** Based on a tally of costs supplied to the author by BMDO.

294 **sent a memo:** Philip Coyle, memo to Jacques Gansler, June 26, 2000.

303 **"Never mind the calculations":** Interview with Hans Mark, September 19, 2000.

303 **"running away from the program":** Interview with Strobe Talbott, July 18, 2000.

CHAPTER 15: COHEN'S LAST STAND

307 **"and other strategic issues":** Interview with Bob Tyrer, February 2, 2000.

310 **a potential $42 million bonus:** John Donnelly and Ann Roosevelt, "Boeing Lost About $20 Million for Missile Defense Woes," *Defense Week,* August 14, 2000.

310 **"management of some elements":** Interview with Lieutenant General Ronald Kadish, January 16, 2000.

311 **"as it can get it":** Interview with Colonel Mary Kaura, January 2, 2001.

316 **"it's not impossible":** Interview with Jacques Gansler, August 30, 2000.

317 **"the schedule will be self-adjusting":** Executive summary, National Missile Defense Independent Review Team, June 13, 2000.

317 **between the two countries:** Howard W. French, "Seoul Leader Arrives in North Korea for Long-Awaited Meeting," *New York Times,* June 13, 2000.

317 **"into a domestic cat":** William Cohen, testimony before the Senate Armed Services Committee, July 25, 2000.

319 **from the U.S. effort:** Steven Lee Myers, "U.S. Missile Plan Could Reportedly Provoke China," *New York Times,* August 10, 2000.

319 **"in the way of trade-offs":** Interview with Sandy Berger, December 6, 2000.

319 **on relations with Russia:** William Drozdiak, "Worries Mount In Europe Over U.S. Missile Defense," *Washington Post,* May 19, 2000.

319 **"tension and world destabilization":** "U.S. Delegation to brief Greenland on antimissile scheme," Agence France Presse, August 21, 2000.

320 **"the threats it perceives":** Tom Buerkle, "U.K. Panel Questions U.S. Missile Shield Plans," *International Herald Tribune,* August 3, 2000.

320 **"the most grave adverse consequences":** Ted Plafker, "China, Russia Unify Against U.S. Missile Shield," *Washington Post,* July 19, 2000.

320 **satellites into space:** Michael Gordon, "North Korea Reported Open to Halting Missile Program," *New York Times,* July 20, 2000.

321 **theater missile defense systems:** "Joint Statement on Cooperation on Strategic Stability," July 21, 2000.

321 **not the only leader:** Marc Lacey, "Putin Bends Clinton's Ear Hoping to Halt Missile Shield," *New York Times,* July 22, 2000.

322 **"confusion was pretty widespread":** Interview with Robert Walpole, June 4, 2001.

323 **moved away from the idea:** "Senate Democrats Urge NMD Delay," *Aerospace Daily,* July 27, 2000.

324 **keeping disagreements to themselves:** Steven Lee Myers, "Russian Resistance Key in Decision to Delay Missile Shield," *New York Times,* September 3, 2000.

326 **visibly tired after an eleven-hour flight:** Ibid.

327 **"to preserve the five-year option":** Clinton interview, July 6, 2001.

327 **"making us less secure":** Ibid.

328 **"proceed with the Shemya radar":** Ibid.

328 **"pretty orderly" by contrast:** Bill Clinton, written answers to questions from author, June 5, 2001.

328 **caught many journalists off-guard:** Jay Branegan, "It Only Looks Like He's Not Doing Anything," *Time,* September 11, 2000.

331 **favorable among Democratic lawmakers:** Eric Schmitt, "President Decides to Put off Work on Missile Shield," *New York Times,* September 2, 2000.

331 **handling of the missile defense program:** Ibid.

332 **applauded the president's action:** Patrick E. Tyler, "European Leaders Praise U.S. Antimissile Decision," *New York Times,* September 2, 2000.

332 **cooperation on theater missile defenses:** "Statement on the Strategic Stability Cooperation Initiative Between the United States of America and the Russian Federation," September 6, 2000.

333 **that interceptors might encounter:** The report was finally sent to Congress in June 2001. See James Dao, "Pentagon Study Casts Doubt on Missile Defense Schedule," *New York Times,* June 25, 2000.

334 **"despite two high-profile test failures":** Ronald Kadish, testimony before the National Security, Veterans' Affairs, and International Relations Subcommittee of the House Government Reform Committee, September 8, 2000.

335 **"ended up wanting to build":** Interview with Lieutenant General Ronald Kadish, November 15, 2000.

335 **"it wasn't there":** Interview with Major General Willie Nance, Jr., May 22, 2001.

336 **invitation to Clinton to visit Pyongyang:** Marc Lacey, "Clinton Trip to North Korea Is Possible This Year," *New York Times,* October 13, 2000.

338 **mood at the Pyongyang meetings:** Michael Gordon, "How Politics Sank Accord on Missiles with North Korea," *New York Times,* March 6, 2001.

338 **several important concessions:** Ibid.

340 **"what they ought to do":** Rice interview, May 25, 2001.

CHAPTER 16: ONCE MORE, WITH FEELING

342 **"and share them with people":** Interview with George Bush, July 2, 2001.

343 **"that sense of disbelief":** Rice interview, May 25, 2001.

344 **military readiness and transformation:** Frank Bruni, "Bush Vows Money and Support for Military," *New York Times,* September 24, 1999.

344 **"even praise him":** Jim Hoagland, "Some Sure Answers from Bush," *Washington Post,* December 21, 1999.

345 **a symbol of Birmingham:** John Lancaster and Terry Neal, "Heavyweight Vulcans Help Bush Forge a Foreign Policy," *Washington Post,* November 19, 1999.

347 **"fresh ways about it":** Interview with Richard Perle, June 5, 2001.

347 **"where we sit today":** Bush interview, July 2, 2001.

348 **cold war arms control treaties:** Alison Mitchell, "Bush Says U.S. Should Reduce Nuclear Arms," *New York Times,* May 24, 2000.

349 **impractical or reckless:** Ceci Connolly, "Gore Warns of Perils in Bush Arms Plan," *Washington Post,* May 28, 2000.

349 **the cold war mold:** "Mr. Bush Talks Arms," *Washington Post,* May 25, 2000; "Bush's Breakthrough," *Wall Street Journal,* May 25, 2000; "Nuclear Swords and Shields," *New York Times,* May 26, 2000.

350 **a system that worked:** ABC-TV, *This Week,* July 16, 2000.

351 **against emerging missile threats:** Donald Rumsfeld, testimony before the Senate Arms Services Committee, January 11, 2001.

351 **withdrawing from the treaty:** Roberto Suro, "Missile Defense Is Still Just a Pie in the Sky," *Washington Post,* February 12, 2001.

352 **walk away from the treaty:** Colin Powell, testimony before the Senate Armed Services Committee, January 17, 2001.

352 **"seen as decoupling":** Rice interview, May 25, 2001.

353 **"examining those elements":** Steven Mufson, "Bush to Pick Up Clinton Talks on North Korean Missiles," *Washington Post,* March 7, 2001.

353 **missile talks anytime soon:** David E. Sanger, "Bush Tells Seoul Talks with North Won't Resume Now," *New York Times,* March 8, 2001.

353 **"too far forward on my skis":** Colin Powell, interview with CNN, May 14, 2001.

354 "conventional military posture": Jane Perlez, "U.S. Will Restart Wide Negotiations with North Korea," *New York Times,* June 7, 2001.

355 prevailed during the cold war: "Rationale and Requirements for U.S. Nuclear Forces and Arms Control," *National Institute for Public Policy* (January 2001).

355 "win the intellectual argument": Interview with Condoleezza Rice, May 25, 2001.

355 than a moral imperative: Michael R. Gordon, "U.S. Tries Defusing Allies' Opposition to Missile Defense," *New York Times,* February 4, 2001.

356 "this made sense": Powell, testimony before the Senate Armed Services Committee, January 17, 2001.

356 than they had been in 1983: Jackson Diehl, "Reaganism II," *Washington Post,* February 5, 2001.

356 the administration of Bush's father: Stephen Hadley, "Global Protection System: Concept and Progress," *Comparative Strategy* (January–March 1993): 3–6.

357 defense system for Europe: Peter Baker and Susan B. Glasser, "Russia Details Anti-Missile Alternative," *Washington Post,* February 21, 2001.

361 "a more rational debate": Lord George Robertson, interview with Defense Writers Group, March 8, 2001.

362 at least a decade away: U.S. Navy and Ballistic Missile Defense Organization, "Naval National Missile Defense: A Potential Expansion of the Land-Based NMD Architecture to Extend Protection," executive summary report, December 8, 2000.

363 "as people had postulated": Interview with Lieutenant General Ronald Kadish, May 24, 2001.

365 trap them in their own rhetoric: James Dao, "Pentagon To Seek Money For Testing Missile Defense," *New York Times*, July 10, 2001.

365 "fielding of an operational system": "A Missile Defense Test For Congress," *New York Times,* July 12, 2001.

365 "devise the system structure": Kadish interview, May 24, 2001.

367 did not fully work: Thomas L. Friedman, "Who's Crazy Here?" *New York Times,* May 15, 2001.

368 "that would abrogate the ABM Treaty": Alison Mitchell, "Senate Democrats Square Off with Bush Over Missile Plan," *New York Times*, May 3, 2001.

369 "the modern architecture of international security": Frank Bruni, "Putin Urges Bush Not To Act Alone On Missile Shield," *New York Times,* June 17, 2001.

369 **Putin gathered with a group:** Patrick Tyler, "Putin Says Russia Would Counter U.S. Shield," *New York Times,* June 19, 2001.

369 **"talk me to death":** Bush interview, July 2, 2001.

370 **"turn it into tomorrow's paranoia":** Leon Fuerth, interviewed on the Charlie Rose show, Public Broadcasting Station, May 2, 2001.

370 **"it's an evolving relationship":** Bush interview, July 2, 2001.

370 **"with India and then with Pakistan":** Clinton interview, July 6, 2001.

371 **"withdraw from it or replace it":** Testimony before the Senate Armed Services Committee, July 12, 2001.

AFTERWORD

377 **blink at certain specific tests:** Bill Keller, "The World According to Powell," *New York Times Magazine,* November 25, 2001.

379 **"defended with increasing passion":** James Schlesinger, "The ABM Treaty's Quiet Demise," *Washington Post,* February 20, 2002.

381 **typical of previous nuclear deals:** Peter Slevin, "Ambitious Nuclear Arms Pact Faces a Senate Examination," *Washington Post,* August 7, 2002.

383 **event of a North Korean threat:** Paul Wolfowitz, "Beyond the ABM Treaty," *Wall Street Journal,* June 24, 2002.

385 **"when it's ready":** Bradley Graham, "Missile Defense Choices Sought," *Washington Post,* September 3, 2002.

387 **"because we overpromised":** Interview with Keith Englander, May 20, 2003.

389 **billions of dollars on inadequate technology:** Bradley Graham, "Missile Defense to Start in 2004," *Washington Post,* December 18, 2002.

389 **to concentrate on building first:** Pat Towell, "Bush's Missile Defense Victory Signifies Changing Times," *Congressional Quarterly,* October 26, 2002.

390 **vowed to do better:** "Test Failure of GMD System Blamed on Problem with Computer Chip," *Aerospace Daily,* March 4, 2003; Tony Capaccio, "Boeing Loses Bonus After Raytheon Warhead Fails in Missile Test," *Bloomberg News Service,* June 9, 2003.

392 **made any such projections meaningless:** Bradley Graham, "Secrecy on Missile Defense Grows," *Washington Post,* June 12, 2002.

393 **antimissile system was over:** Hearing of the Senate Armed Services Committee, March 18, 2003.

397 **"we might as well hang it up":** Interview with Lieutenant General Ronald Kadish, May 24, 2001.

Acknowledgments

Much of the reporting for this book was done while the events portrayed were under way. Early on, I approached a variety of people involved in or likely to be knowledgeable about one aspect or another of the unfolding missile defense story. I frequently went back to a number of them in the months that followed. Many agreed to speak on condition they not be identified as sources and publication would be withheld until they were out of office. But apart from such limits, no one imposed any terms on how the information would be presented in the book. That responsibility is mine alone.

While many sources wish to remain anonymous, there are a number of people whom I can thank for their help. My appreciation goes to both President Clinton and President Bush for granting interviews and providing insight into their own thinking about the missile defense challenge. I am grateful also to Bill Cohen and John Hamre at the Defense Department, Madeleine Albright and Strobe Talbott at the State Department, and Sandy Berger and Jim Steinberg at the NSC, not only for sharing their views but facilitating interviews with others on their staffs. At the Pentagon's Ballistic Missile Defense Organization, Lieutenant General Ron Kadish and Major General Bill Nance allowed exceptional access and delivered an education in missile defense technologies and architectures. Boeing's John Peller, Raytheon's

Bill Carpenter and Chuck LaDue, and a number of others in the defense contracting world provided instructive lessons about the actual business of building an antimissile system. Senators Joseph Biden, Thad Cochran, Jon Kyl, Carl Levin, and Rep. Curt Weldon, along with their staffs, led me through the politics of missile defense. And Larry Welch, Lisbeth Gronlund, David Wright, and the staff at Lincoln Laboratory offered technical expertise and analysis.

For taking extra time to fill in gaps and correct what errors of mine they could, I thank Steve Andreasen, Hans Binnendijk, Bob Bell, Jim Bodner, Steve Cambone, Phil Coyle, Philippe Errera, Jack Gansler, Rich Haver, Mitch Kugler, Glenn Lamartin, David Lyles, Frank Miller, Michael O'Neill, Condoleezza Rice, Walt Slocombe, Glenn Trimmer, Bob Walpole, Ted Warner, and Chris Williams.

Having been drawn to the subject of missile defense as a military affairs correspondent for the *Washington Post*, I thank executive editor Len Downie and managing editor Steve Coll for granting a leave to research and write the book. Special thanks goes to Glenn Frankel and Tom Frail of the paper's Sunday magazine, whose interest in a piece on the July 2000 flight test helped get me going.

The Center for Strategic and International Studies in Washington gave me a place initially to retreat to and write. My agent Esther Newberg quickly recognized the promise in the book. Working with PublicAffairs brought me together again with a former-editor-turned-publisher, Peter Osnos, and introduced me to the very able editing talents of Paul Golob. The rest of the PublicAffairs staff—including managing editor Robert Kimzey and assistants Melanie Peirson Johnstone and David Patterson—walked me patiently and knowledgeably through the publishing process. A special word also to Cindy Buck for her skillful copyediting and to Barbara Werden and Mark McGarry for the book's design.

For looking up countless transcripts, historical tidbits, book references and other invaluable information, I owe a special debt to researcher Lucy Shackelford. Helpful comments on one part or another of the manuscript came from friends Ken Bacon, Bruce Blair, Alan

Cooperman, Merrick Garland, Mike Getler, Jim Hoagland, Fred Ikle and Valerie Strauss.

My appreciation also extends to my brother Russell Graham and my father-in-law Charles Muscatine, who both contributed to improving the manuscript. And to my wife, Lissa Muscatine, whose advice, encouragement, and loving care sustained me all the way along.

Index

PublicAffairs is a publishing house founded in 1997. It is a tribute to the standards, values, and flair of three persons who have served as mentors to countless reporters, writers, editors, and book people of all kinds, including me.

I. F. STONE, proprietor of *I. F. Stone's Weekly*, combined a commitment to the First Amendment with entrepreneurial zeal and reporting skill and became one of the great independent journalists in American history. At the age of eighty, Izzy published *The Trial of Socrates,* which was a national bestseller. He wrote the book after he taught himself ancient Greek.

BENJAMIN C. BRADLEE was for nearly thirty years the charismatic editorial leader of *The Washington Post.* It was Ben who gave the *Post* the range and courage to pursue such historic issues as Watergate. He supported his reporters with a tenacity that made them fearless and it is no accident that so many became authors of influential, best-selling books.

ROBERT L. BERNSTEIN, the chief executive of Random House for more than a quarter century, guided one of the nation's premier publishing houses. Bob was personally responsible for many books of political dissent and argument that challenged tyranny around the globe. He is also the founder and longtime chair of Human Rights Watch, one of the most respected human rights organizations in the world.

For fifty years, the banner of Public Affairs Press was carried by its owner, Morris B. Schnapper, who published Gandhi, Nasser, Toynbee, Truman, and about 1,500 other authors. In 1983, Schnapper was described by *The Washington Post* as "a redoubtable gadfly." His legacy will endure in the books to come.

Peter Osnos, *Publisher*